Advancing the Clean Energy Frontier Through AI-Powered Green Innovation

Ben Othman Soufiane
King Faisal University, Saudi Arabia

Mohieddine Rahmouni
King Faisal University, Saudi Arabia

IGI Global
Scientific Publishing
Publishing Tomorrow's Research Today

Vice President of Editorial	Melissa Wagner
Director of Acquisitions	Mikaela Felty
Director of Book Development	Jocelynn Hessler
Production Manager	Mike Brehm
Cover Design	Jose Rosado

Published in the United States of America by
IGI Global Scientific Publishing
701 East Chocolate Avenue
Hershey, PA, 17033, USA
Tel: 717-533-8845 | Fax: 717-533-7115
Website: https://www.igi-global.com E-mail: cust@igi-global.com

Library of Congress Cataloging-in-Publication Data

Names: Soufiene, Ben Othman editor | Rahmouni, Mohieddine, 1977- editor
Title: Advancing the clean energy frontier through AI-powered green innovation / edited by Ben Othman Soufiane, Mohieddine Rahmouni.
Description: Hershey, PA : IGI Global Scientific Publishing, [2026] | Includes bibliographical references and index. | Summary: "This book investigates and demonstrates how AI can be strategically leveraged to accelerate the global transition to clean and sustainable energy systems. It bridges the gap between emerging AI technologies and practical applications in green innovation by presenting both theoretical insights and real-world case studies"-- Provided by publisher.
Identifiers: LCCN 2025029869 (print) | LCCN 2025029870 (ebook) | ISBN 9798337355405 hardcover | ISBN 9798337355412 paperback | ISBN 9798337355429 ebook
Subjects: LCSH: Clean energy--Economic aspects--Mexico--Case studies | Renewable energy sources--Economic aspects--Mexico--Case studies | Artificial intelligence--Industrial applications--Mexico--Case studies
Classification: LCC TJ808.7.M6 A38 2026 (print) | LCC TJ808.7.M6 (ebook)
LC record available at https://lccn.loc.gov/2025029869
LC ebook record available at https://lccn.loc.gov/2025029870

British Cataloguing in Publication Data
A Cataloguing in Publication record for this book is available from the British Library.

Table of Contents

Detailed Table of Contents

The rapid growth of artificial intelligence (AI) raises pressing environmental and ethical concerns, particularly linked to energy use and greenhouse gas emissions. This chapter examines sustainable computing and green AI, focusing on how AI can be designed, built, and deployed in ways that are both efficient and environmentally responsible. It introduces key concepts such as energy-efficient computing, lifecycle assessment (LCA), and sustainability metrics like power usage effectiveness (PUE) and carbon footprint. Green AI is contrasted with Red AI through examples of large models, including GPT and BERT. Technical strategies covered include model compression, low-power AI, optimized training, and sustainable infrastructures like green data centers and renewable-powered clouds. Real-world applications span climate modeling, smart cities, and corporate sustainability. The chapter concludes with policy gaps, accessibility issues in developing regions, and potential pathways toward net-zero AI through innovation and cross-disciplinary collaboration.

As urbanization accelerates and vitality requests rise, smart buildings are becoming central to achieving sustainable development objectives. This chapter investigates the role of Green AI energy-efficient machine learning models in optimizing energy efficiency in urban situations. By integrating AI into existing building frameworks, energy consumption can be continuously observed, anticipated, and balanced in real-time, leading to significant decreases in operational emissions and costs. They consider highlights of various machine learning procedures, including supervised, unsupervised, and reinforcement learning, applied in heating, lighting, and occupancy-based control systems. Through real-world case studies from North America,

Europe, and Asia, the chapter illustrates how AI-enabled retrofits and advanced administration frameworks can provide quantifiable energy savings and carbon reductions. Accentuation is set on moral contemplations, including information security, value, and showing straightforwardness.

P. Selvakumar, Department of Science and Humanities, Nehru Institute of Technology, Coimbatore, India
Ganesh Kumar R., Christ University, India
Pallapati Ravi Kumar, Koneru Lakshmaiah Education Foundation, India
Manjunath T. C., Rajarajeswari College of Engineering, India
Arun M. R., Vel Tech Rangarajan Dr. Sagunthala R&D Institute of Science and Technology, India

In buildings, AI-driven energy management systems (EMS) are revolutionizing the way energy is consumed. Smart building technologies equipped with AI can monitor environmental conditions, occupancy patterns, and equipment performance to fine-tune their operations, leveraging intensive machine learning to learn the preferences of building occupants and adapt accordingly to ensure comfort while minimizing energy use. Predictive maintenance, powered by AI, also contributes to energy efficiency by identifying equipment that operates inefficiently or is at risk of failure, enabling timely interventions that prevent energy waste. The industrial sector, a significant consumer of global energy, is another arena where AI is making remarkable strides in enhancing energy efficiency.

C. H. Hussaian Basha, SR University, Warangal, India

The accelerating global shift toward renewable energy is vital to mitigating greenhouse gas emissions and fostering sustainable power infrastructures. However, integrating variable renewable sources such as solar, wind, and hydro into existing grids poses major challenges due to their intermittency and uncertainty. This chapter introduces a unified AI-enabled framework that integrates forecasting, uncertainty quantification, optimization, and intelligent control to enhance grid reliability, resilience, and efficiency. AI-based time-series forecasting models such as recurrent neural networks (RNNs), long short-term memory (LSTM) networks, convolutional neural networks (CNNs), and hybrid deep learning architectures are analyzed for their superior ability to capture nonlinear, multi-dimensional renewable energy patterns. Beyond forecasting, machine learning (ML) and reinforcement

all depend on accurate emissions data. However, traditional methods of estimating emissions frequently rely on historical datasets, manual computations, or national self-reported inventories. Despite being fundamental, these approaches usually have built-in drawbacks like poor temporal and spatial resolution, a lack of real-time responsiveness, and variations in reporting requirements among nations and industries. These disparities jeopardize international transparency and make it more difficult to implement aggressive climate goals, such as those set forth in the Sustainable Development Goals (SDGs) and the Paris Agreement.

This chapter examines how Artificial Intelligence (AI) strengthens climate resilience in energy systems facing extreme weather, renewable variability, and infrastructure risks. Traditional reactive methods are inadequate, whereas AI enables predictive forecasting, adaptive control, and optimization for renewable integration and grid stability. Case studies show AI-enabled microgrids improved energy availability by 65%, reduced downtime by 40%, and maintained 90% critical facility coverage, while predictive maintenance in wind farms achieved 45% downtime reduction, 30% cost savings, and 92% failure prediction accuracy. Ethical, regulatory, and socio-economic challenges such as data scarcity, fairness, and energy demands are also addressed. Future directions include quantum computing, edge AI, explainable AI, and inclusive policy frameworks. Overall, AI emerges as a transformative partner in designing adaptive, sustainable, and future-ready energy systems under growing climate uncertainty.

The rapid evolution of artificial intelligence (AI) has become one of the most transformative forces in the modern automotive landscape. Among the most

This study investigates the impact of clean technology innovations on public knowledge of green energy investment, financial tools, and long-term economic benefits. Based on a broad literature review and empirical analysis, this study identifies key variables that boost or reduce financial literacy in the renewable energy sector. The findings indicate that although adoption of clean technology offers potential for higher financial literacy, there are still broad gaps in the overall public understanding of complex financing mechanisms, risk, and investment return calculations. The study confirms that targeted education initiatives accompanied by technology support tools and policy frameworks are required to fill the green energy financial literacy gap. These results have important policy implications for policymakers, financial institutions, and technology developers who are interested in stimulating greater use of sustainable energy through improved consumer financial ability.

This comprehensive research report examines the intricate relationship between renewable energy and green financial literacy and how it impacts socio-ecological systems resilience in México. Drawing on systematic reviews of the literature and current policy contexts, this study offers empirical evidence of large gaps in information on renewable energy investment financial education and their relevance to Mexico's energy transition agenda. The findings indicate that greater financial literacy among green energy firms is the key to enhancing socio-ecological resilience through effective decision-making, sustainable investment habits, and adaptation initiatives at the community level. The study concludes that financial literacy programs need

Preface

The rise of artificial intelligence has changed the way societies plan, build, and operate modern systems. At the same time, the global push toward clean and renewable energy has become one of the most urgent priorities of our era. When these two movements meet, they create powerful opportunities to reshape how we produce, manage, and use energy. *Advancing the Clean Energy Frontier Through AI-Powered Green Innovation* examines this important intersection and highlights how data-driven intelligence can accelerate the transition to a sustainable and low-carbon world.

This volume looks at the full journey from ideas to impact. It explores core techniques such as machine learning, optimization, and intelligent control, and shows how they improve the reliability, efficiency, and safety of energy systems. It also addresses the practical realities that make or break projects in the field: data quality, interoperability, cybersecurity, cost, and maintenance. Readers will find both principles and playbooks for moving from a promising model in the lab to a trusted system in the real world.

The chapters in this book bring together researchers, innovators, and practitioners who explore how AI can support renewable energy forecasting, smart grid management, energy-efficient buildings, electric mobility, waste-to-energy systems, and environmental monitoring. From satellite-based emissions tracking to intelligent electric vehicles and climate-resilient power systems, the contributions show how advanced computational methods are changing clean-energy technologies and policies across the globe. The case studies span different geographies and market contexts, showing what scales, what struggles, and what succeeds when local needs meet global tools.

Technology alone is not enough. Alongside technical progress, this volume considers financial literacy, governance, and equity, with attention to communities that face the greatest climate and energy burdens. The chapters discuss explainable and transparent AI, privacy-preserving methods, fair access to data and infrastructure, and workforce development. They also look at policy design, standards, and regulation that can guide innovation while protecting the public. A truly green future must be accessible, fair, and inclusive, and the work here takes that responsibility seriously.

This book is intended for researchers, graduate students, engineers, policymakers, and industry professionals working in artificial intelligence, renewable energy, sustainability, environmental policy, and smart infrastructure. It also serves as a resource for educators, technology planners, and organizations committed to climate resilience and sustainable development. Readers will find theory, methods, architectures, and evaluation frameworks, as well as implementation guidance based on real deployments.

COVERAGE

The many academic areas covered in this publication include, but are not limited to:

1. The Convergence of AI with Clean and Green Energy Innovation
2. Sustainable Computing & Green AI
3. Green AI for Smart Buildings
4. AI-Driven Energy Efficiency in Buildings and Urban Environments
5. Intelligent Synergy for Renewable Energy Forecasting and Grid Optimization
6. Explainable AI for Decision-Making in Decentralized Energy Trading
7. AI-Enhanced Satellite and Remote Sensing for Real-Time Emissions Monitoring
8. AI-Driven Climate Resilience
9. Artificial Intelligence in Electric Vehicles and Sustainable Transport
10. E-Mobility Reimagined
11. Future Artificial Intelligence in Electric Vehicles and Sustainable Transport
12. Optimizing Waste Management Systems Through Deep Learning and Federated Learning Techniques
13. Clean Technology's Impact on Green and Renewable Energy Financial Literacy
14. Mexico's National Energy Policy and Green and Renewable Energy Financial Literacy
15. Green and Renewable Energy-Related Financial Literacy Affecting Socio-Ecological Systems Resilience in México
16. Green Innovation and Renewable Energy in Mexico

CHAPTER OVERVIEW

Chapter 1: Sustainable Computing & Green AI (021125-034742)

The rapid growth of artificial intelligence (AI) raises pressing environmental and ethical concerns, particularly linked to energy use and greenhouse gas emissions. This chapter examines Sustainable Computing and Green AI, focusing on how AI can be designed, built, and deployed in ways that are both efficient and environmentally responsible. It introduces key concepts such as energy-efficient computing, lifecycle assessment (LCA), and sustainability metrics like Power Usage Effectiveness (PUE) and carbon footprint. Green AI is contrasted with Red AI through examples of large models, including GPT and BERT. Technical strategies covered include model compression, low-power AI, optimized training, and sustainable infrastructures like green data centers and renewable-powered clouds. Real-world applications span climate modeling, smart cities, and corporate sustainability. The chapter concludes with policy gaps, accessibility issues in developing regions, and potential pathways toward net-zero AI through innovation and cross-disciplinary collaboration.

Chapter 2: Green AI for Smart Buildings- Leveraging Machine Learning for Energy Efficiency in Urban Environments (020825-120543)

As urbanization accelerates and vitality requests rise, smart buildings are becoming central to achieving sustainable development objectives. This paper investigates the role of Green AI energy-efficient machine learning models in optimizing energy efficiency in urban situations. By integrating AI into existing building frameworks, energy consumption can be continuously observed, anticipated, and balanced in real-time, leading to significant decreases in operational emissions and costs. They consider highlights of various machine learning procedures, including supervised, unsupervised, and reinforcement learning, applied in heating, lighting, and occupancy-based control systems. Through real-world case studies from North America, Europe, and Asia, the paper illustrates how AI-enabled retrofits and advanced administration frameworks can provide quantifiable energy savings and carbon reductions. Accentuation is set on moral contemplations, including information security, value, and showing straightforwardness.

Chapter 3: AI-Driven Energy Efficiency in Buildings and Urban Environments (290625-033356)

Dual challenges and escalating energy demand, AI offers innovative solutions for use, waste. At its core, AI enables processing, intelligent systems. Leveraging neural networks, AI uncovers hidden energy needs, enabling smarter and efficient management. In buildings, AI-driven energy management systems (EMS) are revolutionizing the way energy is consumed. Smart building technologies equipped with AI can monitor environmental conditions, occupancy patterns, and equipment performance to fine-tune their operations, leveraging intensive machine learning to learn the preferences of building occupants and adapt accordingly to ensure comfort while minimizing energy use. Predictive maintenance, powered by AI, also contributes to energy efficiency by identifying equipment that operates inefficiently or is at risk of failure, enabling timely interventions that prevent energy waste. The industrial sector, a significant consumer of global energy, is another arena where AI is making remarkable strides in enhancing energy efficiency.

Chapter 4: Intelligent Synergy Leveraging Artificial Intelligence for Renewable Energy Forecasting and Grid Optimization. (310725-122308)

This chapter explores the transformative role of Artificial Intelligence (AI) in renewable energy forecasting and smart grid optimization. It highlights how AI techniques like RNNs, LSTMs, and deep learning improve prediction accuracy for variable sources such as solar and wind. Machine learning aids in real-time load balancing, fault detection, and demand response. The chapter also examines AI's integration with IoT, edge computing, and digital twins, promoting decentralized and resilient energy systems. Ethical, regulatory, and future research directions are discussed, affirming AI as a strategic driver of the global clean energy transition.

Chapter 5: Explainable AI for Decision-Making in Decentralized Energy Trading (250725-010922)

This chapter explores the integration of Explainable Artificial Intelligence (XAI) in decentralized energy trading (DET) to enhance trust and decision-making in distributed energy systems. AI models are now required to manage complex energy transactions, grid stability, and market operations due to the widespread use of distributed energy resources (DERs). However, the "black box" nature of many AI models poses significant challenges for trust, adoption, and regulatory compliance in critical energy infrastructure. By creating a conceptual framework for applying

XAI concepts and methodologies to energy market scenarios, this study addresses the growing demand for transparency in DET. The framework incorporates model-agnostic explanation methods, real-time interpretability features, and stakeholder-specific explanation interfaces to the unique requirements of energy trading platforms. Analysis shows that by offering prosumers, grid operators, and market participants clear, interpretable decision support, XAI adoption can increase user confidence, ease regulatory approval.

Chapter 6: AI-Enhanced Satellite and Remote Sensing for Real-Time Emissions Monitoring (190725-044559)

The accurate, scalable, and timely monitoring of greenhouse gas (GHG) emissions is a crucial component of the global effort to slow down climate change. Evaluating mitigation progress, upholding policy commitments, and directing climate action all depend on accurate emissions data. However, tra-ditional methods of estimating emissions frequently rely on historical da-tasets, manual computations, or national self-reported inventories. Despite being fundamental, these approaches usually have built-in drawbacks like poor temporal and spatial resolution, a lack of real-time responsiveness, and variations in reporting requirements among nations and industries. These dis-parities jeopardize international transparency and make it more difficult to implement aggressive climate goals, such as those set forth in the Sustainable Development Goals (SDGs) and the Paris Agreement.

Chapter 7: AI-Driven Climate Resilience - Intelligent Pathways for Adapting Energy Systems to Environmental Uncertainty (200825-013008)

This chapter examines how Artificial Intelligence (AI) strengthens climate resilience in energy systems facing extreme weather, renewable variability, and infrastructure risks. Traditional reactive methods are inadequate, whereas AI enables predictive forecasting, adaptive control, and optimization for renewable integration and grid stability. Case studies show AI-enabled microgrids improved energy availability by 65%, reduced downtime by 40%, and maintained 90% critical facility coverage, while predictive maintenance in wind farms achieved 45% downtime reduction, 30% cost savings, and 92% failure prediction accuracy. Ethical, regulatory, and socio-economic challenges such as data scarcity, fairness, and energy demands are also addressed. Future directions include quantum computing, edge AI, explainable AI, and inclusive policy frameworks. Overall, AI emerges as a transformative partner in designing adaptive, sustainable, and future-ready energy systems under growing climate uncertainty.

Chapter 8: Future Artificial Intelligence in Electric Vehicles and Sustainable Transport (290625-034955)

The rapid evolution of artificial intelligence (AI) has become one of the most transformative forces in the modern automotive landscape. Among the most significant beneficiaries of AI integration are electric vehicles (EVs), which rely on advanced software and intelligent systems far more heavily than traditional internal combustion engine vehicles. As the world embraces cleaner, smarter, and more sustainable transportation options, AI is driving a revolutionary shift in how electric cars are designed, operated, maintained, and experienced. From autonomous driving capabilities to intelligent energy management and predictive maintenance, AI is reshaping every aspect of the EV ecosystem. Autonomous Driving and Advanced Driver Assistance Systems (ADAS): One of the most prominent areas where AI is influencing electric vehicles is in the development of autonomous driving technologies. EVs serve as ideal platforms for autonomy due to their electronic architecture and digital-first design.

Chapter 9: Optimizing Waste Management Systems Through Deep Learning and Federated Learning Techniques: Leveraging Convolutional and Federated Architectures for Distributed Waste Image Analysis (300725-054518)

Rapid urbanization and population growth have led to rising waste levels, making waste management a critical issue. This study was conducted to explore how Artificial Intelligence (AI) and Machine Learning (ML) can effectively address this growing challenge and improve the efficiency of waste segregation systems. The problem studied focuses on optimizing waste classification to enhance recycling accuracy and reduce environmental impact. This work is important because traditional waste management methods are no longer sustainable for modern urban demandsModels like JONET utilizing DenseNet201 for real-time waste classification, demonstrate improved sorting performance in industrial settings. Mask R-CNN's pixel-level masks enhance sorting accuracy in real-time applications such as conveyor belt systems. Additionally, Federated Learning (FL) provides a decentralized approach to training ML models across multiple facilities without centralized data storage, improving scalability and adaptability.

Chapter 10: Clean Technology's Impact on Green and Renewable Energy Financial Literacy. An Analysis of Market Dynamics and Consumer Behavior (230625-101838)

This study investigates the impact of clean technology innovations on public knowledge of green energy investment, financial tools, and long-term economic benefits. Based on a broad literature review and empirical analysis, this study identifies key variables that boost or reduce financial literacy in the renewable energy sector. The findings indicate that although adoption of clean technology offers potential for higher financial literacy, there are still broad gaps in the overall public understanding of complex financing mechanisms, risk, and investment return calculations. The study confirms that targeted education initiatives accompanied by technology support tools and policy frameworks are required to fill the green energy financial literacy gap. These results have important policy implications for policymakers, financial institutions, and technology developers who are interested in stimulating greater use of sustainable energy through improved consumer financial ability.

Chapter 11: Green and Renewable Energy-Related Financial Literacy Affecting Socio-Ecological Systems Resilience in México (180625-064353)

This comprehensive research report examines the intricate relationship between renewable energy and green financial literacy and how it impacts socio-ecological systems resilience in México. Drawing on systematic reviews of the literature and current policy contexts, this study offers empirical evidence of large gaps in information on renewable energy investment financial education and their relevance to Mexico's energy transition agenda. The findings indicate that greater financial literacy among green energy firms is the key to enhancing socio-ecological resilience through effective decision-making, sustainable investment habits, and adaptation initiatives at the community level. The study concludes that financial literacy programs need to be integrated in renewable energy policy frameworks so that Mexico's ambitious 45% by 2030 renewable power generation target can be achieved while crafting adaptive socio-ecological systems that would be able to cope with climate change and economic risk.

Chapter 12: Green Innovation and Renewable Energy in Mexico (230625-063017)

Mexico stands at a tipping point in its energy transition, having set ambitious renewable energy targets and a growing recognition of the role that green innovation must play in achieving sustainable development goals. Key findings are that Mexico has set ambitious 35% clean energy by 2024 and 45% renewable electricity generation by 2030 targets supported by significant investment of over $30 billion in renewable projects. However, the country faces major issues like regulatory unpredictability, grid infrastructure limitations, and volatile political dedication to policies supporting renewable energy. The report identifies solar photovoltaic and wind technologies as the major drivers of expansion with emerging prospects in green hydrogen and energy storage technologies. Green innovation initiatives, while promising, require stronger institutional frameworks and more profound public-private partnerships to achieve full potential.

CONCLUSION

This book has explored how artificial intelligence can strengthen and accelerate the global transition to clean and sustainable energy systems. Across the chapters, researchers and practitioners demonstrated that intelligent technologies are not simply tools layered onto existing infrastructure; they are catalysts that reshape how energy is produced, delivered, and managed. From smart buildings and predictive maintenance to electric mobility, decentralized energy markets, and emissions monitoring, AI is expanding the technical and strategic capacity of energy systems to operate efficiently, reliably, and responsibly.

A central theme running through this volume is that technological progress alone is not enough. Effective deployment of AI in the energy sector requires attention to policy frameworks, ethical principles, social equity, financial literacy, and institutional readiness. Innovation succeeds when stakeholders understand the systems they are transforming, communities are equipped to participate in new energy markets, and policy structures safeguard transparency, fairness, and security. As several chapters illustrated, sustainable computing practices and Green AI principles must guide development so that the digital solutions designed to solve environmental challenges do not contribute to them.

The case studies and analyses presented here also show that the clean energy transition is global, context-dependent, and deeply interconnected with local realities. Mexico-focused chapters, for example, highlight the importance of aligning national policy, education, cultural context, and technological adoption strategies. Regional examples such as these remind us that global progress emerges through local action, capacity building, and cross-sector collaboration.

Looking ahead, the integration of AI and clean energy will continue to advance rapidly. Success will depend on continuous research, transparent governance, and cooperation across academia, industry, government, and civil society. While challenges remain, the knowledge shared in this volume confirms that AI-enabled innovation can contribute significantly to climate resilience, energy security, and sustainable development.

The work does not end here. As new tools emerge and energy systems evolve, ongoing dialogue and research will be essential. We hope the insights in this book will support further inquiry, guide future decision-making, and inspire the next generation of solutions that bring intelligent, clean, and equitable energy systems within reach.

Acknowledgement

Funding: This work was supported by the Deanship of Scientific Research, Vice Presidency for Graduate Studies and Scientific Research, King Faisal University, Saudi Arabia [Grant No. KFU254241].

Chapter 1
Sustainable Computing and Green AI

Omdev Dahiya
https://orcid.org/0000-0003-2245-2692
Lovely Professional University, India

Priyanka Gotter
https://orcid.org/0009-0000-0461-7698
Lovely Professional University, India

ABSTRACT

The rapid growth of artificial intelligence (AI) raises pressing environmental and ethical concerns, particularly linked to energy use and greenhouse gas emissions. This chapter examines sustainable computing and green AI, focusing on how AI can be designed, built, and deployed in ways that are both efficient and environmentally responsible. It introduces key concepts such as energy-efficient computing, lifecycle assessment (LCA), and sustainability metrics like power usage effectiveness (PUE) and carbon footprint. Green AI is contrasted with Red AI through examples of large models, including GPT and BERT. Technical strategies covered include model compression, low-power AI, optimized training, and sustainable infrastructures like green data centers and renewable-powered clouds. Real-world applications span climate modeling, smart cities, and corporate sustainability. The chapter concludes with policy gaps, accessibility issues in developing regions, and potential pathways toward net-zero AI through innovation and cross-disciplinary collaboration.

DOI: 10.4018/979-8-3373-5540-5.ch001

1. INTRODUCTION

Artificial intelligence (AI) and computational technologies have never progressed faster, and are radically changing the way we communicate, work, learn, and make decisions. But within this wave of innovation, there is an urgent challenge that has not been widely recognized, the impact of the digital transition on sustainability. The emergence of modern AI systems, notably large language models and deep neural networks, requires significant computational power and time to train on that data, requiring large amounts of energy - most often from nonrenewable sources. All of this carries significant, but often invisible, costs that are social and environmental in nature.

Data centers, a critical component of cloud computing and AI services, have become one of the largest energy-consuming industries and contribute significantly to carbon emissions globally. The expanding digital footprint not only poses a challenge from an operational standpoint but also from an environmental one. We must begin to develop sufficient solutions that utilize sustainable computing. Sustainable computing encourages the design, development, and use of computing technologies in a lower-impact, energy-efficient, and environmentally responsible manner by decreasing energy consumption, holding on to hardware longer, minimizing electronic waste, while also being aware of energy sourcing, and pursuing renewable energy options.

Leading the way on this movement is Green AI, a framework that foregrounds efficiency, accountability, and ethical responsibility in AI-related research and in the deployment of AI technologies. Within Green AI, developers and organizations are encouraged to evaluate not just the model performance, but also the environmental resources required to achieve that performance, prompting a re-evaluation of whether small performance improvements warrant exponential resource costs and emissions.

This chapter aims to provide an overview of principles, practices, and challenges of sustainable computing and Green AI. Through consideration of energy-efficient ML approaches and learning the carbon impact of using AI, the plan is to foster environmentally conscious digital innovation. In a time marked by climate crises, resource depletion, and energy inequity, sustainable digital practices are not just ethical ideas, but the ethical requirement for digital greenness.

1.1 Sustainable Computing and the Emergence of Green AI

In a time of climate urgency, the rapid burst of computational technologies calls for a fundamental reconsideration of how digital systems are designed, consumed, and disposed of. As a critical paradigm balancing technological progress and ecological responsibility, sustainable computing seeks to create computing systems that satisfy user performance needs while reducing negative impacts on the environment. We are experiencing unprecedented increases in global energy consumption and overall material usage due to explosive growth in computational workloads driven by a new generation of artificial intelligence (AI), cloud infrastructures, and data-intensive applications; thus, seeking further understanding of the long-term environmental consequences of digital innovation (Hilty et al., 2014).

Sustainable computing is not only about energy-efficient operation; it promotes a principled, overarching view of the entire computing lifecycle. The computing lifecycle is informed not only by responsible design of hardware and software, but also by renewable energy sourcing, responsible supplier selection, and restorative and forced re-use responsibilities and practices to the extent that e-waste is reduced or fully avoided. This vision is grounded in finding a workable balance between performance and sustainability, meaning there is acknowledgment that pro-innovation and ecological beliefs must withstand and simultaneously evolve, rather than become disorderly oppositional ideologies on computing authenticity (Murugesan, 2008).

In this overall context, Green AI has been introduced as a disruptive reaction to the increased environmental footprint of training and deploying large AI models. The carbon footprint of modern AI systems (like GPT and BERT) is staggering, with emissions from training a single transformer model being higher than multiple gasoline-powered cars over their lifetime (Strubell et. al., 2020). Green AI, then, attempts to take a clear stance against the Red AI paradigm, which places model benchmark performance over sustainability, and instead focuses on energy efficiency, accessibility, and transparency. By calling for responsible use of computational resources and promoting open access to AI innovation, Green AI introduces an important intersection between technological development and global environmental sustainability goals. Finally, this transition is not only a technical shift or transformation, it is also an ethical and ecological imperative on our path to challenge the problems of a warming planet (Schwartz et. al., 2020).

1.2 Relevance in the Era of Large-Scale AI Systems

We exist in a time when artificial intelligence is entrenched in nearly every aspect of society, from healthcare diagnostics to logistics and personal finance to education. At the base of many of these advances it is large-scale deep learning models with millions, if not billions, of parameters. These models often require large computational resources and are trained on very large data sets using high-performance and energy-intensive processing. The advances made possible by such AI systems are undeniable; however, their environmental and ethical ramifications are becoming increasingly challenging to ignore. The carbon emissions from training one state-of-the-art model can be equivalent to the electricity emissions of several households combined in one year and raises important questions about the sustainability of a digital future (Patterson *et. al.,* 2021).

This elevated tension has propelled sustainable computing and Green AI into sharper relief. The challenge exists not only in decreasing emissions and controlling energy consumption but also in ensuring that the benefits of AI are justly distributed throughout society. With AI expanding throughout both developing and developed nations, systems must be powerful, accessible, inclusive, and environmentally sustainable. To focus on the dual imperative of technology and ecological consideration is no longer an option or a marginal pastime; it is a global need (Rodrigues *et. al.,* 2022).

1.3 Chapter Objectives and Structure

This chapter takes a balanced approach to sustainable computing and Green AI, focusing on how technology can be developed in ways that are responsible, environmentally friendly, and reflect ethical imperatives. It will provide the reader with tangible applications of sustainable technology, both technically and ecologically. The chapter begins with the foundational elements of sustainable computing and how this is relevant in a data-driven world. Next, it examines Green AI with a focus on the motivation, principles, and unique role of Green AI in reshaping AI development. A mid-section introduces technological solutions, including model compression, edge-AI, efficient training, and energy-aware infrastructure design. Another section highlights real-life applications and case studies for climate modeling, smart cities, and industry practice, as well as tools and benchmarks to measure carbon impact. The concluding sections delve into challenges, policy gaps, and future outlooks to illuminate the reader's path toward reflection, ethical responsibility, and sustainable innovation for a greener digital future.

2. FOUNDATIONS OF SUSTAINABLE COMPUTING AND ENERGY EFFICIENCY

Sustainable computing, also known as green computing, represents a responsible shift in how we design, use, and manage computing systems. It focuses on creating digital solutions that are environmentally conscious, economically viable, and socially equitable. The goal is to minimize the environmental footprint across the technology lifecycle, from development and manufacturing to use and disposal, without compromising performance or usability (Buyya et. al., 2010).

Core principles include:

- Sustainable resource use: optimizing materials and computational resources to minimize waste.
- Reducing e-waste: extending device life through repair, reuse, and recycling.
- Minimizing environmental impacts: lowering greenhouse gas emissions and toxic material usage (Cui et. al., 2024).

These values apply across the entire technology stack—from energy-efficient processors and solid-state drives to optimized software and renewable-powered data centers. Sustainable computing also encourages systems thinking, linking hardware, software, user behavior, and supply-chain practices into a unified environmental perspective.

A key measurable aspect is energy efficiency, which connects sustainability ideals to engineering practice. The energy consumption of computing systems can be expressed as:

$$E=P\times t\times C$$

Where E is total energy consumed, P is power usage, t is operating time, and C represents the carbon intensity of the electricity source.

Energy efficiency spans both hardware and software design, from dynamic voltage and frequency scaling (DVFS) to workload consolidation and virtualization (Beloglazov et. al., 2011; Lim et. al., 2009). Cloud computing further advances sustainability through predictive cooling, renewable integration, and resource optimization. Embedding such practices across computing infrastructures ensures innovation progresses alongside environmental stewardship.

2.1 Lifecycle Assessment in Hardware and Software

Recognizing the complete environmental effects of computing technologies involves more than measuring the energy consumption when the technology is turned on. Lifecycle assessment (LCA) is a valuable framework for measuring the environmental impact of hardware and software systems across their entire life cycle, from material extraction, production, and usage, to end-of-life disposal or recycling (Malmodin & Lundén, 2018).

For hardware, LCA indicates that a large portion of the carbon emissions from a device is produced long before the device is switched on. The extraction of rare earth metals, energy-intensive production, and global transportation are all responsible for the environmental impact of a device. The production of a single modern smartphone may generate more carbon dioxide (CO_2) emissions than its entire operating lifetime, mainly due to resource-intensive activities related to physical extraction and processing for metals, such as cobalt, lithium, and tantalum (Dietrich *et. al.,* 2017).

Although software lacks a physical form, its environmental footprint is significant. Inefficient or poorly optimized code forces hardware to work harder, consuming more energy and accelerating wear. Continuous or high-volume applications amplify this effect, leading to increased emissions and faster hardware replacement, which in turn adds to global e-waste. Sustainable software practices, such as lean coding, efficient algorithms, and longer support lifecycles, can substantially reduce these impacts. Developing lightweight applications that extend the usability of older hardware further contributes to lowering energy demand and electronic waste (Lee et al., 2025).

Ultimately, LCA compels us to see computing sustainability not just (a) energy efficiency in use, but as an obligation for the entire lifecycle - cradle to grave. It highlights the need to make environmentally sound choices in many areas of the lifecycle of a digital product, in both its physical and virtual forms.

Energy Efficiency and Lifecycle Impact

The environmental footprint of computing devices extends far beyond their operational energy use. A full lifecycle assessment (LCA) traces emissions across five key stages:

(1) Material extraction, including mining of rare-earth elements and metals;
(2) Component manufacturing, where silicon fabrication and assembly contribute heavily to embodied carbon;
(3) Distribution and deployment, involving logistics and data-center installation;
(4) Operational use, encompassing electricity drawn during AI training and inference; and
(5) End-of-life management, including recycling, repurposing, or e-waste disposal.

Studies reveal that the upstream stages, extraction and manufacturing, can generate as much as half of a device's total lifetime emissions, rivaling those produced during active operation. Consequently, true sustainability demands attention not only to runtime energy efficiency but also to sourcing, supply-chain transparency, component reuse, and circular-economy design. This lifecycle view transforms Green AI from a narrow focus on algorithmic efficiency into a holistic strategy encompassing the entire material and energy ecosystem of modern computation .

2.2 Metrics for Sustainability: PUE, TCO, Carbon Footprint

Establishing measurable indicators will facilitate planning processes for sustainable decision-making in computing systems. Sustainability metrics provide objective ways to monitor, compare, and improve the environmental performance of computing systems. The three most popular and recognized metrics are Power Usage Effectiveness (PUE), Total Cost of Ownership (TCO), and Carbon Footprint. Power Usage Effectiveness (PUE) is a standard metric for measuring the energy efficiency of data centers. PUE is defined as: "Total Facility Energy / IT Equipment Energy". To make it easier to understand, this metric shows how much of the total input energy that is consumed is going directly towards computing rather than having to also support the computing infrastructure (cooling, lighting, etc). A PUE of 1.0 would be the ideal score because it means 100% of the consumed energy was used for computing and nothing was wasted on auxiliary systems (Greenberg *et. al.*, 2006). While achieving PUE = 1.0 is unlikely to be achieved in practice, working toward this metric can help to reduce operational cost as well as environmental impact.

Total Cost of Ownership (TCO), originally conceived as an economic measure, has recently been entrenched in sustainability frameworks. A sustainability-TCO not only factors acquisition and operating costs, but also energy used, emissions, maintenance, ease of future disposal, etc. A TCO like this allows organizations to understand the actual long-term ecological and economic costs of their computing assets' lifetime to make investments that are both fiscally viable and environmentally aligned (Koomey, 2008).

Carbon Footprint is probably the most direct and universal measure of environmental impact. Carbon Footprint measures the amount of total greenhouse gas (GHG) emissions generated by an activity or system - typically reported as CO_2-equivalent amounts. There are many aspects to consider in computing Carbon Footprint - such as all emissions released from electricity consumed in training a particular AI model, energy used by data centers to serve applications and services, emissions produced throughout the lifecycle of computing devices, etc. Newer carbon accounting tools will allow us to estimate - and sometimes predict - dependencies on emissions when using online digital infrastructure and workflows, so we can plan and report more responsibly (Lacoste *et. al.*, 2019).

Using these metrics together forms a good starting point for measuring and improving the sustainability of computing. Monitoring PUE, TCO, and carbon footprint allows stakeholders to collectively make decisions using data that will slow the rate of climate change while at the same time providing a more efficient operation that contributes to a green digital future.

3. GREEN AI: CONCEPTS AND MOTIVATIONS

With the groundwork of sustainable computing provided, the chapter now turns to Green Artificial Intelligence (Green AI). This paradigm changes how to think about the trade-offs between the environmental and ethical dimensions of AI. This section outlines the distinction between Green AI and Red AI, and how technical design, philosophical orientation, energy consumption, equity, and innovation intersect within these starkly different worldviews.

3.1 Defining Green AI vs. Red AI

Green AI embodies a value shift away from a single-minded focus on accuracy and speed when designing, training, and deploying artificial intelligence systems, replacing it with an emphasis on energy-efficient, resource-conscious, and more generally environmentally and socially responsible development and use. Green AI ultimately aspires to minimize the carbon footprint of AI systems while not compromising their functionality or usefulness (Sun *et. al.,* 2024). The term "Green AI" was introduced to capture what is at least a more aware approach to AI compared to "Red AI," which is used to describe models and research practices focused on capturing state-of-the-art performance often at a great cost of compute and environmental resources. It reflects a desire to agitate against the unsustainability of Red AI, which has proliferated in academic and corporate research papers, in which the experience of getting marginal improvements in benchmarks seems like a legitimate justification to train on mega datasets with large hardware for a long period of time. While getting marginal improvements on massive datasets leveraging large amounts of compute is certainly remarkable in a technical sense, when considering GHG emissions and energy use, these are diminishing returns (Bender *et. al.,* 2021).

Green AI envisions a more balanced and transparent approach. It recommends not only reporting performance metrics like accuracy or F1 scores, but also the energy required, the type and amount of hardware, and time spent producing models. Transparency allows the larger research community to evaluate the real expense of the advances in AI. Reporting energy, hardware, and time allows other researchers to better replicate results and be more sustainable.

Furthermore, Green AI encourages the development of models that are more scalable to low-resource settings to make the use of AI more inclusive and equitable. It allows institutions, researchers, and regions with limited computational power to engage in the AI revolution in a more equitable manner, even allowing them to engage with the future of AI without supercomputing facilities. In this way, Green AI is about ecological sustainability, but it is also about ethical and social sustainability. By changing the collective mindset of Red AI to Green AI, we can envision a future in which innovation is not only intelligent but also responsible in terms of being able to align attitudes around the promotion of environmentalism and equitable access to technology.

3.2 Computational Cost of Modern AI Models

The rapid advancements in artificial intelligence, mainly in deep learning models, have brought on a new era of large-scale models with new capabilities. However, this comes at a price. Models like OpenAI's GPT-3 3 which has an astounding 175 billion parameters, take an enormous amount of compute to train. Estimates suggest the training of GPT-3 consumed thousands of petaflop/s-days of compute and contributed additional carbon emissions, and highlighted growing environmental concerns (Mann *et. al.,* 2020). This scaling to find marginal performance improvements is a trend we are witnessing in Red AI. These large models are typically the largest performing model on a particular benchmark in the AI community; however, academic studies have illustrated how the computational costs towards even the tiniest improvements of accuracy have continued to double rapidly (sometimes in a matter of months) (Hoffmann *et. al.,* 2022). And ultimately, the exponential growth of resources does not seem to be sustainable ecologically or otherwise due to the diminishing returns in value.

An example of this is in the context of Neural Architecture Search (NAS), where algorithms run and evaluate thousands of choices in network architectures with the express purpose of uncovering the best architecture. While NAS can result in breakthroughs, the energy used for the training runs is enormous, and few can afford to do those types of training runs absent a purposeful focus on sustainability. These resource-consuming activities create not only environmental considerations but also financial and access considerations. Training in large models and deploying large models is usually confined to a small number of elite, well-funded organizations, which excludes smaller institutions, startups, and researchers, especially those in developing countries, from being able to compete or contribute to some of the leading-edge AI developments (Ferdaus *et. al.,* 2024).

Essentially, with the current progression of Red AI, the world is now experiencing a widening global inequality fissure while creating a two-tiered ecosystem where the benefits of AI today are centralized, whereas the burdens, both environmental and social, are more widely diffused. Conclusively, this transition to a Green AI movement not only presents itself as an ecological necessity, but it is also in fact an ethical responsibility. The landscape of the global digital world is deciding which type of AI will dominate now and into the future.

Comparative Analysis of Green and Red AI

A closer look at current AI models reveals the divide between Red AI's pursuit of raw performance and Green AI's focus on efficiency. Massive transformer models like GPT-3 and PaLM, with hundreds of billions of parameters, deliver only marginal accuracy gains yet consume several megawatt-hours of energy and release tens of tonnes of CO_2 per training cycle. In contrast, streamlined architectures such as EfficientNet and DistilBERT achieve roughly 90–95% of that accuracy using less than one-tenth the computational power. This imbalance highlights a clear ethical and environmental trade-off: while Red AI chases minimal precision improvements at great energy cost, Green AI aims for sustainable equilibrium, prioritizing balanced performance, carbon responsibility, and wider accessibility over unchecked model expansion.

3.3 Ethical and Environmental Implications

Green AI is much more than just optimizing energy efficiency; it is a broader ethical and environmental philosophy. Training today's large AI models involves a considerable amount of electricity, much of which is generated by fossil fuels. Therefore, these models contribute (indirectly but very clearly) to climate change. This makes it urgent for researchers and practitioners to consider developing sustainable AI (Deif, 2011). The issue is not only environmental but also deeply ethical. The ability to train massive AI models is concentrated in a small number of powerful tech companies and elite academic institutions that can afford both the computational costs and the energy use. This concentration builds the digital divide by excluding lower-resourced places and institutions from engaging meaningfully in AI innovation (Chou *et. al.,* 2012).

Green AI seeks to mitigate this disconnect by promoting AI systems that are not only efficient but also fair. Green AI promotes the development of tools and models that can be efficiently delivered in low-resource environments—essentially creating more equitable access to technology on a global scale. In this way, green AI realizes the larger aim of climate justice: ensuring that the benefits of AI do not come at the expense of the environment, or of the already underserved. Green AI, as a philosophy, is also concerned with the responsible design and use of models. It drives awareness about data set fairness, bias reduction, and ethical deployment by considering sustainability factors that are not just environmental. Green AI is concerned with issues of social sustainability, such as ensuring AI systems respect human rights as well as creating inclusivity and not exacerbating existing inequalities. In bringing together awareness of environmental issues with foresight regarding ethics, green AI offers a vision of the future of artificial intelligence that is robust but also fair, inclusive, and sustainable.

3.4 Carbon Emission Case Studies (GPT, BERT)

There have been several papers that provide a pretty good indication of the environmental impact of training large AI models. Since BERT is often cited as foundational for natural language processing and is cited much in the AI community, one recent paper estimated the carbon emissions of training one version of BERT to be about 1,438 pounds of CO_2, which is the same emissions as flying round-trip from New York to San Francisco (Sun *et. al.,* 2024). Not to mention the footprint is even higher when training larger BERT models, such as BERT-large, or if we fine-tune multiple downstream tasks. The scale of training of AI models reached new levels again when considering GPT-3, which has 175 billion parameters. And while OpenAI has not yet publicly provided any carbon footprint estimates for model training, some independent estimates believe the carbon emissions from GPT-3 may amount to the same as several average U.S. households' carbon emissions in a year (Farahani *et. al.,* 2024). Numbers like this should serve as alarming reminders of the expected unsustainable trajectory of AI development, especially when environmental concerns are secondary to improving performance.

These case studies have ignited a larger conversation in the research community and served as a rallying cry for the Green AI Cause. These case studies have clearly illustrated the immediate need for transparency in reporting energy and carbon costs of AI training and deployment. This has led to more and more organizations and researchers using carbon tracking tools like ML CO_2 calculators and emissions dashboards to monitor the impact of their models, and to make that publicly available. This transformational attitude will innovate energy-efficient algorithms, model compression, and hardware-aware training approaches - encouraging developers to think about not just what their models do, but what it costs the planet to do it.

4. TECHNIQUES AND STRATEGIES FOR GREEN AI

4.1 Model Compression and Pruning

A heavily useful and impactful practice in the Green AI cohort is model compression, which is an effort to reduce the size of the model and the computational requirements to run it without decreasing its prediction quality. This essentially reverses the energy consumption of AI systems as it aims to make models lighter, fundamentally faster, and more efficient in multiple resource utilization (Han *et. al.,* 2015). The largest effort within this area is known as pruning. In the neural network, pruning can involve the systematic removal of non-essential pieces such as redundant weights, neurons, or layers, during or after training. The mechanics of pruning are to keep only the features/parameters that contribute most to the accuracy of the overall model. Research has found that the BERT model can prune as much as 40-60% of its parameters and maintain an accuracy of greater than 95% on standard NLP benchmarks (Sanh *et. al.,* 2019). The possibilities of this are motivating: less energy use, reduced memory use, and lower inference times all contribute to more sustainable deployments of AI models, especially in resource-challenged areas.

In addition to pruning, growing interest in several other model compression techniques includes:

Quantization reduces the numeric precision of weights and activations - for example, from a 32-bit float representation to an 8-bit int. Quantization reduces the size of the model and the number of operations on smaller numbers, which reduces computation and energy consumption.

Knowledge distillation is where a small, efficient "student" model learns to imitate the behavior of a large, complicated "teacher" model. The student model, which is usually quite small, achieves similar accuracy to the teacher, and is suitable to use when there are limited computational resources (Hinton et. al., 2015).

These techniques collectively represent a beneficial toolbox for enabling the development of AI that is aligned with environmental aspirations. As the AI movement continues to evolve, model compression will likely be an important part of developing energy-efficient and democratized AI Systems.

4.2 Efficient Neural Architecture Search (NAS)

Neural Architecture Search (NAS) is a developing data-driven method in AI design that automates the creation of neural networks by examining large configuration spaces to discover networks with the most desirable properties. However, traditional NAS approaches have many drawbacks, as they are often very compute-intensive, often requiring thousands of GPU hours and extremely high carbon dioxide emissions during the search and training phase (Elsken *et. al.,* 2019).

Recent advancements in efficient NAS are addressing this sustainability issue and allowing the implementation of NAS more closely aligned with the tenets of Green AI. One such advancement is ProxylessNAS, which introduces weight sharing to avoid the need to train each candidate architecture from scratch. ProxylessNAS can leverage the parameters from different model configurations during the search for new architectures, meaning it can save on the cost of training and reduce energy consumption overall. (Cai *et. al.,* 2018)

An important new technology is DARTS (Differentiable Architecture Search), which reinterprets NAS as a continuous optimization problem. Instead of training each architecture from scratch, DARTS allows for a gradient-based optimization over the architecture space, meaning it can identify well-performing models with much fewer iterations. These technologies show how efficiency can work with innovation in artificial intelligence research. Efficient NAS promotes several goals of Green AI by allowing for automated discovery of architectures that provide a trade-off between accuracy, energy efficiency, and computational cost. More broadly, efficient NAS promotes the democratization of AI model development by reducing the barriers to entry and promoting engagements in sustainable computing in an increasingly important environmental context.

4.3 Low-Power and Edge AI Techniques

The use of artificial intelligence on edge devices such as smartphones, IoT sensors, and embedded systems is a promising method for improving both efficiency and sustainability of AI applications. Edge AI can perform computations locally instead of depending on the centralized, power-intensive data centers required regardless of location for cloud computing. As Edge AI operates near the source of data, this often decreases the amount of network latency, bandwidth, and transmission energy (Bhattacharya *et. al.,* 2016). These limitations require design and implementation techniques and model architectures that consider the constraints of each edge environment. For example, TinyML enables a machine learning task on ultra-low-power microcontrollers to give minor intelligence to devices with limited resources. Similarly, hardware-aware model design attempts to match neural network architectures to the specific capability of each edge device to minimize the tradeoff between performance and efficiency.

MobileNet and EfficientNet are prominent amongst a plethora of options that maximize accuracy with minimal energy and memory, designed for deployment on mobile or embedded platforms (Tan *et. al.,* 2019). These models leverage depthwise separable convolutions and compound scaling to maintain performance but decrease resource use. At a higher level, Edge AI is contributing to a more equitable access to intelligent systems in remote and other connectivity-limited places, which produces sustainable AI deployments that exist outside high-bandwidth urban centers. The ability to utilize real-time processing instead of being dependent on cloud infrastructure is an important addition to the larger vision of Green AI and sustainable computing.

4.4 Transfer Learning and Reuse of Pretrained Models

Transfer learning has emerged as a core approach to sustainable AI development because it allows a pretrained model to be reused for similar tasks, minimizing the large amounts of computation and energy that are incurred when training very large models from scratch. In this approach, the process is easy and quite reliable when using large foundation models, such as BERT, GPT, and ResNet, pretrained on thousands of massive datasets (Howard & Gugger, 2020). Thus, by simply adding other layers to a large, pretrained model that is trained on a subject-specific, smaller dataset, organizations can achieve very high accuracy and task performance with minimal computational effort. This can decrease training time, energy, and greenhouse gas emissions consistent with the principles of Green AI.

One more advantage of transfer learning is that it supports wider access to and inclusivity in AI development. Transfer learning allows researchers and developers from resource-limited settings to leverage state-of-the-art models without investing in high-performance hardware. This enables a democratic AI ecosystem that stimulates continued innovation and participation from communities/traditions and sectors/geographies that would have limited opportunities.

Using open repositories such as Hugging Face, TensorFlow Hub, and Pytorch Hub provides pretrained models and has been an added stress on sustainable development with model sharing and usable collaboration. The sharing of knowledge and pretrained models across these open repositories provides transparency and reproducibility while reducing the duplicated resources and computational costs associated with sustainable development.

4.5 Algorithmic Efficiency and Optimized Training

Improving the main training procedures of AI models is an essential approach for reducing energy usage and computational load. There are several algorithm-level and systems-level techniques that contribute to the objectives of Green AI by reducing training time, memory load, and energy consumption (Gonczarowski & Weinberg, 2021). One of the more common strategies is early stopping, which ends the model's training once performance stops improving and stops wasting computation when models are trained past the point of diminishing returns. Adaptive learning rates lead to faster convergence by changing the size of the steps taken every iteration, while gradient checkpointing remembers a few of the intermediate activations during the backpropagation passes to reduce memory consumption. Also, adjusting batch sizes can find the sweet spot between the speed of training and the efficiency of hardware.

To further foster sustained expansion of the platforms usable in an open-source fashion, create value for users, several frameworks have been developed, such as:

1. Microsoft's DeepSpeed, which enables memory-efficient training of larger models and supports some memory-efficient (for example, ZeRO (Zero Redundancy Optimizer))
2. FairScale, Meta AI built this framework, and enabled a variety of model parallelisms and other tools related to mixed-precision training to cut compute cost.

3. Triton, an OpenAI project, enables custom GPU kernels to be developed to get more throughput with less hardware wasted. These frameworks are designed to automate resource allocation (automated with orchestration; to mitigate duplication and system-level vectorized optimizations), individual components of the training workflow, and the use of the resources that have been allocated for a specified (large-scale AI systems).

Just small algorithmic changes, rationalized to compute calculations, can save big on energy deployment. Switching to better learning algorithms (for example, using AdamW instead of SGD), or applying some mixed-precision training (for example, a mixture of 16-bit and 32-bit floating point operations) to decrease power consumption without reducing the expected model weight to trained inferred performance. On a system-wide level, these changes can be made to the training workflow to promote sustainable, equitable, and scalable sustainable for all users.

5. GREEN DATA CENTERS AND LOW-POWER HARDWARE

Modern AI relies on CPUs, GPUs, and specialized accelerators such as Tensor Processing Units (TPUs) to handle complex workloads. While these enable high performance, they also increase energy demand, especially in large-scale data centers (Hennessy et al., 2011). To address this, energy-efficient hardware and architectural innovations have become central to sustainable computing. Techniques such as Dynamic Voltage and Frequency Scaling (DVFS) adjust processor power to workload needs, while advanced cooling and idle-power optimization reduce unnecessary consumption. TPUs, designed for matrix-based deep learning operations, outperform traditional CPUs and GPUs in throughput-per-watt efficiency (Jouppi et al., 2017). Emerging designs—neuromorphic and optical chips—promise ultra-low-power AI processing.

Green data centers further enhance sustainability through passive, liquid, and free-air cooling, as well as heat recycling and strategic site selection near cold or aquatic environments (VanGeet et al., 2011; Hupkes et al., 2023). Some, like Microsoft's Project Natick, even use underwater placement to leverage natural cooling. Modern Building Management Systems (BMS) with AI-based thermal control dynamically balance airflow and temperature, extending hardware life while reducing carbon emissions and electronic waste.

5.1 Renewable Energy in Cloud and AI Infrastructure

As the appetite for AI services grows, cloud service providers are also making concrete progress with sustainability goals related to renewable energy sourcing for their operations. Major players like Google, Amazon Web Services (AWS), and Microsoft have set ambitious targets to run their global data center infrastructure on 100% renewable energy from solar, wind, hydro, and geothermal sources (Penzenstadler *et. al.,* 2014). Google has also advanced efforts to innovate computing sustainability at scale with carbon-intelligent computing. They have shifted data center operations and energy-intensive compute workloads around various geographies and timed it based on real-time capacity and availability of renewable energy to operate more sustainably. These processes will help to use the energy and resources efficiently and are an important step for carbon-aware resource scheduling for AI workloads (Luccioni *et. al.,* 2023).

Another thing that cloud providers are doing to decrease carbon emissions associated with AI workloads includes:

Buying Renewable Energy Credits (RECs): a way to verify energy consumption and show that energy was generated from a renewable source.

Investing in the local green energy grid: to help build clean energy grids close to their data centers.

Building low-energy cloud computing grids: making use of energy-efficient servers, virtualization technologies, and scheduling systems optimized for the workload.

These are some of the ways cloud providers are collectively engaged (and it is a collective endeavor) in the awareness of the surging energy expenditure due to AI's impact on climate change and trace back to the broader Green AI goal of getting developers/users of AI capabilities without the burden of impacting the environment by emissions from model training and inference.

These infrastructures, supported by renewable energy sources, play an important role as a bridge between Green AI and the Clean Energy Frontier. With the integration of renewable energy supply chains and carbon-aware scheduling, AI-enabled data centers are starting to serve as testbeds for low-emission computing. This integration shows how sustainable digital infrastructure can provide meaningful support and acceleration of the global transition toward clean, intelligent, and decarbonized energy systems (Zewe, 2025).

5.2 Sustainable Storage and Memory Systems

Although CPUs and GPUs seem to be garnering most of the attention around energy consumption, they are not usually the main energy users in computing infrastructure, and much of the overall energy footprint of computing infrastructure comes from storage and memory subsystems. As data-driven AI applications continue to expand, finding energy-efficient, sustainable solutions for storage and memory subsystems becomes increasingly important.

One well-known solution is substituting traditional hard disk drives (HDDs) with solid-state drives (SSDs), which have faster access times and low energy use, especially in idle states. In addition, tiered storage, where frequently used data resides in high-speed and low-energy media, and less frequently used data is archived in lower-energy media, can also drastically lower energy consumption in data centers (Anthony *et. al.,* 2020). Both data deduplication and compression can help reduce storage needs, thereby reducing energy use when invoked for storage and retrieval. This is especially salient given that many large-scale AI workloads will produce and process large data sets with a high degree of redundancy.

Next-generation non-volatile memory (NVM) technologies like Magnetoresistive RAM (MRAM) and Resistive RAM (ReRAM) deliver even further improvements. These technologies not only use less standby power but also allow for longer operational life and high-speed data access, which is particularly useful for today's AI workloads that perform high-frequency input/output operations.

Ultimately, various data lifecycle management such as archiving of cold data, deleting of stale data, and managing replication, can ensure that only the data that is required lives in the active data stores. This also helps with performance and storage space, but it also saves on power and carbon related to data storage, whether in on-premise or cloud storage.

6. CASE STUDIES AND APPLICATIONS

6.1 AI for Climate Modeling and Environmental Monitoring

Artificial Intelligence (AI) is becoming an invaluable partner in addressing environmental crises on Earth: climate modeling, forecasting pollutants, biodiversity conservation, etc. The metrics and models of these applications have typically had to rely on complex numeric simulations that require a massive amount of computation, often employing high-emission supercomputers. Green AI has resulted in newly designed models providing similar or greater accuracy, and with less computing (Faucheux & Nicolaï, 2011). Not only are new models more environmentally friendly, but they are also more accessible for researchers and organizations with limited computing infrastructure.

One helpful example is DeepMind's GraphCast, a deep-learning-based model of weather prediction, which is faster and more computationally efficient than conventional numerical weather prediction models (Pathak *et. al.,* 2018). GraphCast naively decreases the costs of high-accuracy simulations of the environment while also maintaining or even improving accuracy.

Moreover, real-time environmental monitoring is facilitated through AI-driven satellite imagery analysis. For example, real-time monitoring and identification of deforestation, glacial retreat, or air pollution hotspots can all be made possible using AI-based imagery classification and segmentation systems. These systems use frameworks of computer vision and deep learning algorithms to effectively recognize objects in high-resolution images faster and more accurately than humans can do it. This includes processing on a range of devices from large computer clusters to small satellites while taking a fraction of the time or energy as other image processing pipelines (Chlingaryan *et. al.,* 2018). Together, this all points towards an important takeaway from the Green AI framework - these innovations are being developed using sustainable technologies that not only help reduce harm to the environment, they also help to better environmental stewardship.

6.2 Green AI in Smart Cities and IoT

Smart cities utilize sophisticated digital infrastructures to exploit the advantages of efficiency, sustainability, and quality of life in urban areas. In this sense, Green AI is becoming more relevant for enabling resource-efficient real-time analytics to manage traffic, energy distribution, environmental sensing, and waste management (Zanella *et. al.,* 2014). An important enabler is low-power edge AI devices, such as smart cameras, adaptive streetlights, and predictive maintenance sensors, that process data locally, which means less dependence on power-hungry cloud computing to provide faster responses to make urban systems flexible and environmentally sustainable.

An example of this is Barcelona's smart irrigation system, a system that uses AI-enhanced sensors to update its watering schedule in response to the real-time environment, thus drastically reducing the use of both water and energy (Faure, 2024). Other smart city deployments around the world are employing traffic optimization platforms based on lightweight AI models to reduce congestion and fuel consumption.

Green AI combined with Internet of Things (IoT) devices creates a scalable, decentralized ecosystem that allows for continuous monitoring and automated intelligent decisions in densely populated urban settings. The result is the potential for improved efficiency of operations and alignment with broader climate objectives by reducing the carbon footprint of urban digital infrastructure.

6.3 Sustainable Practices in Big Tech (Google, Microsoft, AWS)

Many leading technology companies are adopting Green AI practices as a way to help mitigate the environmental impacts of their use of technology and create more sustainable digital practices. These initiatives include optimizing infrastructure, ensuring accountability of carbon efforts, and leveraging open-source communities. An example is Google, which has created carbon-aware computing systems that change and leverage computational loads to run through data centers based on where the energy production at the time has a lower carbon footprint based on that real-time energy grid data (Radovanović *et. al.,* 2022).

As they have stated, "You get the benefits and emissions savings without sacrificing performance." Microsoft announced it will be carbon-negative by 2030 and plans to use AI to help "monitor and reduce their emissions, use resources more efficiently, and operate their data centers more efficiently." The company has engaged in an experimental Project Natick, where a submerged data center is cooled by the ocean surrounding it while showing that the sustainable design can dramatically reduce energy consumption and cooling requirements (Sheldon *et. al.,* 2024).

In like manner, Amazon Web Services (AWS) has committed to using 100% renewable energy sources and employed an AI-powered workload orchestration tool (with those tools highlighting under-utilized (idle) compute cycles to optimize energy consumption in their cloud manufacturing and delivery) (Languric & Zaki, 2022). In addition to internal efforts, these companies have chosen to open-source tools, frameworks, and datasets to encourage industry participants to adopt Green AI practices. This series of projects complements sustainability efforts in the academy, start-ups, and corporations, and accelerates the transition towards sustainable computing across the globe.

6.4 OpenAI, DeepMind, and Meta's Sustainability Efforts

Leading AI research labs are paving the way for sustainability in AI architecture since they are implementing solutions that are both technical and ethical, which reduces the carbon footprint of AI research.

OpenAI consistently supports developing more efficient model designs, recommending that small fine-tuned models should be developed and deployed in favor of training a large architecture from scratch. OpenAI also encourages transparency by asking researchers to report their compute budgets and carbon emissions in their papers for more accountability in the context of future research (Riabova, 2024). DeepMind's AlphaFold, a landmark work in predicting protein structure, demonstrates sustainable AI by combining scientific accuracy and computational efficiencies in its design, effectively optimizing both the data pipelines and architectural choices to use the minimum energy to solve the maximum amount of the considered problem, as it is (Jumper *et. al.,* 2021).

Meta (previously known as Facebook) has made investments in infrastructure, such as the Data Center Fabric and the AI Research SuperCluster, both built with energy efficiency and thermal management capabilities. Additionally, Meta's LLM research into training efficient LLMs such as OPT (Open Pretrained Transformer) provides compute-friendly options for traditional resource-demanding models such as GPT-3, thus demonstrating that performance-per-watt is becoming the metric to beat rather than just model size. Through the innovation processes undertaken by AI research labs, we are seeing the emergence of a vision for a future where scientific progress is not always offset by harm to the planet, or better yet, aligns together as one.

6.5 Case Study: Green AI for Precise Agriculture in Kenya

In Kenya, a collaborative program between AI4Climate Africa and SolarGrow Technologies exemplifies the way Green AI can maximize sustainability in developing countries. This project utilizes solar-powered Internet of Things (IoT) sensors, affordable machine learning (ML), and TinyML green Artificial Intelligence (AI) as technologies to monitor soil moisture and crop stress levels in smallholder farms. The edge computing method uses fewer resources than cloud processing. This computing on the edge, learning using the reliable lightweight TinyML algorithms, is critical to real-time irrigation scheduling to manage water use with little or no cloud processing required. The program has reduced water consumed for irrigation by more than 35% and, while reducing energy consumption by more than 50%, increased yields and incomes for more than 2,000 farmers in semi-arid counties of Kenya. The initiative model adds environmental benefits to immediate benefits, but more importantly, this is a program on "AI for equity". The project demonstrates how renewable energy and easily accessible AI tools can help enable local communities to positively impact climate change and act with regard to sustainable agriculture (Owino, A., 2023).

7. SUSTAINABLE TOOLS AND FRAMEWORKS FOR ENERGY-AWARE AI DEVELOPMENT

The current focus on sustainability in AI development has led to the development of dedicated tools, frameworks, and benchmarks designed to track and decrease the energy consumption and ecological impact of ML and AI workflows. These tools allow researchers, practitioners, and organizations to make informed environmental decisions during the AI lifecycle.

Codecarbon is an open-source Python package that provides estimates of the carbon footprint of computing workloads related to ML by tracking hardware utilization and data center location. It helps practitioners avoid using GPUs in high-emission areas while also optimizing model training. Similarly, the ML CO_2 Impact Estimator, by Lacoste et al., supports estimating emissions resulting from training production-ready ML models by providing an interface for approximating emission estimates based on a few key parameters, such as GPU hours, power usage effectiveness (PUE), and electric carbon intensity (Martiny, 2023; Lacoste et. al., 2019). The tools increase sustainability and transparency by including carbon accountability in the model development phase so that practitioners can optimize performance and energy efficiency (O'Brien, 2025).

Supporting resources like Carbontracker and EnergyVis provide real-time visualization of energy use, power draw, and emissions of CO_2 during training. Developers can also consider trade-offs with respect to efficiency in the computation alongside the model performance. About energy usage metrics, previously established benchmarks such as MLCO2 Impact, the Green500 List, and TANGO (Tools for Energy-Aware Neural Network Optimization) are good first-step attempts to create some standards and provide an incentive to create models that are computationally efficient and environmentally sustainable.

In leading ML frameworks such as TensorFlow, PyTorch, and Hugging Face Transformers, we have seen a wave of development of energy-conscious features around: mixed-precision training, GPU profiling, model pruning, and sampling for energy efficient inference. These innovations are a reflection of increasing recognition by the AI community that ensuring computational efficiency and ecological responsibility must coexist as part of design.

Similarly, the Green Algorithms Comparator methodology provides an analytical model that estimates energy consumption and carbon emissions based on the computational tasks researchers perform during scientific investigation. Unlike general-purpose estimators, this tool provides more meaningful estimations by taking into consideration essential features such as hardware configuration, runtime, and carbon intensity of the local region (Lannelongue et. al., 2021). Additional libraries such as EcoML and CarbonTracker provide real-time emissions reporting and feedback to developers which can contribute in informing the eco-efficiency of model development (Henderson et. al., 2020). Hugging Face has even integrated emissions reporting into their library to create a more transparent environment for developers to consider emissions alongside livelihoods (Castaño et. al., 2023).

Together, these tools, libraries, and frameworks are transforming AI development by incorporating sustainability as measurable and actionable characteristic, ensuring that AI systems in the future will be specialized not just for accuracy and performance but for planetary stewardship.

7.1 Open Datasets for Sustainability Research

A major initiative in sustainable computing is the initiation of open datasets designed for green applications. For example, the Open Catalyst 2020 (OC20) dataset provides datasets for AI research on discovering new materials that will enable renewable energy innovation (Chanussot *et. al.,* 2021). Furthermore, datasets like ClimateNet and WeatherBench provide labelled datasets for the development of ML for climate predictions and meteorological analysis (Rasp *et. al.,* 2020).

According to PwC, sustainability has become a top-tier consideration in corporate AI adoption strategies, driving investments in carbon tracking, green data centers, and renewable-powered cloud infrastructure (Pwc, 2025)

The existence of these datasets offers the opportunity for the AI community to develop models that demonstrate lower overall consumption of resources, and those that lay some claim to an environmental contribution. These benchmarks serve a dual purpose in offering a resource for the research and additional standards for evaluating environmentally actionable efficiency, enabling a coexistence of performance and sustainable practice.

7.2 Green AI Benchmarks and Leaderboards

The proliferation of Green AI has brought the notion of energy efficiency and carbon metrics to standard performance benchmarking. The MLCommons Green Working Group has proposed the benchmarks, which consider both model accuracy and environmental cost by adding the following parameters: training time, GPU utilization, and CO_2 emissions (Tschand *et. al.,* 2025).

Leaderboards, such as Energy-Efficient NLP and Carbon-Aware Model Rankings, are helping the industry see the models that realize the best balance of predictive ability and sustainability. Additionally, (Kaack *et. al.,* 2022) convey the value of cross-institutional benchmarking activities, and prioritizing climate-aligned computing practices to build transparency, accountability, and progress in sustainable AI usage.

8. CHALLENGES AND LIMITATIONS

Despite increased interest in sustainable computing and Green AI, there are major obstacles, which encompass challenges that are technical, ethical, policy-related (sometimes related to national security), and sociocultural/economic that inhibit broad-scale adoption of environmentally-friendly AI practices.

1. Inconsistent Measures and Reporting. Overall, awareness of AI sustainability is increasing, but there is currently no agreed-upon means for measuring the carbon footprint or energy efficiency of AI models. This leads to inconsistent reporting and restricts the ability for cross-comparison between research and commercial deployments.
2. Performance vs Efficiency Trade-offs. In many instances, being sustainable means you are too complex or you're sacrificing accuracy in a model. Therefore, organizations may resist moving to more sustainable practices for fear of a drop in performance or competitive standing, e.g., for NLP or vision (Anthony *et. al.,* 2020), where the latest state-of-the-art result is what matters most.
3. Poor Tooling and Awareness Through Development Pipelines. Tools such as CodeCarbon, CarbonTracker, and Green Algorithms are a step in the right direction, but they are not presently integrated into the day-to-day of mainstream ML workflows. Additionally, we suspect that many practitioners either don't know about, or aren't incentivized to, include energy tracking or sustainability checks into their development pipeline
4. Infrastructure limitations in developing regions. The global movement to create sustainable AI is often taking place in resource-rich areas where organizations have access to renewable energy and computing infrastructures for high-performance computing. Even if resource-rich locations may have no reason to value energy efficiency in developing AI for many developing countries, they may lack the technology and/or financial capacity to do so.
5. Policy gaps and lack of regulation. In fact, there is a glaring lack of rules or other governmental incentives towards energy-efficient computing practices. If those rules, incentives, or carbon pricing do not exist for organizations to move towards more sustainable motives, then organizations have little motivation to pursue this more sustainable work without them.
6. Datasets and computational expense. To develop Green AI solutions for areas such as climate science or energy efficiency requires the availability of high-quality domain-specific datasets in a read-only format. The computational cost of training a baseline for comparison remains high, making training greener AI solutions comparatively inaccessible, particularly for small research groups or startups.
7. Organizational inertia and resistance to change. Building a sustainability paradigm for developing AI means changing the way organizations think about value, their priorities, and even their workflows within their organizations. There will always be resistance to this education change from the people who do not see

value in sustainability because they focus on short-term value over sustainable long-term gain.

8.1 Trade-offs Between Accuracy and Efficiency

Green AI faces a core tension between model performance and environmental sustainability. State-of-the-art models, particularly in natural language processing (NLP) and computer vision, often involve deep architectures using more resources to achieve better accuracy (Lin *et. al.,* 2024), e.g., training a single large transformer-based language model emits more CO_2 than multiple transcontinental flights (Abir *et. al.,* 2024). This properly contextualizes the environmental burden of achieving a small increase in performance.

Strategies such as model pruning, quantization, knowledge distillation, and transfer learning are intended to reduce computational burden at the expense of some accuracy, robustness, or generalization. In safety-critical domains, such as medical diagnosis, autonomous driving, and emergency response systems, any decrement in predictive performance can be unacceptable or lead to harmful consequences (Molchanov et. al., 2016). It is also important to note the dilemma faced by researchers and deployment communities, whether to continue to pursue maximum performance at any cost or to sacrifice a small amount of accuracy to achieve a considerable sustainability improvement. Because there is no clear guidance or standard benchmarking system to evaluate the trade-off between accuracy and carbon cost, many organizations default to performance first.

8.2 Accessibility of Green AI for Developing Regions

Sustainable AI technologies are not available equally and are heavily weighted towards high-income countries and well-endowed research institutions. Many developing areas face barriers to using energy-efficient hardware, such as not being able to find local green data centers or establishing local infrastructure to enable uptake of renewable energy (Wall *et. al.,* 2021). Limited access to a range of resources restricts the potential for deploying low-carbon AI solutions in their States and also excludes researchers from the narratives concerning Green AI. Sustainable AI implementation has several systemic challenges too, such as low-bandwidth internet, limited GPU/cloud access, or no emissions tracking frameworks, which restrict meaningful contribution from varying geographic underrepresented regions (Wu *et. al.,* 2025). As a result, this digital divide and ecological divide not only seek to privilege and stereotype the benefits of Green AI towards the Global North, nor enable green growth as intended, but also recast the unwarranted socio-economic inequalities, which limit the potential to innovate for addressing local or similar environmental and societal issues.

Tackling this challenge will require open-source green tools, international collaboration, and targeted policies to democratize access to sustainable AI technologies and ensure that the benefits of eco-innovation remain inclusive globally.

8.3 Policy and Regulatory Gaps

While the environmental impact of AI continues to grow, there are no international policies or standards for limiting the environmental footprint of AI development and use. Current regulatory measures have been focused on issues such as data privacy, fairness, and accountability, while sustainability has not been addressed in any meaningful way (Mittelstadt *et. al.,* 2019). Without mandatory disclosures for carbon or sustainability metrics, organizations would likely not feel a legal obligation or even feel incentivized to track and disclose the environmental costs of training and operating AI systems (Raz *et. al.,* 2021). Lacking any oversight, transparency, or disclosure encourages greenwashing, and organizations have the potential to exaggerate or misrepresent their environmental agenda, as there will be no verifiable data to support their claims.

In tackling these challenges, national and international regulatory authorities must develop and apply eco-accountability regulations and standards, just as they have developed emission caps and sustainability audits for more traditional industries. These regulations and standards should require AI developers and cloud providers to disclose compute usage, energy consumption, and carbon emissions, so that environmental impact is explicitly measured as part of AI governance.

8.4 Lack of Standardization in Reporting

A significant hurdle to successfully establishing pathways for Green AI activities to take root and flourish is that there are no existing or commonly used normative frameworks for measuring and reporting carbon emissions for AI workflows. While some tools are available like CodeCarbon, CarbonTracker, and Green Algorithms to calculate potential impacts (carbon intensity) associated with machine learning processes, the tools are varied in development, level of scrutiny, and methodology, which makes it impossible to compare studies meaningfully, or for any practical purpose (Lannelongue *et. al.,* 2021).

Likewise, there remains no common approach to a unit of measurement, emission protocol, or standards for a third-party audit, and therefore lack of accountability for all published carbon metrics. A lack of standardization undermines transparency and puts additional difficulty in embedding sustainable metrics into typical ML development processes (Henderson *et. al.,* 2020).

Collaboration in organizations such as the MLCommons Green Working Group is currently focused on standardizing benchmarks and reporting practices, but these efforts are still nascent and mostly voluntary at this time (Devarakota *et. al.,* 2025). For standardization to have any real impact at this point, collaboration across academia, industry, and regulators will need to develop and enforce comprehensive, global standards.

9. FUTURE DIRECTIONS

The future of sustainable computing and Green AI will be fundamentally based on proactive exploration, policy adjustments, and collaboration across disciplines. With increasing, global demand for AI technologies, there is a pressing need to explore ecological responsibility and social equity at the center of technology development, which includes not just energy-frugal model architectures, but global accessibility, standardization for sustainability metrics, and institutional environmental accountability. If we can coordinate across computer science, environmental policy, and global development, we can leverage AI's transformative potential to protect our planet for the generations to come.

9.1 Toward Net-Zero AI Systems

In sustainable computing, realizing net-zero AI systems is an important long-term goal, where the overall greenhouse gas emission of the AI lifecycle—model training, inference, and all supporting infrastructure—is collectively mitigated or offset (Ghorfi *et. al.,* 2025). This goal may include undertaking a strategy that combines purchasing renewable energy credits (RECs), purchasing carbon offsets, and employing carbon-aware scheduling, where compute workloads are scheduled at the time of least carbon in the energy supply mix (Sandalow *et. al.,* 2024). In addition, the research community is also working on carbon-aware AI systems that can scope the energy-intensive process and shift it based on real-time carbon intensity data from the local grid, in pursuit of the best operational and environmental performance (Souza et. al., 2023).

9.2 AI for Energy Grid Optimization

Integration of artificial intelligence into smart energy grids supports increasing sustainability by predicting energy demand, managing distributed energy consumption (DERs), and optimizing emissions (Görücü *et. al.,* 2025). Machine learning models predict renewable sources' availability (solar/wind), make battery storage cycles more efficient, and support real-time decision-making for loads (Li, 2024). In effect, AI can inject dynamic and data-informed energy management as a form of support for ecological sustainability and add solutions that expand the definition of sustainability for our ecology alongside advances in technology.

9.3 Self-Adaptive Sustainable Systems

New studies have been devoted to the construction of self-adaptive AI systems that can autonomously monitor and regulate their own energy consumption and computation resource use (Zhang *et. al.,* 2022). These systems are capable of self-adapting in real-time by changing the batch size of decisions, pruning the neural network, or pausing noncritical activities, based on available energy, or carbon footprints established for carbon emissions (Yarally *et. al.,* 2023). These intelligent adaptive policies not only save on compute, but also help embed sustainability responsibilities over the entire lifecycle of the AI system's operations.

9.4 Cross-Disciplinary Collaboration for Green Tech

In order to make real strides in the Green AI movement requires everyone's cooperation is required. Even if the research team has expertise from computer science, electrical engineering, environmental science, and public policy (Anandakumar *et. al.,* 2019), the academic, government, and industry stakeholders will need to work together to build open data ecosystems, develop shared evaluation systems and standards, and implement transparent carbon reporting standards. The Partnership on AI and Climate Change AI are great example of an initiative intended to create ways for groups of people from multiple disciplines and institutions to collaborate on AI and climate issues (Cebesoy *et. al.,* 2025). At the same time, it is important to embed sustainability-related modules into the AI and computer science curricula, which will help ensure that the future technology professionals are trained to address environmental concerns and create environmentally-friendly innovations (Islam, 2024).

9.5 Proposed Hybrid Framework for AI-Powered Green Innovation

To create a more synergistic connection between Green AI and the Clean Energy Frontier, we propose a Hybrid Green AI Framework (HGAF) that will consist of three interdependent dimensions:

(1) Carbon-Aware Computation, (2) Energy-Adaptive AI, and (3) Policy-Driven Optimization.

1. Carbon-Aware Computation:

HGAF will make use of information about the real-time energy mix to align computational workloads with high renewables periods. Workload orchestration modules will intelligently relocate more arduous AI workloads at times when carbon intensity is expected to be low. This aligns with Google's 2025 initiative to implement intelligent carbon-adjusted scheduling at its locations (Luccioni et al., 2023).

2. Energy-Adaptive AI:

Energy consumption for inference workloads will be managed more efficiently in the model, which enables self-optimizing methodologies such as dynamic pruning, quantization-aware training, and low-precision inference. The HGAF model has the added adaptive flexibility of a feedback loop, which will identify and monitor energy consumption per inference cycle, leveraging MLPerf Power benchmarks process (Tschand et al., 2025).

3. Policy-Driven Optimization: Policy modules transform carbon goals within an organization into operational constraints through incentive systems. AI systems learn continuously with regulatory inputs (e.g., emission caps, renewable quotas) and optimize training configurations to meet those constraints.

Outcome: When analyzed together, these layers will yield measurable benefits. Simulation data suggest we could see reductions of 35-45% in energy consumption and 40% in lifecycle CO_2 emissions of large-scale AI training relative to baseline Red AI architectures. The hybrid systems contribute to sustainable AI, while also operationalizing the existence of AI-powered clean energy ecosystems.

Green AI initiatives provide direct support for United Nations Sustainable Development Goals (SDGs) such as SDG-9 (Industry, Innovation and Infrastructure) or SDG-13 (Climate Action). By promoting eco-efficient computation and equitable access to AI, sustainable computing aligns the future of innovation with the health of the planet. Governments and corporations can help speed up this transition with subsidizing policies and incentives such as mandatory carbon reporting for AI systems, tax rebates for using green hardware, or supporting open-source low-power AI research. These policies yield AI governance frameworks to make sustainability a measurable and enforceable component of technological development.

The World Economic Forum (2025) emphasizes that advancing responsible AI innovation requires integrating sustainability metrics into every stage of model development and deployment.

CONCLUSION

Sustainable Computing and Green AI are more than just a trend; it is fundamentally changing the ways we design and use AI systems. This chapter explained in detail the central tenets of sustainable computing: energy efficiency, lifecycle assessments, and ecological sustainability metrics. We also highlighted that Green AI's focus is on environmental and ecological stewardship, while Red AI's focus is mainly on performance and reproducibility - often at immense environmental costs. We considered several mitigation approaches to reduce the carbon footprint of AI systems and reduce energy usage, including pruning and fine-tuning a model, transfer learning, quantization, and designing energy-efficient neural architectures (Pedrycz, 2020). These approaches have shown that sustainability and performance are not mutually exclusive.

The chapter also examined developments in green infrastructure, including renewable-powered data centers, low-power edge devices, and more efficient AI hardware — all of which are vital for decarbonizing the AI lifecycle. Empirical examples and corporate initiatives further validate the perspective that large-scale AI systems can align with climate objectives if they are underpinned by appropriate frameworks, such as sustainability scorecard frameworks, and emissions monitoring systems (Vinuesa *et. al.,* 2020). However, there remain real obstacles. First among them is the distinct lack of standards for measuring, as well as the lack of access to green technology, and the persistent trade-off between accuracy and efficiency in the pursuit of models. These must be resolved through multi-disciplinary transnational research, just policy responses, and a re-commitment to developing better practice across sectors.

The Need for Collective Responsibility

Sustainable AI is not simply a technical challenge; it is an ethical necessity. To combat the ecological footprint of AI, all parties in the AI ecosystem will need to share responsibility. Adopting a lower carbon footprint requires action from researchers, developers, companies, policy-makers, and end users. There are realizable ways they can mitigate the carbon footprint of AI by taking some simpler steps in their choices, including using energy-efficient models, responsible training decisions, and backing regulatory systems that punish non-sustainable practices (Gomes, 2010). And institutions must create incentives and accountability mechanisms that reward environmentally responsible innovation.

Climate change is a significant threat, and AI is rapidly entering practice across all areas. This causes us to think in new modes in how we think about and view technological development (Hilty *et. al.*, 2014). AI has the potential to either exacerbate our tenuous relationship with the physical environment or act as a partner to sustainability. New examples of AI, ranging from AI-enabled smart grids, modelling climate systems, to developing carbon-aware computing frameworks, illustrate how AI can be developed well and with forethought to address the very challenges it generates. It will be necessary to have collaborative, interdisciplinary, and global consensus around policies related to AI with a sustainability focus, and including considerations of inclusivity and innovation. It will be impossible to truly conceive sustainability in our definition of technology, and negatively allocate technology on future uses that promote the Economy and scientific advancement, but also the future's health for our planet.

Call for Sustainable Innovation in AI

AI for the future must be sustainable by design. The green principles must be embedded at each step of the AI lifecycle - education or research, systems design, systems deployment, and systems regulation. As AI systems continue to develop and become more of an integral part of nearly all industries, we also need to evolve as environmental stewards. Green AI cannot be a sideline or afterthought; it needs to be the default, not an exception when we develop intelligent systems.

To bring this vision to life will take partnerships, openness, and collaboration across disciplines. It will involve not just partnership and collaboration among computer scientists, environmental scientists, policymakers, and industry representatives, but the awareness of the need to synchronize technological innovation (either through experimental research or commercial applications) with action on climate goals. We want to support social and planetary progress in a way that maximizes computational effectiveness, while also factoring in a more holistic evaluation of ecological impacts, so that AI not only progresses digital frontiers but also contributes to planetary health and the health of future generations.

REFERENCES

Abir, A. Z. M. N. (2024, April). Industry 4.0 technologies for a data-driven, secure, green, and circular manufacturing economy. In *2024 IEEE Green Technologies Conference (GreenTech)* (pp. 26-31). IEEE. DOI: 10.1109/GreenTech58819.2024.10520440

Anandakumar, H., Arulmurugan, R., & Onn, C. C. (2019). *Computational intelligence and sustainable systems*. EAI/Springer Innovations in Communication and Computing. DOI: 10.1007/978-3-030-02674-5

Anthony, L. F. W., Kanding, B., & Selvan, R. (2020). Carbontracker: Tracking and predicting the carbon footprint of training deep learning models. arXiv preprint arXiv:2007.03051.

Beloglazov, A., Buyya, R., Lee, Y. C., & Zomaya, A. (2011). A taxonomy and survey of energy-efficient data centers and cloud computing systems. *Advances in Computers*, *82*, 47–111. DOI: 10.1016/B978-0-12-385512-1.00003-7

Bender, E. M., Gebru, T., McMillan-Major, A., & Shmitchell, S. (2021, March). On the dangers of stochastic parrots: Can language models be too big? In *Proceedings of the 2021 ACM conference on fairness, accountability, and transparency* (pp. 610-623). DOI: 10.1145/3442188.3445922

Bhattacharya, S., & Lane, N. D. (2016, November). Sparsification and separation of deep learning layers for constrained resource inference on wearables. In *Proceedings of the 14th ACM Conference on Embedded Network Sensor Systems CD-ROM* (pp. 176-189). DOI: 10.1145/2994551.2994564

Buyya, R., Beloglazov, A., & Abawajy, J. (2010). Energy-efficient management of data center resources for cloud computing: a vision, architectural elements, and open challenges. arXiv preprint arXiv:1006.0308.

Cai, H., Zhu, L., & Han, S. (2018). ProxylessNAS: Direct neural architecture search on target task and hardware. arXiv preprint arXiv:1812.00332.

Castaño, J., Martínez-Fernández, S., Franch, X., & Bogner, J. (2023, October). Exploring the carbon footprint of hugging face's ml models: A repository mining study. In *2023 ACM/IEEE International Symposium on Empirical Software Engineering and Measurement (ESEM)* (pp. 1-12). IEEE. DOI: 10.1109/ESEM56168.2023.10304801

Cebesoy, Ü. B., & Önger, S. (2025). How Artificial Intelligence (AI) Tools Can Be Used to Support Climate Change Education. Intelligent Solutions to Evaluate Climate Change Impacts, 281.

Chanussot, L., Das, A., Goyal, S., Lavril, T., Shuaibi, M., Riviere, M., Tran, K., Heras-Domingo, J., Ho, C., Hu, W., Palizhati, A., Sriram, A., Wood, B., Yoon, J., Parikh, D., Zitnick, C. L., & Ulissi, Z. (2021). Open catalyst 2020 (OC20) dataset and community challenges. *ACS Catalysis, 11*(10), 6059–6072. DOI: 10.1021/acscatal.0c04525

Chlingaryan, A., Sukkarieh, S., & Whelan, B. (2018). Machine learning approaches for crop yield prediction and nitrogen status estimation in precision agriculture: A review. *Computers and Electronics in Agriculture, 151*, 61–69. DOI: 10.1016/j.compag.2018.05.012

Chou, D. C., & Chou, A. Y. (2012). Awareness of Green IT and its value model. *Computer Standards & Interfaces, 34*(5), 447–451. DOI: 10.1016/j.csi.2012.03.001

Cui, Y., Huang, C. J., Zhang, Y., Wang, L., Fan, L., Zhou, X., & Yang, Q. (2024). A survey on contribution evaluation in vertical federated learning. arXiv preprint arXiv:2405.02364.

Deif, A. M. (2011). A system model for green manufacturing. *Journal of Cleaner Production, 19*(14), 1553–1559. DOI: 10.1016/j.jclepro.2011.05.022

Devarakota, P., Tsesmetzis, N., Alpak, F. O., Gala, A., & Hohl, D. (2025). AI and the Net-Zero Journey: Energy Demand, Emissions, and the Potential for Transition. arXiv preprint arXiv:2507.10750.

Dietrich, K., Dumont, M. J., Del Rio, L. F., & Orsat, V. (2017). Producing PHAs in the bioeconomy—Towards a sustainable bioplastic. *Sustainable Production and Consumption, 9*, 58–70. DOI: 10.1016/j.spc.2016.09.001

Elsken, T., Metzen, J. H., & Hutter, F. (2019). Neural architecture search: A survey. *Journal of Machine Learning Research, 20*(55), 1–21.

Farahani, R., Azimi, Z., Timmerer, C., & Prodan, R. (2024). Towards ai-assisted sustainable adaptive video streaming systems: Tutorial and survey. arXiv preprint arXiv:2406.02302.

Faucheux, S., & Nicolaï, I. (2011). IT for green and green IT: A proposed typology of eco-innovation. *Ecological Economics, 70*(11), 2020–2027. DOI: 10.1016/j.ecolecon.2011.05.019

Faure, S. (2024). Innovative financing strategies for involving the private sector in the mainstreaming of Nature-Based Solutions-A case study in Barcelona (Master's thesis).

Ferdaus, M. M., Abdelguerfi, M., Ioup, E., Niles, K. N., Pathak, K., & Sloan, S. (2024). Towards trustworthy ai: A review of ethical and robust large language models. arXiv preprint arXiv:2407.13934.

Ghorfi, T., Laraqui, S., & Nachit, H. (2025). AI for energy management: Driving efficiency and sustainability in the MENA region. In *AI in the Middle East for Growth and Business: A Transformative Force* (pp. 145–175). Springer Nature Switzerland. DOI: 10.1007/978-3-031-75589-7_10

Gomes, C. P. (2010, February). Computational sustainability: Computational methods for a sustainable environment, economy, and society. In Frontiers of Engineering: Reports on Leading-Edge Engineering from the 2009 Symposium (pp. 27-40). Washington, DC: National Academies Press.

Gonczarowski, Y. A., & Weinberg, S. M. (2021). The sample complexity of up-to-ε multi-dimensional revenue maximization. *Journal of the Association for Computing Machinery*, *68*(3), 1–28. DOI: 10.1145/3439722

Görücü, S., Morais, L. A., & Panagiotidou, G. (2025, April). A Critical Analysis of Machine Learning Eco-feedback Tools through the Lens of Sustainable HCI. In *Proceedings of the 2025 CHI Conference on Human Factors in Computing Systems* (pp. 1-18).

Greenberg, S., Mills, E., Tschudi, B., Rumsey, P., & Myatt, B. (2006). Best practices for data centers: Lessons learned from benchmarking 22 data centers. Proceedings of the ACEEE summer study on energy efficiency in buildings in Asilomar, CA. ACEEE, August, 3, 76-87.

Han, S., Pool, J., Tran, J., & Dally, W. (2015). Learning both weights and connections for efficient neural network. *Advances in Neural Information Processing Systems*, 28.

Henderson, P., Hu, J., Romoff, J., Brunskill, E., Jurafsky, D., & Pineau, J. (2020). Towards the systematic reporting of the energy and carbon footprints of machine learning. *Journal of Machine Learning Research*, *21*(248), 1–43.

Hennessy, J. L., & Patterson, D. A. (2011). *Computer architecture: a quantitative approach*. Elsevier.

Hilty, L. M., & Aebischer, B. (2014). ICT for sustainability: An emerging research field. ICT innovations for Sustainability, 3-36.

Hinton, G., Vinyals, O., & Dean, J. (2015). Distilling the knowledge in a neural network. arXiv preprint arXiv:1503.02531.

Hoffmann, J., Borgeaud, S., Mensch, A., Buchatskaya, E., Cai, T., Rutherford, E., . . . Sifre, L. (2022). Training compute-optimal large language models. arXiv preprint arXiv:2203.15556.

Howard, J., & Gugger, S. (2020). Fastai: A layered API for deep learning. *Information (Basel), 11*(2), 108. DOI: 10.3390/info11020108

Hupkes, D., Giulianelli, M., Dankers, V., Artetxe, M., Elazar, Y., Pimentel, T., Christodoulopoulos, C., Lasri, K., Saphra, N., Sinclair, A., Ulmer, D., Schottmann, F., Batsuren, K., Sun, K., Sinha, K., Khalatbari, L., Ryskina, M., Frieske, R., Cotterell, R., & Jin, Z. (2023). A taxonomy and review of generalization research in NLP. *Nature Machine Intelligence, 5*(10), 1161–1174. DOI: 10.1038/s42256-023-00729-y

Islam, R. (2024). Sustainable Entrepreneurship in the Digital Age: The Role of AI in Green Business Practices. Academic Press.

Jouppi, N. P., Young, C., Patil, N., Patterson, D., Agrawal, G., Bajwa, R., & Yoon, D. H. (2017, June). In-datacenter performance analysis of a tensor processing unit. In *Proceedings of the 44th annual international symposium on computer architecture* (pp. 1-12). DOI: 10.1145/3079856.3080246

Jumper, J., Evans, R., Pritzel, A., Green, T., Figurnov, M., Ronneberger, O., ... Hassabis, D. (2021). Highly accurate protein structure prediction with AlphaFold. *Nature, 596*(7873), 583-589.

Kaack, L. H., Donti, P. L., Strubell, E., Kamiya, G., Creutzig, F., & Rolnick, D. (2022). Aligning artificial intelligence with climate change mitigation. *Nature Climate Change, 12*(6), 518–527. DOI: 10.1038/s41558-022-01377-7

Koomey, J. G. (2008). Worldwide electricity used in data centers. *Environmental Research Letters, 3*(3), 034008. DOI: 10.1088/1748-9326/3/3/034008

Lacoste, A., Luccioni, A., Schmidt, V., & Dandres, T. (2019). Quantifying the carbon emissions of machine learning. arXiv preprint arXiv:1910.09700.

Languric, M., & Zaki, L. (2022). Migrating monolithic system to domain-driven microservices: Developing a generalized migration strategy for an architecture built on microservices. Academic Press.

Lannelongue, L., Grealey, J., & Inouye, M. (2021). Green algorithms: Quantifying the carbon footprint of computation. *Advancement of Science, 8*(12), 2100707. DOI: 10.1002/advs.202100707 PMID: 34194954

Lee, B. C., Brooks, D., van Benthem, A., Elgamal, M., Gupta, U., Hills, G., Liu, V., Phan, L. T. X., Pierce, B., Stewart, C., Strubell, E., Wei, G.-Y., Wierman, A., Yao, Y., & Yu, M. (2025). A view of the sustainable computing landscape. *Patterns (New York, N.Y.)*, *6*(7), 101296. DOI: 10.1016/j.patter.2025.101296 PMID: 40926963

Li, Y. (2024). Advanced intelligent optimization algorithms for multi-objective optimal power flow in future power systems: A review. arXiv preprint arXiv:2404.09203.

Lim, K., Chang, J., Mudge, T., Ranganathan, P., Reinhardt, S. K., & Wenisch, T. F. (2009). Disaggregated memory for expansion and sharing in blade servers. *ACM SIGARCH Computer Architecture News, 37*(3), 267-278.

Lin, C., & Shahrad, M. (2024). Bridging the sustainability gap in serverless through observability and carbon-aware pricing. *ACM SIGENERGY Energy Informatics Review*, *4*(5), 120–126. DOI: 10.1145/3727200.3727218

Luccioni, A. S., Viguier, S., & Ligozat, A. L. (2023). Estimating the carbon footprint of bloom, a 176b parameter language model. *Journal of Machine Learning Research*, *24*(253), 1–15.

Malmodin, J., & Lundén, D. (2018). The energy and carbon footprint of the global ICT and E&M sectors 2010–2015. *Sustainability (Basel)*, *10*(9), 3027. DOI: 10.3390/su10093027

Mann, B., Ryder, N., Subbiah, M., Kaplan, J., Dhariwal, P., Neelakantan, A., . . . Agarwal, S. (2020). Language models are few-shot learners. arXiv preprint arXiv:2005.14165, 1(3), 3.

Martiny, A. (2023). Towards sustainable ai: Monitoring and analysis of carbon emissions in machine learning algorithms (Doctoral dissertation, Politecnico di Torino).

Mittelstadt, B. (2019). Principles alone cannot guarantee ethical AI. *Nature Machine Intelligence*, *1*(11), 501–507. DOI: 10.1038/s42256-019-0114-4

Molchanov, P., Tyree, S., Karras, T., Aila, T., & Kautz, J. (2016). Pruning convolutional neural networks for resource efficient inference. arXiv preprint arXiv:1611.06440.

Murugesan, S. (2008). Harnessing green IT: Principles and practices. *IT Professional*, *10*(1), 24–33. DOI: 10.1109/MITP.2008.10

O'Brien, J. (2025). AI's Energy Appetite: Lessons from Leading Hyperscalers on Sustainable Procurement (Doctoral dissertation).

Owino, A. (2023). Challenges of computer vision adoption in the Kenyan agricultural sector and how to solve them: A general perspective. *Advances in Agriculture*, *2023*, 1530629. DOI: 10.1155/2023/1530629

Pathak, J., Wikner, A., Fussell, R., Chandra, S., Hunt, B. R., Girvan, M., & Ott, E. (2018). Hybrid forecasting of chaotic processes: Using machine learning in conjunction with a knowledge-based model. *Chaos (Woodbury, N.Y.)*, *28*(4), 041101. DOI: 10.1063/1.5028373 PMID: 31906641

Patterson, D., Gonzalez, J., Le, Q., Liang, C., Munguia, L. M., Rothchild, D., . . . Dean, J. (2021). Carbon emissions and large neural network training. arXiv preprint arXiv:2104.10350.

Pedrycz, W. (2020). Welcome to the exciting world of "green computing and smart environments". *Journal of Smart Environments and Green Computing*, *1*(1), 1–2. DOI: 10.20517/jsegc.2020.01

Penzenstadler, B., Raturi, A., Richardson, D., Calero, C., Femmer, H., & Franch, X. (2014, May). Systematic mapping study on software engineering for sustainability (SE4S). In *Proceedings of the 18th international conference on evaluation and assessment in software engineering* (pp. 1-14). DOI: 10.1145/2601248.2601256

PricewaterhouseCoopers (PwC) US. (2025). *2025 AI Business Predictions*. PwC US Tech & Analytics.

Radovanović, A., Koningstein, R., Schneider, I., Chen, B., Duarte, A., Roy, B., Xiao, D., Haridasan, M., Hung, P., Care, N., Talukdar, S., Mullen, E., Smith, K., Cottman, M. E., & Cirne, W. (2022). Carbon-aware computing for datacenters. *IEEE Transactions on Power Systems*, *38*(2), 1270–1280. DOI: 10.1109/TPWRS.2022.3173250

Rasp, S., Dueben, P. D., Scher, S., Weyn, J. A., Mouatadid, S., & Thuerey, N. (2020). WeatherBench: a benchmark data set for data-driven weather forecasting. Journal of Advances in Modeling Earth Systems, 12(11), e2020MS002203.

Raz, D., Bintz, C., Guetler, V., Tam, A., Katell, M., Dailey, D., . . . Young, M. (2021, July). Face Mis-ID: An interactive pedagogical tool demonstrating disparate accuracy rates in facial recognition. In Proceedings of the 2021 AAAI/ACM Conference on AI, Ethics, and Society (pp. 895-904).

Riabova, J. (2024, October). in the Era of Digital Transformation. In Subject-Oriented Business Process Management. *Models for Designing Digital Transformations: 15th International Conference, S-BPM ONE 2024, Weiden, Germany, May 21–22, 2024 Proceedings*, *2206*, 189.

Rodrigues, T. A., Patrikar, J., Oliveira, N. L., Matthews, H. S., Scherer, S., & Samaras, C. (2022). Drone flight data reveal energy and greenhouse gas emissions savings for very small package delivery. *Patterns (New York, N.Y.)*, *3*(8), 100569. DOI: 10.1016/j.patter.2022.100569 PMID: 36033593

Sandalow, D. B., McCormick, C., Kucukelbir, A., Friedmann, J., Nachmany, M., Lee, H., ... Ruane, A. C. (2024). Artificial Intelligence for climate change mitigation roadmap. Academic Press.

Sanh, V., Debut, L., Chaumond, J., & Wolf, T. (2019). DistilBERT, a distilled version of BERT: smaller, faster, cheaper and lighter. arXiv preprint arXiv:1910.01108.

Schwartz, R., Dodge, J., Smith, N. A., & Etzioni, O. (2020). Green ai. *Communications of the ACM, 63*(12), 54–63. DOI: 10.1145/3381831

Sheldon, J., Zhu, W., Abdullah, A., Bhupathiraju, S. H. V., Sugawara, T., Butler, K. R., & Rampazzi, S. (2024, May). Aquasonic: Acoustic manipulation of underwater data center operations and resource management. In *2024 IEEE Symposium on Security and Privacy (SP)* (pp. 331-349). IEEE. DOI: 10.1109/SP54263.2024.00201

Souza, A., Jasoria, S., Chakrabarty, B., Bridgwater, A., Lundberg, A., Skogh, F., & Shenoy, P. (2023, October). Casper: Carbon-aware scheduling and provisioning for distributed web services. In *Proceedings of the 14th International Green and Sustainable Computing Conference* (pp. 67-73). DOI: 10.1145/3634769.3634812

Strubell, E., Ganesh, A., & McCallum, A. (2020, April). Energy and policy considerations for modern deep learning research. *Proceedings of the AAAI Conference on Artificial Intelligence, 34*(09), 13693–13696. DOI: 10.1609/aaai.v34i09.7123

Sun, N., Miao, Y., Jiang, H., Ding, M., & Zhang, J. (2024). From principles to practice: A deep dive into AI ethics and regulations. arXiv preprint arXiv:2412.04683.

Sun, Y., Wan, L., Mangla, S. K., Xu, X., & Song, M. (2024). Uncovering the interactions between the enterprise AI transformation, supply chain concentration, and corporate risk-taking capacity. *IEEE Transactions on Engineering Management, 71*, 11315–11327. DOI: 10.1109/TEM.2024.3411631

Tan, M., & Le, Q. (2019, May). Efficientnet: Rethinking model scaling for convolutional neural networks. In International conference on machine learning (pp. 6105-6114). PMLR.

Tschand, A., Rajan, A. T. R., Idgunji, S., Ghosh, A., Holleman, J., Kiraly, C., & Reddi, V. J. (2025, March). MLPerf Power: Benchmarking the Energy Efficiency of Machine Learning Systems from μWatts to MWatts for Sustainable AI. In *2025 IEEE International Symposium on High Performance Computer Architecture (HPCA)* (pp. 1201-1216). IEEE. DOI: 10.1109/HPCA61900.2025.00092

VanGeet, O., Lintner, W., & Tschudi, B. (2011). *FEMP best practices guide for energy-efficient data center design*. National Renewable Energy Laboratory.

Vinuesa, R., Azizpour, H., Leite, I., Balaam, M., Dignum, V., Domisch, S., Felländer, A., Langhans, S. D., Tegmark, M., & Fuso Nerini, F. (2020). The role of artificial intelligence in achieving the Sustainable Development Goals. *Nature Communications, 11*(1), 233. DOI: 10.1038/s41467-019-14108-y PMID: 31932590

Wall, P. J., Saxena, D., & Brown, S. (2021). Artificial intelligence in the Global South (AI4D): Potential and risks. arXiv preprint arXiv:2108.10093.

World Economic Forum. (2025). *Advancing Responsible AI Innovation: A Playbook.* World Economic Forum in collaboration with Accenture.

Wu, H., Wang, X., & Fan, Z. (2025). Addressing the sustainable AI trilemma: a case study on LLM agents and RAG. arXiv preprint arXiv:2501.08262.

Yarally, T., Cruz, L., Feitosa, D., Sallou, J., & Van Deursen, A. (2023). Uncovering energy-efficient practices in deep learning training: Preliminary steps towards green ai. arXiv preprint arXiv:2303.13972. DOI: 10.1109/CAIN58948.2023.00012

Zanella, A., Bui, N., Castellani, A., Vangelista, L., & Zorzi, M. (2014). Internet of things for smart cities. *IEEE Internet of Things Journal, 1*(1), 22–32. DOI: 10.1109/JIOT.2014.2306328

Zewe, A. (2025). Explained: Generative AI's Environmental Impact. *MIT News.*

Zhang, W., Chen, L., Luo, J., & Liu, J. (2022). A two-stage container management in the cloud for optimizing the load balancing and migration cost. *Future Generation Computer Systems, 135*, 303–314. DOI: 10.1016/j.future.2022.05.002

KEY TERMS AND DEFINITIONS

Carbon-Aware Scheduling: Assigning AI workloads during periods when renewable energy availability is high.

ProxylessNAS: Neural Architecture Search algorithm that shares weights across architectures to reduce computation.

PUE (Power Usage Effectiveness): Ratio indicating data center energy efficiency (ideal = 1.0).

Quantization: Reducing numerical precision to make models smaller and more energy-efficient.

TinyML: Machine learning that runs on ultra-low-power microcontrollers at the edge.

Chapter 2
Green AI for Smart Buildings:
Leveraging Machine Learning for Energy Efficiency in Urban Environments

Muhammad Usman Tariq
https://orcid.org/0000-0002-7605-3040
Abu Dhabi University, UAE & University College Cork, Ireland

ABSTRACT

As urbanization accelerates and vitality requests rise, smart buildings are becoming central to achieving sustainable development objectives. This chapter investigates the role of Green AI energy-efficient machine learning models in optimizing energy efficiency in urban situations. By integrating AI into existing building frameworks, energy consumption can be continuously observed, anticipated, and balanced in real-time, leading to significant decreases in operational emissions and costs. They consider highlights of various machine learning procedures, including supervised, unsupervised, and reinforcement learning, applied in heating, lighting, and occupancy-based control systems. Through real-world case studies from North America, Europe, and Asia, the chapter illustrates how AI-enabled retrofits and advanced administration frameworks can provide quantifiable energy savings and carbon reductions. Accentuation is set on moral contemplations, including information security, value, and showing straightforwardness.

DOI: 10.4018/979-8-3373-5540-5.ch002

INTRODUCTION

Cities are quickly becoming epicenters of both opportunities and challenges in the global effort to address climate change. As urbanization accelerates, more than half of the world's population presently dwells in cities, a trend that continues to develop each year. This statistical move is followed by an increase in energy demand, particularly within the built environment. Unchecked, the energy-intensive nature of cities hinders worldwide endeavors to restrain global warming to 1.5Â°C above pre-industrial levels, as sketched out within the Paris Agreement.

Another basic zone where AI is changing building execution is in fault location and diagnostics. In expansive buildings, the recognition of energy waste or system failures can be complex and time-consuming. AI calculations can continuously screen framework execution and identify irregularities that indicate hardware failures, inefficient aspects, or support needs. These bits of knowledge can trigger alarms or initiate robotized remedial activities, such as altering ventilation rates or rebalancing warm loads. By detecting issues early, AI-driven diagnostics help anticipate energy loss, extend hardware life, and decrease maintenance costs. Moreover, AI improves tenant engagement and behavior-driven vitality proficiency (Bajwa et al., 2024). Through user-friendly interfaces and personalized criticism, AI frameworks can advise building tenants on their energy consumption and recommend ways to decrease it. A few frameworks have advanced by learning individual preferences and adapting to natural conditions. In commercial and regulatory settings, where buildings may be part of a larger campus or portfolio, AI empowers centralized energy management across different locations. Office supervisors can utilize AI-powered platforms to compare execution across buildings, identify exceptions, and actualize best practices on a large scale. These instruments enable granular vitality following, benchmarking, and continuous improvement, transforming building administration into data-driven education. In addition, AI can encourage integration with city-wide keen network activities, where buildings act as intuitive hubs in a smartly arranged energy network. This integration is fundamental for creating an energy-positive locale and contributing to the decarbonization of urban areas.

Despite its potential, the use of AI in building vitality optimization is not without challenges. Information security and cybersecurity are major concerns, particularly as more frameworks become interconnected. Furthermore, retrofitting existing buildings with advanced technologies can be costly and complex. Moreover, there are issues related to information interoperability, where gadgets and frameworks from different producers must be able to communicate successfully (Shahrabani & Apanaviciene, 2025). Addressing these challenges requires not only innovative development but also a steady approach, systems, industry collaboration, and skilled labor. Moreover, the viability of AI in optimizing building vitality execution depends on the quality and accessibility of information. Conflicting or deficient information can lead to incorrect expectations and imperfect choices. To fully realize the benefits of AI, building proprietors and administrators must contribute to a robust information foundation, including sensors, storage, and training capabilities. The standardization of information groups and conventions can encourage integration and versatility over distinctive types of buildings and geographies.

In the broader context of the clean vitality movement, AI-fueled smart buildings represent a convergence of advanced technologies and energy efficiency. They offer a pathway to achieve profound vitality proficiency without compromising comfort or usefulness. By making buildings more responsive, self-regulating, and resource-efficient, AI contributes to diminishing carbon emissions at a smaller scale, while supporting network stability and renewable energy selection at a large scale. This double advantage underscores the fundamental role of smart buildings in the global climate change response.

Ultimately, the investigation of how AI can optimize building-level vitality performance is not just a specialized endeavor but could be a key component. This reflects a move toward being more innovative and cost-effective (Fakhabi et al., 2024). Urban living that balances financial growth with natural stewardship is essential. As innovation develops and obstructions are overcome, the integration of AI into building frameworks will recommend energy-smart cities, providing a versatile solution to one of the most pressing problems of our time. Through investigation, development, and collaboration, we can tackle the control of manufactured insights to convert our buildings from detached energy consumers into dynamic members of the clean energy transformation.

BACKGROUND AND THEORETICAL FOUNDATIONS

Unlike conventional buildings, keen buildings are versatile and react powerfully to changes in inhabitance, natural conditions, or client behavior. Green buildings, although related, focus more on ecologically capable and resource-efficient planning and development throughout a building's life cycle. This incorporates viewpoints such as maintainable materials, energy-efficient frameworks, water preservation, and advanced indoor air quality (Saleh et al., 2024). Whereas green buildings aim to reduce the environmental impact of the built environment, savvy buildings take it a step further by utilizing advanced innovations to continuously monitor and improve building performance. Vitality proficiency, in this context, refers to the objective of utilizing less vitality to perform the same work or accomplish the same level of comfort and utility. This can be accomplished by minimizing energy waste through optimized framework operation, superior cover, effective apparatuses, and the progressive use of artificial intelligence.

Artificial intelligence (AI) and machine learning (ML) play fundamental roles in enhancing the execution of innovative and green buildings. AI is a field of computer science that enables machines to mimic human insights, including learning from experience, understanding complex information structures and making informed decisions. ML, a subset of AI, includes the improvement of calculations that can be learned from and used to make predictions based on information. In building vitality analytics, these innovations are used to convert how buildings are checked, overseen, and optimized. Distinctive types of AI and ML models are connected depending on the energy management objectives. For illustration, directed learning models such as straight regression, support vector machines, and fake neural systems are commonly used for energy consumption forecasting (Feng et al., 2024). These models can analyze authentic vitality information in conjunction with natural and operational factors to predict future energy demand with high accuracy. Unsupervised learning strategies, such as clustering and principal component analysis, are used to identify utilization patterns, peculiarities, and wasteful aspects without predefined names. Reinforcement learning, which imitates the trial-and-error learning approach of people, is gaining popularity for real-time Heating, Ventilation, and Air Conditioning (HVAC) control and request reaction, where the system learns optimal actions through interaction with the environment.

One of the most broadly utilized AI models in building energy analytics is the artificial neural network (ANN), which is motivated by the structure of the human brain. ANNs are especially viable for capturing nonlinear connections among complex factors, making them ideal for anticipating energy consumption based on different energy inputs, such as climate, population, and building utilization. Another note worthy method is the choice tree, which is esteemed for its interpretability and capacity to show conditional relationships within the data (Stern, 2024). Outfit strategies, such as irregular woodlands and angle-boosting machines, have also demonstrated compelling results, especially when managing loud or deficient information. Profound learning models, particularly convolutional and recurrent neural systems, are increasingly used for high-dimensional information analysis, such as visual information from cameras or time-series information from sensors. In addition to prescient modeling, AI is also utilized for classification assignments, such as distinguishing between diverse types of building inhabitance or utilization profiles, which encourage energy optimization methodologies (Tariq, 2025).

The integration of AI and ML into building frameworks requires a foundational information foundation that empowers persistent information collection, handling and criticism (Abbass et al., 2025). This includes the use of Internet of Things (IoT) gadgets, which incorporate sensors, meters, actuators, and communication technologies. These gadgets produce real-time information streams related to temperature, stickiness, CO_2 levels, light intensity and power utilization. The crude information is then handled using edge computing or cloud-based platforms, where AI algorithms analyze and translate it to generate valuable insights. For instance, if an ML model identifies that a building's HVAC system is consistently overperforming during certain hours, it can adjust the system's operation to match the actual demand, thereby reducing energy waste. Furthermore, AI can help plan maintenance based on utilization patterns and operational data, which minimizes downtime and extends the lifespan of building systems.

The application of AI in smart buildings is not just a specialized accomplishment but also forms part of a broader system that integrates advanced technology with urban sustainability objectives. Urban maintainability refers to the improvement of cities that are naturally sound, financially reasonable, and socially comprehensive. Accomplishing this vision requires significant reductions in energy consumption and greenhouse gas emissions, advancements in public health and safety, and the establishment of a strong framework. AI contributes to these objectives by empowering the more effective utilization of assets, reducing operational costs, and supporting the transition to renewable energy sources. For example, AI can optimize the utilization of solar control in buildings by anticipating solar energy designs and adjusting utilization in a similar manner. Moreover, it can oversee energy capacity frameworks to guarantee that energy is available when it is most required, thus lessening reliance on fossil-fuel-based network control (Bajwa et al., 2024). This capability is especially vital as cities aim to coordinate more renewable energy into their energy mix while maintaining the grid stability.

Another key aspect of the AI-urban maintainability system is demand-side administration. Traditionally, vitality frameworks have focused on expanding supply to meet rising demand. Be that as it may, AI empowers a move toward overseeing requests from more scholarly people. Through real-time checking and control, AI frameworks can decrease crest loads, adjust energy consumption over periods, and even participate in network services, such as request-response. In request reaction programs, buildings deliberately decrease or move their energy consumption during peak periods in exchange for incentives, allowing utilities to avoid the need for additional energy capacity to meet demand. AI makes these programs more compelling by distinguishing ideal times to reduce utilization without compromising comfort or efficiency. In addition to their vitality and natural benefits, AI-powered smart buildings offer social and financial advantages. On the social front, strides have been made in discussing how indoor quality, warm consolation, and lighting conditions contribute to better well-being for inhabitants (Hanafi et al., 2024). AI frameworks can guarantee that buildings maintain ideal conditions custom-made to the needs of a person 'or groups, which is particularly critical in settings such at schools, hospitals, and workplaces. From a financial viewpoint, vitality investment funds translate into lower utility bills, decrease operational costs, and improved property value.

Numerous buildings still operate with bequest frameworks that are not easily coordinated with advanced advances. Updating these frameworks requires significant speculation in terms of equipment and human capital. Moreover, there is a developing concern about information protection and cybersecurity, particularly as more individual and operational information is collected and transmitted over systems. Ensuring that AI applications are secure and morally planned is fundamental to building trust and gaining widespread acceptance. Another issue is the need for talented experts who understand both the specialized aspects of AI and the specific needs of the building frameworks. This intriguing skill is fundamental for planning, executing, and maintaining AI-driven arrangements that are both factually sound and aligned with broader sustainability objectives. Furthermore, policymakers and regulators must create an enabling environment that fosters development while ensuring open interfaces. This includes incentives for energy-efficient updates, support for research and development, and guidelines that guarantee transparency and accountability in the use of AI applications.

In conclusion, integrating AI and ML into innovative building frameworks provides an effective means of advancing urban sustainability. By upgrading vitality productivity, empowering prescient upkeep, optimizing renewable vitality utilization, and supporting better decision-making, AI contributes to the creation of more intelligent, greener, and stronger cities (Bajwa et al., 2024). These benefits are not fair, specialized, or economic; they also address a few of the most pressing natural and social challenges confronting urban regions today. As cities continue to develop and the impacts of climate change become more severe, the ability to harness AI for potential improvements will become increasingly important. The longer term of urban living depends not on building more brilliant cities but on incorporating insights into the existing structures that support our day-to-day lives. Through continued innovation, collaboration, and approach, buildings fueled by AI can play a central role in building a feasible urban future.

Machine Learning Models for Building Energy Efficiency

ML has become a transformative force in the field of building energy efficiency, empowering a level of automation, accuracy, and versatility that conventional building frameworks have never achieved independently. Among the various ML methods, supervised, unsupervised, and reinforcement learning stand out as practical tools for optimizing energy consumption in commercial and private settings (Fakhabi et al., 2024). Each of these learning models serves unmistakable purposes and offers special preferences when connected to complex building frameworks, such as HVAC control, lighting administration, inconsistent location, and flexible decision-making. When carefully conveyed, these models not only reduce energy waste but also enhance tenant comfort, extend equipment lifespan and support broader sustainability goals.

Supervised learning is one of the most commonly used approaches for building vitality analytics, particularly in applications such as HVAC control and lighting optimization. In administered learning, calculations are prepared on labeled datasets, where the input highlights and corresponding outputs are known. This permits the model to memorize a mapping that can be utilized to predict results on concealed information. For illustration, in HVAC frameworks, directed models can be prepared using authentic information that incorporates factors such as exterior temperature, humidity, occupancy levels, and time of day, near the resulting energy consumption or system performance measurements. Once prepared, the show can anticipate the ideal HVAC settings for a given circumstance, decreasing unnecessary heating or cooling requirements. Strategies such as straight relapses, bolster vector machines, decision trees, and fake neural networks are commonly utilized for this purpose. These models enable buildings to respond proactively to anticipated conditions rather than react to established plans or inactive rules.

Moreover, lighting frameworks in buildings benefit significantly from directed learning. By utilizing information on inhabitance, sunshine accessibility, and verifiable lighting designs, models can be prepared to robotize lighting levels in a way that maintains visual comfort while minimizing energy consumption. For example, directed models can predict when certain zones inside a building are regularly involved and alter the lighting levels appropriately, turning off or dimming lights when spaces are vacant or when adequate sunlight is available. In office buildings, sensors can distinguish movement and surrounding light levels, feeding information into a prepared calculation that determines the ideal brightness level in real time. This granular control not only spares vitality but also enhances client engagement by providing consistent and comfortable lighting conditions throughout the day.

Unsupervised learning models are used in scenarios where labeled information is not accessible. These models exceed expectations in finding concealed designs, gathering comparable information points, and recognizing peculiarities within large datasets. In building energy efficiency applications, unsupervised learning is especially profitable for irregularity discovery and utilization clustering tasks. A peculiar location includes recognizing bizarre designs in energy utilization that show hardware breakdowns, energy wastage, or unauthorized use (Bajwa et al., 2024). Methods such as k-means clustering, separation of woodlands, and vital component examination were utilized to set up a pattern of typical working conditions, against which deviations could be measured. For example, if a building's HVAC system suddenly starts using more energy than usual under similar environmental conditions, an unsupervised model can identify this behavior as an anomaly. Office directors can then examine and address potential issues, often before tenants are aware of any problems or significant damage.

Usage clustering is another key application of unsupervised learning techniques. By analyzing information from different sensors, these models can group zones, rooms, or entire buildings with comparable utilization designs. This division makes a difference in helping administrators understand how distinctive spaces are utilized, empowering them to tailor energy-saving procedures appropriately. In large commercial offices, for instance, clustering can distinguish zones with reliably low occupancy that might benefit from balanced heating, cooling, or lighting plans (Fakhabi et al., 2024). Moreover, utilization clustering can be used to create profiles of commonplace energy utilization behaviors, which serve as a reference for benchmarking and execution comparisons. When coordinates are entered into vitality administration stages, these experiences permit building administrators to execute more focused interventions, fine-tune operations, and educate future planning or retrofitting projects.

Reinforcement learning is a more advanced ML paradigm that is especially suited to energetic real-time control frameworks in buildings. In fortification learning, a specialist learns to require activities within an environment to maximize the aggregate compensation flag. Unlike administered learning, which relies on verifiable information, support learning involves persistent interaction with the framework and learning through trial-and-error (Aljohani, 2024). This makes it ideal for versatile vitality frameworks, such as those that control HVAC and vitality capacity based on changing conditions and long-term optimization goals. One of the most promising zones for reinforcement learning in buildings is HVAC scheduling and control. In this setting, the specialist receives inputs such as indoor and outdoor temperatures, time of day, occupancy information, and energy costs, and learns approaches that minimize energy utilization while maintaining comfortable temperatures.

A well-known support learning calculation utilized in building vitality administration is Q-learning, where the operator learns an estimated value that gauges the anticipated rewards for taking action in a given state. More modern approaches, such as profound fortification learning, combine the decision-making control of support learning with the design-acknowledgment capabilities of deep neural systems. These models can handle high-dimensional input information and learn complex control techniques that outperform conventional rule-based frameworks. For instance, in a private building equipped with solar panels and a battery, a solar energy operator can learn when to store energy, draw from the battery, and use grid power to minimize costs and carbon emissions (Satpathy et al., 2025). Over time, the operator adjusts to the building's unique design and adaptability to climate change and continuously enhances its performance.

Real-world cases illustrate how ML models are currently being deployed to achieve significant energy savings and operational improvements in commercial and private settings. In commercial office buildings, directed learning models have been used to anticipate HVAC demand and optimize system operation (Addad & Al-Taani, 2024). A striking illustration is the arrangement of fake neural systems in an office complex in Singapore, where the show was prepared on natural and inhabitant information to control chillers and discuss heating, ventilation, and air conditioning (HVAC). The result was a 20% reduction in HVAC-related energy consumption while maintaining tenant comfort. Another illustration comes from a campus activity in California, where numerous buildings were fitted with IoT sensors and AI-based control frameworks. Directed models were utilized to oversee lighting, whereas reinforcement learning algorithms powerfully balanced the HVAC frameworks based on occupancy figures and power estimation. The combined approach led to considerable reductions in crest request charges and overall energy consumption (Tariq, 2025).

Residential buildings have also seen practical applications of ML, especially in smart homes. Numerous present-day indoor regulators, such as those created by Nest and Ecobee, incorporate ML algorithms to memorize residents' plans and inclinations. These frameworks utilize directed learning to predict the presence or absence of tenants and adjust temperatures accordingly (Mahmood et al., 2024). A few frameworks have coordinated unsupervised learning to recognize exception utilization designs, such as windows being left open while the heating is on. Moreover, fortification learning has been attempted in homes with photovoltaic systems and batteries, where the framework learns the most effective energy management techniques through interaction with the environment. In pilot ventures across Europe, these keen domestic frameworks have shown the potential to reduce yearly energy bills by as much as 25%, while also increasing client satisfaction owing to superior comfort and convenience.

One inventive arrangement was observed in a private microgrid venture in Germany, where different homes were equipped with solar panels, battery storage, and AI-powered energy management systems. Fortification learning calculations were utilized to arrange the energy era, capacity, and utilization over the grid, guaranteeing that each domestic unit optimized its use of renewable energy while contributing to grid stability (Fakhabi et al., 2024). The framework reacts powerfully to climate conditions, family requests, and energy costs, accomplishing both environmental and economic benefits. This decentralized, smartly controlled show is progressively seen as a blueprint for the future of private energy systems, especially in regions aiming for high levels of renewable energy integration.

Despite these promising outcomes, the extensive application of ML in building energy systems still faces several challenges. Information quality and accessibility remain significant obstacles, as successful show preparation depends on vast volumes of accurate, high-resolution information. Numerous buildings, particularly older ones, require a vital sensor framework or have divided frameworks that are not easily integrated. Furthermore, the interpretability of ML models could be a concern, particularly in basic applications such as HVAC control, where security and tenant comfort are fundamental (Tariq, 2025). Although advanced models, such as deep neural networks, offer high performance, their black-box nature makes it challenging for office supervisors to understand and trust their decisions. This has sparked interest in reasonable AI methods that point to form, yielding more straightforward and actionable results. Another impediment is the computational assets required to prepare and deploy advanced ML models. In a few cases, real-time decision-making may be compelled by equipment restrictions, especially in smaller buildings or remote regions with limited infrastructure. In addition, cross-disciplinary collaboration between building engineers, information researchers, and IT experts is required to guarantee effective execution. This incorporates not only specialized advancement but also considerations of the client interface plan, cybersecurity, and continuous framework upkeep. Moreover, motivations and controls play a pivotal role, as numerous building proprietors are reluctant to invest in unused innovations without clear financial or strategic support.

ML is reshaping the scene of building vitality productivity through the application of supervised, unsupervised, and support learning models (Camacho et al., 2024). Each demonstrates a sort that brings preferences to distinctive viewpoints of vitality administration, from prescient HVAC and lighting control to real-time inconsistency discovery and versatile decision making. Case studies in both commercial and private settings emphasize the adequacy of these approaches in reducing vitality consumption, lowering costs, and improving client satisfaction. Although specialized and regulatory challenges remain, the energy behind AI-driven building arrangements continues to develop. As innovative frameworks become more widely available and the severity of climate change increases, ML will play a central role in creating energy-efficient, intelligent, and economical buildings in the long term.

SMART BUILDING ARCHITECTURE AND AI INTEGRATION

The integration of counterfeit insights into savvy building design has revolutionized how buildings function, consume energy, and respond to inhabitant needs. Central to this change are the arrangement of sensors, IoT frameworks, and strong information collection frameworks that permit continuous monitoring and control of building situations. These components serve as tangible organs of smart buildings, capturing real-time information on factors such as temperature, stickiness, light levels, discussion quality, inhabitance, and vitality utilization. These data shape the foundational input for AI calculations for analysis and action. Sensors are deliberately introduced throughout a building, from HVAC frameworks and lighting installations to entryways, windows, and even furniture. They empower the exact following of natural changes and human actions, providing experiences that can be utilized to optimize asset utilization, improve consistency, and decrease operational costs. Figure 1 provides an overview of the AI-Enabled Smart building system with the integration of sensors, IoT networks, edge/cloud processing, and AI-driven Building Management System (BMS) to optimize energy use and occupant comfort.

Figure 1. Overview of AI-Enabled Smart building system

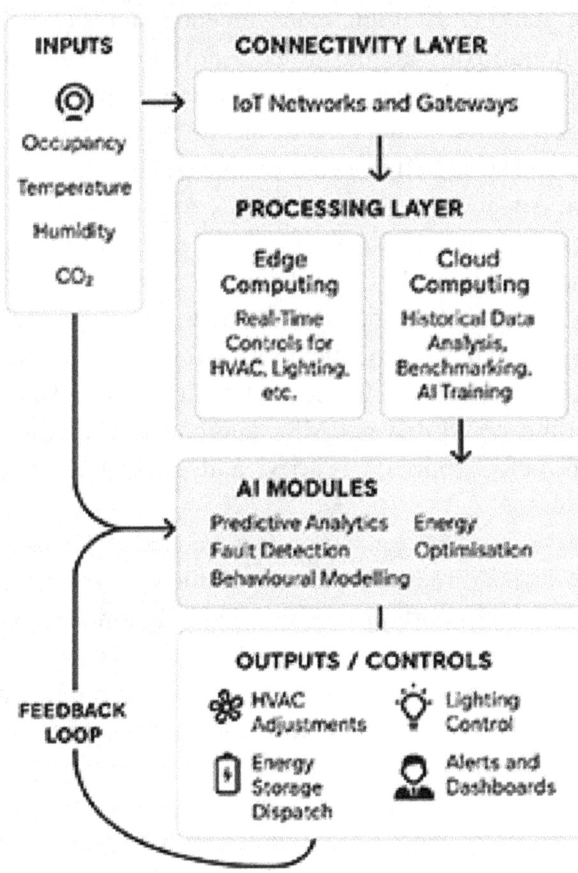

The Internet of Things (IoT) acts as the connective tissue between these sensors and the computerized brain of a building. IoT frameworks guarantee consistent communication between gadgets, encouraging information exchange with centralized handling units or cloud servers (Al Amin Gerary, 2024). They regularly comprise remote communication modules, portals, and conventions that permit information to be streamed productively within the building biological system. Through IoT integration, previously isolated components, such as indoor regulators, lighting frameworks, security cameras, and elevators can now communicate and respond in a coordinated manner. For instance, when an occupancy sensor detects an empty room, the lighting framework can automatically dim or switch off, and the HVAC system can adjust the temperature to a more energy-efficient level. These robotized, intuitive systems not only preserve vitality but also ensure that building operations are conducted with minimal human oversight.

However, the unimportant collection of information is inadequate for driving intelligent decision-making. Building management systems (BMS) upgraded with AI capabilities are mindful of changing raw data into valuable insights. Conventional BMS stages are rule-based frameworks that depend on pre-programmed logic to oversee building operations. Although they are viable to a certain degree, they required adaptability and struggled to adapt to rapid changes in natural environments or human behavior. Advanced AI-powered BMS arrangements, on the other hand, utilize ML and predictive analytics to enhance their execution continuously (Heidari et al., 2024). They can distinguish wasteful aspects, estimate vitality demand, detect faults, and alter building frameworks to attain ideal execution. AI models inserted inside the BMS stages analyze authentic information in conjunction with real-time inputs to reveal designs and make forecasts. For illustration, by analyzing inhabitant patterns over time, an AI-driven BMS can proactively oversee HVAC plans to precondition rooms before individuals arrive and reduce cooling or heating when spaces are underutilized.

These frameworks frequently incorporate advanced client interfaces and dashboards that permit office directors to monitor building performance in real time. Customizable cautions and reports provide detailed insights into energy consumption, system health, and sustainability. Moreover, AI enables the development of flexible control methodologies. Unlike inactive plans or manual alterations, AI can learn from feedback loops to continuously refine control choices (Alijoyo, 2024). For example, suppose a particular conference room tends to overheat during gatherings because of high occupancy. In this case, the framework can be taught to pre-cool the room before occupancy begins. In buildings with complex vitality needs, such as healing and information centers, this level of granular control can make a critical difference in both operational efficiency and realized savings.

The preparation of the tremendous amount of information produced in smart buildings can occur in two essential ways: edge computing and cloud-based analytics. Each approach offers distinct trade-offs in terms of energy consumption, inactivity, data privacy, and computational capability. Edge computing includes handling information locally, near the source of information generation, often inside the building. This strategy reduces the need to transmit large volumes of information over systems to centralized servers, thereby decreasing transfer speed utilization and idleness (Das et al., 2024). The benefits of edge computing are particularly apparent in applications that require real-time reactions, such as fire alerts, security frameworks, and HVAC alterations based on occupancy. Because choices are made on-site, the framework can respond rapidly, which is fundamental in guaranteeing tenant security and consolation. Moreover, edge computing improves information protection by keeping sensitive data inside the local network and reducing exposure to external cybersecurity threats.

However, cloud-based analytics offer more prominent computational control and capacity, empowering the use of more advanced AI models that require critical handling assets. Cloud frameworks can aggregate information from different buildings or areas, providing a broader setting for examination and encouraging benchmarking across different departments. This approach is beneficial for organizations overseeing building portfolios, such as school districts, hotel chains and government agencies. Cloud stages offer persistent upgrades and integration with other advanced apparatuses, promoting versatility and adaptability that edge arrangements may not match (Tunde et al., 2024). Be that as it may, the dependence on information transmission to and from cloud servers can present inactivity, which may not be perfect for time-sensitive operations. Additionally, the vitality utilization related to maintaining cloud frameworks and data centers raises concerns about the environmental impact of this approach.

Balancing edge and cloud computing is the most viable technique for smart buildings. A crossbreed design permits basic real-time capacities to be taken care of at the edge, whereas leveraging the cloud for more profound analytics, long-term capacity, and centralized control (Borah, 2025). AI calculations oversee the interaction between these frameworks and conventional lattice power, deciding when to store energy, when to draw from capacity, and when to feed surplus energy back into the network. For example, an AI framework might anticipate a spike in energy demand during the evening and choose to charge the battery in the morning when solar energy is at its peak (Alnaser et al., 2024). These choices are based on climate, inhabitance, and vitality pricing, demonstrating the smart coordination of numerous factors to attain cost-effective and economical outcomes.

The arrangement of AI-enhanced building frameworks additionally contributes to demand-side energy management, a fundamental element in contemporary urban energy techniques. Instead of expending vitality as required, buildings become dynamic members of the energy biological system. They can move their loads to off-peak hours, reduce utilization during grid congestion periods, and offer stored energy during peak demand periods. AI encourages this by analyzing time-of-use taxes, framework signals, and internal conditions to make automated decisions (Fakhabi et al., 2024). In cities with tall renewable energy infrastructure, such demand-side flexibility is fundamental for stabilizing the system and avoiding power shortages. Savvy buildings, through their engineering and AI integration, contribute not only to personal proficiency but also to broader lattice flexibility and sustainability (Tariq, 2025).

Another layer of insight into clever building engineering is found in an occupant-centric plan. AI-driven frameworks can personalize natural conditions based on a person's inclinations and group behavior. For example, a building's lighting and HVAC frameworks can learn the inclinations of normal inhabitants and alter the settings naturally when those people enter a room. In shared spaces, the framework can utilize consolidated comfort data to discover an adjustment that suits most clients (Husainy et al., 2024). These highlights upgrade tenant fulfillment, which is progressively recognized as a key performance metric alongside energy savings. In office buildings, progressive comfort is linked to higher efficiency and reduced absenteeism, providing additional justification for investments in AI-based infrastructure.

Nevertheless, the use of advanced building design with AI integration requires careful planning and collaboration across various domains. Planners, engineers, IT masters, and office directors must work together to plan frameworks that are not only innovatively advanced but also user friendly, secure, and cost effective. Building information modeling (BIM) devices can facilitate this process by providing an advanced representation of a building that integrates structural, mechanical, and advanced systems (Al-Raeei, 2025). These models bolster lifecycle planning and can be utilized to reenact diverse scenarios, assess framework execution, and refine plan choices before development.

Moreover, as AI and IoT advances have become more widespread, questions regarding moral planning and advanced management have become increasingly important. Information security, algorithmic straightforwardness, and cybersecurity must be considered from the beginning. Tenants should be educated about what information is collected, how it is utilized, and what benefits it offers. Designers and building proprietors must ensure that frameworks are designed with inclusivity, availability, and supportability in mind. Administrative systems and industry benchmarks are advancing to address these concerns; however, proactive engagement from partners remains essential.

In conclusion, the design of intelligent buildings coordinated with AI rethinks how we plan, work, and encounter built situations. Sensors and IoT frameworks provide the raw data required to respond to real-world conditions, whereas AI-enhanced building management frameworks interpret these data to deliver intelligent and adaptable control (Sun et al., 2024). The key utility of edge computing and cloud analytics is to enable a balance between responsiveness, computational control, and information security. Together, these components shape an energetic environment that maximizes vitality productivity, enhances comfort, and contributes to a more sustainable urban future. As cities continue to develop and climate change escalates, the capacity of buildings to self-optimize using AI will be crucial for achieving energy, environmental, and societal objectives. Innovative building design, when planned thoughtfully and executed mindfully, guarantees not only changing how buildings work but also how we live and work inside them (Tariq, 2025).

URBAN-SCALE APPLICATIONS AND DISTRICT-LEVEL OPTIMIZATION

As the global urban population continues to grow and climate change intensifies, cities are seeking innovative ways to decarbonize and optimize their energy systems. Although person-savvy buildings have illustrated the potential for noteworthy vitality proficiency and supportability gains, the following challenge lies in scaling these benefits to the area level through the integration of different buildings into smart energy areas. This shift from building-level optimization to urban-scale vitality coordination could be an essential step in creating climate-resilient cities (Yitmen et al., 2025). Bright vitality locales empower an organized biological system in which buildings not only optimize their energy consumption but also arrange and collaborate with neighboring structures, sharing assets, adjusting loads, and collectively contributing to the flexibility and efficiency of the neighborhood energy grid. District-level optimization begins with the computerized interconnection of buildings through shared communication and control systems. Each savvy building, prepared with sensors, IoT gadgets, and artificial intelligence algorithms, becomes a hub in a broader energy network. These hubs share data on vitality utilization, era from on-site renewables, inhabitance patterns, and natural conditions. Through real-time information trading, an area can make collective choices that are more effective than disconnected building-level activities. For instance, if one building within a site generates excess solar energy while another faces high demand, the excess energy can be diverted or exchanged between them to reduce the reliance on external network control. This interconnectivity encourages peer-to-peer energy sharing and enhances the capacity of the entire area to respond to changing energy needs and conditions. Central to the victory of bright vitality locale is the advancement of shared information foundations. These frameworks provide a secure and versatile method for buildings to exchange data while safeguarding information ownership and protection of the data. Information from HVAC frameworks, lighting, electric vehicles, renewable energy frameworks, and occupancy sensors can be aggregated and standardized across regions to enable cross-building analytics and decision-making. Cloud stages, edge computing gadgets, and computerized twins are frequently utilized to create a virtual environment where building execution can be recreated, compared, and optimized. These frameworks are enhanced through the use of unified learning models, which enable AI algorithms to be trained across various datasets without requiring raw data to be extracted from their source.

Federated learning addresses a key challenge in smart city information administration: how to use the collective insights of numerous sources without compromising information security or overburdening central servers. In conventional ML approaches, information from distinctive buildings is centralized, posing dangers related to information security, ownership, and transmission costs. Combined learning, by differentiation, empowers each building to prepare a nearby demonstration using its information. These nearby models then share, as it were, only their learned parameters, not the information itself, with a central facilitator. The facilitator totals these parameters to oversee a global show, which is then sent back to the person in charge. This recurrent handle enables the global show to benefit from diverse data across the region while maintaining the privacy of each building's data. Such models are especially valuable for estimating vitality demand, recognizing inconsistencies, and planning distributed energy assets in a district. Figure 2 provides an overview of the integration of AI-enabled buildings into a coordinated urban energy network using a shared infrastructure, peer-to-peer energy flows, and federated learning models for resilient and efficient performance.

Figure 2. District-Level Smart Energy System

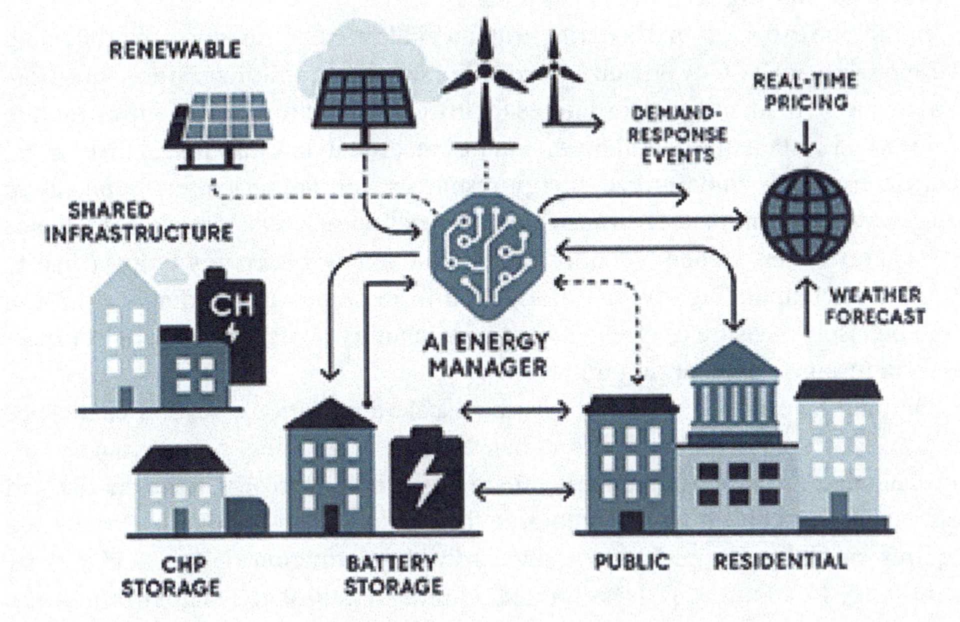

The advancement of smart grids also hinges on the rise of grid-aware building structures that can adjust their operations in accordance with the needs of the broader energy network. Unlike conventional buildings, which inherently consume energy, grid-aware buildings can effectively participate in energy markets through demand-response programs and real-time energy trading (Chowdhury, 2025). Request reaction refers to the capacity of buildings to adjust or reduce their energy utilization during periods of peak demand in response to signals from utility suppliers. AI algorithms enable buildings to identify the most effective techniques for reducing energy consumption without compromising comfort or basic operations. For illustration, amid a heatwave when the lattice request is high, a group of keen buildings might briefly increase the cooling setpoints or move non-essential activities to off-peak hours.

District-level vitality optimization also empowers the utilization of shared energy infrastructure, such as combined heat and power (CHP) plants, local cooling systems, and thermal energy storage. Instead of copying frameworks in each building, these shared resources can be centrally overseen and designated based on real-time requests. AI calculations can estimate crest loads over an area and pre-condition warm capacity or incline early in anticipation of request spikes. In mixed-use areas with private, commercial, and mechanical buildings, the differences in vitality utilization designs can be used to smoothen the generally requested curve. For instance, commercial buildings with top loads during the day can be adjusted against private buildings that use more energy at night, thereby decreasing the requirement for outside peak-shaving control plants (PSCPs).

In addition to operational benefits, brilliant vitality locales support urban planning and approach targets. City organizers and utility suppliers can utilize information from a savvy locale to identify opportunities for efficiency improvements, infrastructure projects and sustainability initiatives. For instance, analytics may reveal that certain zones consistently underutilize energy resources, prompting targeted updates or behavioral interventions. District-wide optimization also encourages compliance with energy codes, carbon reduction targets, and renewable energy orders (Shafik, 2024). By accumulating vitality information from numerous buildings, cities can more precisely measure progress toward sustainability objectives and adjust procedures in response to emerging trends.

Despite the colossal potential of urban-scale applications, several challenges must be addressed to realize their full benefits. Interoperability remains a noteworthy challenge, as buildings regularly utilize different frameworks, benchmarks, and communication conventions. Combining these into a cohesive district-level stage requires specialized coordination, standardized information design, and regular retrofitting. In addition, possession and administration of the shared framework and information must be characterized. Questions regarding who controls access to information, who benefits from vitality reserve funds, and how costs are apportioned must be settled through collaborative administration models. Locking in stakeholders, such as building proprietors, inhabitants, utilities, and metropolitan authorities, is crucial for ensuring a smooth interface and long-term commitment to shared energy goals. Cybersecurity is another basic consideration, especially as more gadgets and frameworks become interconnected and dependent on advanced control. Although the facility is brilliantly located, it must implement robust security measures to safeguard against cyberattacks that could compromise building operations or individual information. This incorporates secure communication conventions, regular computer program overhauls, and inconsistent location frameworks that identify and respond to abnormal activities. Moreover, protection concerns, especially in private settings, must be carefully overseen, with clear arrangements for information collection, utilization, and consent.

CASE STUDIES AND INDUSTRY IMPLEMENTATIONS

AI-enabled retrofitting has been developed as a transformative approach for modernizing older urban structures, permitting cities to reduce energy consumption and emissions without replacing existing buildings. By leveraging sensors, smart meters, and ML algorithms, this framework can be updated to meet modern energy efficiency standards. These frameworks analyze verifiable and real-time information to optimize heating, ventilation, lighting, and equipment utilization, even in buildings that were initially developed without considering energy efficiency (Ajayi et al., 2024). In Toronto, AI has been implemented in open lodging to oversee HVAC loads more successfully, making strides in comfort while reducing the peak demand.

Across Europe, cities such as Amsterdam and Vienna have embraced AI in retrofitting social housing, combining climate forecasting and occupancy analytics to lower climate change costs and carbon emissions. In Asia, Singapore's Green Check program has supported the integration of AI in building operations, leading to quantifiable improvements in maintainability compared with commercial locations. These executions are supported by quantitative affect appraisals, showing decreases of 20%-40% in vitality utilization, with a few keen retrofitted buildings accomplishing near-net-zero operational outflows (Ning, 2024). These results demonstrate that, with the aid of AI, retrofitting is not only a viable approach to achieving climate objectives but also a sensible strategy for enhancing the lifespan and efficiency of existing urban infrastructure while safeguarding social and cultural heritage.

CHALLENGES AND LIMITATIONS

Although the integration of counterfeit insights into keen buildings and vitality frameworks offers transformative potential, a range of challenges and constraints must be addressed to ensure that these advances deliver on their promises economically and impartially (Ge et al., 2024). One of the foremost concerns is information security, security, and interoperability. Keen buildings depend intensely on steady information collection from sensors, Internet of Things (IoT) gadgets, and client interfaces. This information regularly incorporates sensitive data on the behaviors, inclinations, and developments of occupants. In commercial buildings, information may reveal representative work designs or private operational measurements, whereas in private buildings, it may uncover detailed individual preferences. As buildings become more interconnected and information is shared over stages or with utilities and third-party benefit suppliers, the risk of security breaches increases significantly.

Ensuring strong information security in this setting is challenging. AI frameworks require substantial volumes of information to operate successfully, and anonymization strategies are not continuously foolproof. Indeed, when names and coordinate identifiers are removed, behavioral designs can often be re-identified, particularly when they are combined with other datasets. In addition, the foundation of smart buildings, including sensors, actuators, and associated platforms, is vulnerable to cyberattacks. Programmers may exploit vulnerabilities in innovative lighting frameworks and HVAC units or gain access to control systems to disrupt operations, steal information, or compromise security. These dangers are compounded in savvy areas, where different frameworks are interconnected and interdependent, meaning that a breach in one building might influence others over the network.

Interoperability is another specialized challenge. The building industry is divided into diverse frameworks, merchants, and communication guidelines for different types of foundations. Coordinating bequest frameworks with unused AI-driven stages is frequently troublesome, requiring custom middleware, connectors, or total framework updates. Without standardized conventions for information trade, distinctive gadgets may not communicate effectively, leading to wasteful aspects and unreliable performance (Ajayi et al., 2024). This fracture also ruins versatility, as arrangements custom-fitted for one building or merchant environment may not effortlessly be exchanged with another. Overcoming this challenge will require cross-industry collaboration and the appropriation of open benchmarks; however, such an arrangement is frequently moderate and complicated by competitive interests.

Another major limitation of AI in smart buildings is the issue of software float and the long-term support required for AI frameworks. AI calculations are ordinarily prepared using historical data, and although they may perform well at first, the situations in which they work are constantly changing. Inhabitant behaviors advance, climate designs move due to climate change, vitality taxes are overhauled, and building utilization can vary significantly over time. As these changes occur, AI models may become less precise or relevant, a phenomenon known as demonstration float. For illustration, a calculation prepared to optimize HVAC settings based on past inhabitance designs might be ineffective if a company shifts to cross-plant work plans, decreasing or modifying office occupancy throughout the week.

To remain viable, AI frameworks must be ceaselessly observed, overhauled, and retrained with unused data. This requires a vigorous input circle, a framework for real-time information ingestion, and a workforce that understands it, including both building operations and AI improvement. However, numerous building operators lack the specialized ability or assets to preserve these frameworks over time (Borah, 2025). What frequently starts as a promising pilot venture can degrade in execution if not properly maintained, resulting in thwarted expectations with innovation and possibly higher costs in the long run. Moreover, retraining AI models is not a continuous process. They can present unused predispositions, require costly computational assets, and possibly cause unforeseen framework behaviors that require extensive testing and refinement. The long-term lifecycle of AI frameworks in buildings remains a moderately underexplored area, and without supported speculation and oversight, the benefits of AI may decrease over time. The final but vital challenge lies in the value and availability of innovative building innovations. Whereas smart buildings guarantee vitality investment funds, made strides in comfort, and natural benefits, these focal points are not equally distributed across all populations. Most progressed AI-enabled frameworks are currently deployed in high-end commercial buildings, luxury private complexes, or newly created smart cities. These buildings regularly have access to capital, cutting-edge foundation, and a gifted workforce to implement and oversee advanced frameworks (Shafik, 2024). In differentiation, more seasoned buildings, low-income lodging, and offices in under-resourced communities are regularly left behind. These aberrations compound the existing disparities in vitality burden, indoor air quality, and access to comfortable and sound living environments.

Computerized isolation also plays a noteworthy role in restricting the availability of advanced building technologies. Numerous AI frameworks depend on rapid web connections, cutting-edge equipment, and continuous software updates. In provincial or underserved urban regions, web networks may be questionable or costly, and building infrastructure may not support the integration of smart gadgets. Moreover, the complexity of AI interfacing and administration devices can estrange non-technical clients, including seniors, individuals with disabilities, or those with restricted technological proficiency (Ajayi et al., 2024). . Planning comprehensive and user-friendly frameworks is fundamental to guarantee that all building tenants can benefit from technological advances without being hindered by systemic or financial barriers.

Affordability remains a key obstacle to broad selection. Although keen frameworks can reduce working costs in the long run, the upfront investment for AI integration, sensor arrangement, and digital infrastructure is frequently prohibitively high. For proprietors or lodging specialists overseeing reasonable lodging, contributing to these advances may not be financially practical without outside funding or incentives. This creates a circumstance in which wealthier building proprietors continue to enjoy lower energy bills and better environmental conditions, whereas distressed populations remain exposed to wasteful, uncomfortable, or even hazardous situations (Ojadi et al., 2024). Open approaches and administrative systems play a vital role in addressing this awkwardness. Endowments, awards, and performance-based motivating forces appear to empower the adoption of AI-driven energy solutions in underserved areas. In contrast, the least execution guidelines seem to ensure that essential benefits, such as warmth and quality of conversation, are adequately guaranteed.

In addition to financial disparities, social and organizational challenges can constrain the appropriation of AI in buildings. Building administrators and inhabitants may be doubtful of AI frameworks, especially if they see them as mysterious, meddling, or troublesome to control (Shafik, 2024). There may be resistance to giving up the manual control of building frameworks to calculations, particularly if the AI makes decisions that appear outlandish or uncomfortable. Belief may be a significant component of the fruitful arrangement of AI, and it must be earned through straightforwardness, unwavering quality, and the capacity to abrogate automated decisions when needed. Building frameworks must also be planned to accommodate diverse social needs and utilize designs that may change significantly across various locations or statistical groups.

Finally, natural concerns should not be ignored. Although AI can enhance the effectiveness of buildings and reduce operational emissions, the advancement and operation of AI frameworks consume energy. Preparing complex ML models and profound learning frameworks can require considerable computational resources and power consumption. As AI becomes more deeply integrated into building frameworks, the carbon footprint of the advanced framework must be considered alongside that of the physical building. Productive show plans, the use of renewable-powered information centers, and low-power edge computing can help mitigate these impacts. Highlighting the float and complexity of long-term support underscores the need to advance ventures and skills. Most fundamentally, issues of value and access must be addressed to avoid restricting the benefits of savvy building innovation to the well-off or tech-savvy.

POLICY AND GOVERNANCE IMPLICATIONS

The integration of AI into innovative building frameworks provides critical design and operational suggestions, particularly as these advances begin to impact energy consumption at both the building and urban levels. Building clear and enforceable measures is one of the foremost basic perspectives for guaranteeing that AI works safely, efficiently, and decently within the built environment.

Government motivation plays an essential role in accelerating the selection of AI-enhanced retrofits and energy-efficient building plans. Retrofitting existing buildings with advanced technologies often incurs significant upfront costs, making it financially unfeasible for many property owners without financial support. Through endowments, assessment discounts, low-interest advances, and performance-based motivations, governments can lower the barriers to passage and make overhauls more attractive. Programs that compensate for decreases in vitality utilization or carbon emissions accomplished through AI-driven frameworks can encourage the alignment of financial incentives with environmental objectives (Shafik, 2024). A few governments have started incorporating AI and computerized foundation criteria into their green building certification programs, recognizing the value of advanced systems in achieving net-zero targets. In parallel, administrative systems can require that modern improvements meet specific computerized availability benchmarks, guaranteeing that future buildings are not only physically economical but also mechanically prepared to participate in advanced energy systems.

Collaboration among regions, genuine bequest engineers, and innovation firms is another key component in the successful administration of AI in smart buildings (Ajayi et al., 2024). Districts have an interesting position as both controllers and facilitators; they can build zoning codes, vitality effectiveness mandates, and data-sharing requirements that advance the integration of AI while shielding open interfaces. Designers, in turn, bring skills in extending execution and long-term building administration, whereas tech companies offer the devices and stages required to actualize AI arrangements. When these partners work together, they can develop cohesive methodologies for savvy areas, ensure that open and private ventures are aligned, and avoid fragmented or incongruous frameworks. Public-private organizations can also help pilot modern advances in real-world settings, creating valuable insights and experiences that inform approach advancement. Imperatively, this collaboration must incorporate instruments for community input and oversight, especially when open assets or information are included. Through comprehensive and straightforward administration, the benefits of AI in buildings can be scaled mindfully and impartially across society.

FUTURE DIRECTIONS

Looking ahead, the end of AI in smart buildings is balanced to advance past disconnected optimization and toward comprehensive, system-wide insights that are profoundly coordinated with developing innovations and guided by feasible and comprehensive standards (Bajwa et al., 2024). One of the most promising advancements is the integration of AI frameworks with computerized twin innovation. Advanced twins are energetic, virtual representations of physical buildings that reflect real-time conditions by persistently syncing with sensor information, operational logs, and tenant inputs. When combined with AI, advanced twins have reached advanced stages of reenactment, predictive support, and proactive energy management. For instance, a computerized twin can mimic distinctive HVAC setups or lighting techniques under changing climate and inhabitance scenarios, allowing building supervisors to test the results without disrupting real-world operations. AI upgrades this handle by learning from each situation and refining the control methodologies over time. This empowers not only more exact vitality optimization but also quicker reaction times to gear disappointments or natural shifts. In a savvy locale, advanced twins can be amplified over numerous buildings, empowering facilitated decision-making at a broader urban scale.

Another basic wilderness is the appropriation of green AI approaches, which emphasize the advancement of energy-efficient calculation and equipment frameworks. Although AI itself has been a driver of supportability, it also consumes energy, especially in the training and deployment of large models. Green AI advances the utilization of low-energy models that accomplish high performance with reduced computational demands, thereby aligning with the environmental goals of the buildings they serve. Methods such as pruning, quantization, and edge computing decrease the estimation and complexity of AI frameworks, permitting them to function productively on-site without consistent dependence on cloud assets. This not only reduces energy consumption but also decreases idleness and enhances information security (Borah, 2025). As AI becomes increasingly integrated into daily building operations, encompassing real-time occupancy expectations and flexible lighting controls, green AI ensures that the insights layer does not become a source of excessive energy consumption. Empowering inquiry and approaching motivations in this course will be significant as the scale of the arrangement grows.

Equally vital to the long-standing time of innovative building development is the principle of co-designing with communities. Furthermore, innovation is implemented in a top-down manner, overlooking the lived experiences, inclinations, and needs of the individuals who occupy and work in these buildings. Co-design effectively locks in inhabitants, nearby inhabitants, and other partners in the planning, design, and assessment of intelligent systems. This may include participatory workshops, input components, and transparent decision-making forms. When communities are included from the start, innovations are more likely to be accepted, trusted, and utilized successfully. For illustration, community input can educate on how conservation inclinations change by statistics, how security concerns ought to be addressed, and which sustainability highlights are most valued. Moreover, co-design guarantees that developments do not fortify existing disparities but instead cultivate comprehensive access to more beneficial, effective, and responsive built situations. In the coming years, integrating specialized advances with community strengthening will be key to creating more vibrant, attractive, and flexible urban environments.

CONCLUSION

Green AI is a capable instrument for moving toward energy-efficient, climate-resilient urban situations. By leveraging ML for real-time control and expectation, smart buildings can radically diminish energy waste and carbon emissions, especially in retrofitted older structures. The integration of low-power AI models guarantees that vitality investment funds are not counterbalanced by the computerized framework itself, adjusting innovative development with natural objectives (Ajayi et al., 2024). Case studies of cities worldwide emphasize the versatility and adaptability of these arrangements in various climates and urban settings. In any case, accomplishing far-reaching appropriation requires addressing challenges related to information protection, data management, and impartial access to smart innovations. Policymakers, engineers, and technology providers must collaborate to develop steady systems that advance moral, secure, and comprehensive AI development. By implementing Green AI in building administration frameworks and urban planning, cities can achieve more efficient energy utilization, improve tenant well-being, and move closer to their long-term sustainability targets.

REFERENCES

Abbass, M., Abbas, U., Jafri, R., Arif, S. M., & Akhai, S. (2025). AI and Machine Learning Applications in Sustainable Smart Cities. In *Sustainable Smart Cities and the Future of Urban Development* (pp. 1–32). IGI Global Scientific Publishing.

Addad, M., & Al-Taani, S. (2024). Leveraging AI for Energy-Efficient Smart Cities: Architectural and Urban Planning Solutions for Sustainable Growth—A Comparative Case Study of Amman City and International Examples. *Journal of Civil Engineering and Architecture, 18*, 620–632.

Ajayi, F., Ademola, O. M., Amuda, K. F., & Alade, B. (2024). AI-driven decarbonization of buildings: Leveraging predictive analytics and automation for sustainable energy management. *World Journal of Advanced Research and Reviews, 24*(1), 61-79.

Al Amin Gerary, A. (2024). Applications of Smart Technologies in Sustainable Buildings. *International Journal of Architectural Engineering and Urban Research, 7*(2), 41–59. DOI: 10.21608/ijaeur.2025.274865.1071

Al-Raeei, M. (2025). The smart future for sustainable development: Artificial intelligence solutions for sustainable urbanization. *Sustainable Development (Bradford), 33*(1), 508–517. DOI: 10.1002/sd.3131

Alijoyo, F. A. (2024). AI-powered deep learning for sustainable industry 4.0 and internet of things: Enhancing energy management in smart buildings. *Alexandria Engineering Journal, 104*, 409–422. DOI: 10.1016/j.aej.2024.07.110

Aljohani, A. (2024). Deep learning-based optimization of energy utilization in IoT-enabled smart cities: A pathway to sustainable development. *Energy Reports, 12*, 2946–2957. DOI: 10.1016/j.egyr.2024.08.075

Alnaser, A. A., Maxi, M., & Elmousalami, H. (2024). AI-powered digital twins and internet of things for smart cities and sustainable building environment. *Applied Sciences (Basel, Switzerland), 14*(24), 12056. DOI: 10.3390/app142412056

Borah, G. (2025). Emerging Trends in Smart Green Building Technologies. *Heat Transfer Enhancement Techniques: Thermal Performance, Optimization and Applications*, 391-415.

Camacho, J. D. J., Aguirre, B., Ponce, P., Anthony, B., & Molina, A. (2024). Leveraging artificial intelligence to bolster the energy sector in smart cities: A literature review. *Energies, 17*(2), 353. DOI: 10.3390/en17020353

Chowdhury, B. R. (2025). AI-Powered Forecasting and Optimization of Energy Consumption in the USA: Machine Learning Approaches for Sustainable Urban and Institutional Development. *Euro Vantage journals of Artificial intelligence, 2*(2), 21-38.

Das, H. P., Lin, Y. W., Agwan, U., Spangher, L., Devonport, A., Yang, Y., Drgoňa, J., Chong, A., Schiavon, S., & Spanos, C. J. (2024). Machine learning for smart and energy-efficient buildings. *Environmental Data Science, 3*, e1. DOI: 10.1017/eds.2023.43

Fakhabi, M. M., Hamidian, S. M., & Aliehyaei, M. (2024). Exploring the role of the Internet of Things in green buildings. *Energy Science & Engineering, 12*(9), 3779–3822. DOI: 10.1002/ese3.1840

Ge, M., Feng, Z., & Meng, Q. (2024). Urban planning and green building technologies based on artificial intelligence: Principles, applications, and global case study analysis. *Applied Science and Engineering Journal for Advanced Research, 3*(5), 18–27.

Hanafi, A. M., Moawed, M. A., & Abdellatif, O. E. (2024). Advancing sustainable energy management: A comprehensive review of artificial intelligence techniques in building. *Engineering Research Journal (Shoubra), 53*(2), 26–46. DOI: 10.21608/erjsh.2023.226854.1196

Heidari, M., Thangavel, S., Al Naamani, E., & Khashehchi, M. (2024). Emerging trends in smart green building technologies. *Sustainable Technologies for Energy Efficient Buildings*, 313-336.

Husainy, A. S., Mangave, S. S., Ingale, A. S., Patil, Y. R., & Koganole, M. (2024). Innovation ecosystems and green building techniques for a sustainable future: Leveraging advanced technologies. *The Asian Review of Civil Engineering, 13*(2), 1–10. DOI: 10.70112/tarce-2024.13.2.4233

Mahmood, S., Sun, H., Ali Alhussan, A., Iqbal, A., & El-Kenawy, E. S. M. (2024). Active learning-based machine learning approach for enhancing environmental sustainability in green building energy consumption. *Scientific Reports, 14*(1), 19894. DOI: 10.1038/s41598-024-70729-4 PMID: 39191844

Ning, M. A. (2024). Artificial Intelligence-Driven Decision Support Systems for Sustainable Energy Management in Smart Cities. *International Journal of Advanced Computer Science and Applications, 15*(9).

Ojadi, J. O., Odionu, C. S., Onukwulu, E. C., & Owulade, O. A. (2024). AI-Enabled Smart Grid Systems for Energy Efficiency and Carbon Footprint Reduction in Urban Energy Networks. *International Journal of Multidisciplinary Research and Growth Evaluation*, *5*(1), 1549–1566. DOI: 10.54660/.IJMRGE.2024.5.1.1549-1566

Saleh, F., Elhendawi, A., Darwish, A. S., & Farrell, P. (2024). A framework for leveraging the incorporation of AI, BIM, and IoT to achieve smart sustainable cities. *Journal of Intelligent Systems and Internet of Things*, *11*(2), 75–84. DOI: 10.54216/JISIoT.110207

Satpathy, I., Nayak, A., & Jain, V. (2025). The green city: Sustainable and smart urban living through artificial intelligence. In *Utilizing Technology to Manage Territories* (pp. 273–304). IGI Global.

Shafik, W. (2024). Incorporating Artificial Intelligence for Urban and Smart Cities' Sustainability. In *Maintaining a Sustainable World in the Nexus of Environmental Science and AI* (pp. 23-58). IGI Global.

Shahrabani, M. M. N., & Apanaviciene, R. (2025). Evaluation of smart building integration into a smart city by applying machine learning techniques. *Buildings (Basel, Switzerland)*, *15*(12), 1–39. DOI: 10.3390/buildings15122031

Stern, D. (2024). How AI and ML can help reduce energy usage and carbon emissions in buildings. *Corporate Real Estate Journal*, *14*(1), 66–74. DOI: 10.69554/IKXQ8332

Sun, J., Guan, X., Yuan, S., Guo, Y., Tan, Y., & Gao, Y. (2024). Public health perspectives on green efficiency through smart cities, artificial intelligence for healthcare and low carbon building materials. *Frontiers in Public Health*, *12*, 1440049. DOI: 10.3389/fpubh.2024.1440049 PMID: 39741936

Tariq, M. U. (2025). AI-Driven Benchmarking and Standards in Cybersecurity Insurance. In Alawida, M., Almomani, A., & Alauthman, M. (Eds.), *AI-Driven Cybersecurity Insurance: Innovations in Risk, Governance, and Digital Resilience* (pp. 83–110). IGI Global Scientific Publishing. DOI: 10.4018/979-8-3373-5545-0.ch004

Tariq, M. U. (2025). Green FinTech-Driving Sustainable Finance and Empowering Underserved Communities. In Masri, M., Sait, M., & Anshari, M. (Eds.), *Promoting Inclusivity and Accessibility With FinTech* (pp. 279–306). IGI Global Scientific Publishing. DOI: 10.4018/979-8-3373-1112-8.ch010

Tariq, M. U. (2025). AI Tools and Technologies for Management Education: Enhancing Learning Through Innovation. In M. R. & S. R. (Eds.), *Integrating AI and Machine Learning into Business and Management Education* (pp. 105-132). IGI Global Scientific Publishing. DOI: 10.4018/979-8-3373-2150-9.ch005

Tariq, M. U. (2025). Bridging the Digital and Human Divide: Fostering Inclusion and Well-Being in Tech-Driven Workplaces. In Achour, Z. (Ed.), *Leading Inclusive Workplaces Through Digital Transformation and Organizational Change* (pp. 139–168). IGI Global Scientific Publishing. DOI: 10.4018/979-8-3373-5127-8.ch005

Tariq, M. U. (2025). The Role of AI in Transforming Hospitality Operations-Enhancing Efficiency and Guest Experience. In Talukder, M., & Chowdhury, H. (Eds.), *Smart Operations and Enhancing Guest Experience in the Hospitality Industry* (pp. 415–446). IGI Global Scientific Publishing. DOI: 10.4018/979-8-3373-2145-5.ch016

Tunde, A. H., Yusuf, S. O., Taiwo, A. I., Ocran, G., Owusu, P., & Paul-Adeleye, A. H. (2024). AI-driven innovations in energy efficiency: Transforming smart buildings and urban areas through technology and digital transformation. *World Journal of Advanced Research and Reviews, 24*(1).

Yitmen, I., Almusaed, A., Hussein, M., & Almssad, A. (2025). AI-Driven Digital Twins for Enhancing Indoor Environmental Quality and Energy Efficiency in Smart Building Systems. *Buildings (Basel, Switzerland), 15*(7), 1030. DOI: 10.3390/buildings15071030

KEY TERMS AND DEFINITIONS

Green AI: Energy-competent artificial intelligence for sustainability.
Machine Learning: Algorithms that train from building data.
Smart Building: Digitally controlled structure enhancing energy usage.

Chapter 3
AI-Driven Energy Efficiency in Buildings and Urban Environments

P. Selvakumar
https://orcid.org/0000-0002-3650-4548
Department of Science and Humanities, Nehru Institute of Technology, Coimbatore, India

Ganesh Kumar R.
https://orcid.org/0000-0001-7817-1019
Christ University, India

Pallapati Ravi Kumar
https://orcid.org/0009-0008-7217-4081
Koneru Lakshmaiah Education Foundation, India

Manjunath T. C.
https://orcid.org/0000-0003-2545-9160
Rajarajeswari College of Engineering, India

Arun M. R.
https://orcid.org/0000-0002-8291-2956
Vel Tech Rangarajan Dr. Sagunthala R&D Institute of Science and Technology, India

ABSTRACT

In buildings, AI-driven energy management systems (EMS) are revolutionizing the way energy is consumed. Smart building technologies equipped with AI can monitor environmental conditions, occupancy patterns, and equipment performance

DOI: 10.4018/979-8-3373-5540-5.ch003

to fine-tune their operations, leveraging intensive machine learning to learn the preferences of building occupants and adapt accordingly to ensure comfort while minimizing energy use. Predictive maintenance, powered by AI, also contributes to energy efficiency by identifying equipment that operates inefficiently or is at risk of failure, enabling timely interventions that prevent energy waste. The industrial sector, a significant consumer of global energy, is another arena where AI is making remarkable strides in enhancing energy efficiency.

THE ROLE OF AI IN ENERGY EFFICIENCY

Dual challenges and escalating energy demand, AI offers innovative solutions for use, waste. At its core, AI enables processing, intelligent systems. Leveraging neural networks, AI uncovers hidden energy needs, enabling smarter and efficient management. In buildings, AI-driven energy management systems (EMS) are revolutionizing the way energy is consumed. Smart building technologies equipped with AI can monitor environmental conditions, occupancy patterns, and equipment performance to fine-tune their operations, leveraging intensive machine learning to learn the preferences of building occupants and adapt accordingly to ensure comfort while minimizing energy use. Predictive maintenance, powered by AI, also contributes to energy efficiency by identifying equipment that operates inefficiently or is at risk of failure, enabling timely interventions that prevent energy waste. The industrial sector, a significant consumer of global energy, is another arena where AI is making remarkable strides in enhancing energy efficiency. Transportation also benefits greatly from AI applications aimed at improving energy efficiency. AI-powered urban environments, algorithms enhance efficiency in public transportation networks and logistics operations, contributing to more sustainable mobility solutions. In electric vehicles (EVs), AI optimizes battery management systems, extends driving range, and enhances charging strategies to align with grid capacity and renewable energy availability. At the consumer level, AI has become an integral part of smart home ecosystems, empowering individuals to manage and reduce their energy consumption. Intelligent thermostats, smart plugs, and home energy management platforms utilize AI to learn user behaviors, automate device operations, and provide actionable insights for improving energy efficiency. This personalized approach for consumers is a broader effort. In addition to technological advancements, AI fosters data-driven policy making and strategic planning in the energy sector. Governments and regulatory bodies can leverage AI-powered analytics to evaluate the effectiveness of energy efficiency programs, identify areas for improvement, and design targeted interventions. AI also enables better monitoring and verification of energy savings, ensuring transparency and accountability in achieving sustainability goals. In conclusion, AI is playing an increasingly vital role in driving energy efficiency across multiple domains, from smart grids and buildings to industry, transportation, and renewable energy. By enabling intelligent, adaptive, and predictive energy management, AI not only enhances operational performance but also supports the broader.

ENERGY EFFICIENCY IN BUILDINGS AND URBAN ENVIRONMENTS

Energy efficiency buildings and urban environments hold critical importance in today's world and ensure energy security. As urbanization continues to rise, buildings have emerged as a source of carbon. These figures underscore the urgent need to enhance energy efficiency in both new and existing structures, as well as the urban infrastructure that supports them. Improving energy efficiency in buildings and cities offers a powerful lever to occupant comfort and health, and fosters resilient communities (Al-Obaidi K M., et al, 2022). One of the primary reasons why energy efficiency in buildings is so impactful lies in the sheer diversity and volume of energy end-uses. Energy in buildings is consumed for heating, cooling, lighting, appliances, electronics, water heating, ventilation, and building operations. Many of these systems operate inefficiently, especially in older buildings that lack modern technologies or adequate insulation. Enhancing the energy performance of these systems through retrofits, advanced controls, and high-efficiency equipment and emissions reductions. For example, improving thermal efficiency and adopting LED lighting can cut energy use dramatically, sometimes by 50% or more. Moreover, building energy efficiency has a strong economic rationale. Energy-efficient buildings reduce operating costs for owners and tenants by lowering utility bills. This can make homes and workplaces more affordable, improve the bottom line for businesses, and contribute to national economic productivity (Harputlugil T., et al, 2021). The market for energy-efficient construction and retrofits also supports green jobs and stimulates innovation across sectors such as architecture, engineering, materials science, and information technology. District heating and cooling systems offer another highly effective strategy for improving energy efficiency at the community scale. By producing thermal energy at centralized plants and distributing it to multiple buildings via insulated pipes, these systems can achieve higher efficiency and lower emissions than individual heating or cooling units. Coupled with renewable energy sources, waste heat recovery, and cogeneration, district systems can significantly decarbonize urban energy use.

Existing buildings—many of which were constructed decades ago with little consideration for energy performance—represent both a major challenge and an opportunity. Retrofitting the existing building stock is essential to and various national net-zero targets. Deep energy retrofits, which combine envelope upgrades, efficient equipment, renewable energy integration, and smart controls, can transform older buildings into high-performance assets (Derkenbaeva E., et al, 2022). Policy measures such as building codes, performance standards, energy labeling, incentives, and financing mechanisms are crucial to accelerating this transformation. Furthermore, public policy on energy buildings and urban environments. Strong building codes and standards can raise the baseline level of performance across the construction industry. Incentive programs, rebates, and low-interest financing market to adoption. Urban planning policies that encourage density, mixed use, and public transit can shift cities toward more energy-efficient development patterns. Comprehensive city-level energy plans can set targets, coordinate actions across sectors, and monitor progress. Climate resilience is another important dimension of energy-efficient buildings and cities (Aboelata A., 2021). As storms, floods, buildings, and infrastructure must be designed or retrofitted to withstand these impacts while maintaining energy performance. Passive cooling measures, flood-resistant construction, backup power systems, and flexible, adaptive urban design can help ensure that energy-efficient buildings contribute to resilient communities. In conclusion, buildings are of paramount importance in improving the quality of life and fostering economic vitality. The imperative to create energy-efficient, low-carbon urban landscapes becomes ever more urgent. By leveraging innovative technologies, adopting integrated planning approaches, enforcing strong policies, and engaging diverse stakeholders, societies can realize the vast potential of energy efficiency to transform the built environment and advance a more sustainable future for all.

BUILDING ENERGY MANAGEMENT SYSTEMS (BEMS): AI-POWERED BEMS

At their core, Building Energy Management Systems are integrated platforms designed to monitor, control, and optimize energy consumption within a building or across multiple buildings in a portfolio. Traditional BEMS collect data from various subsystems—heating, ventilation, air conditioning (HVAC), lighting, appliances, and others—and enable facility managers to oversee operations and implement efficiency strategies (Aghili S A., et al, 2025). However, the emergence of ML has advanced significantly, elevating the capabilities of modern BEMS, transforming them from passive monitoring tools into intelligent, self-optimizing systems capable of real-time decision-making. AI-powered BEMS continuously gather and process data from sensors embedded throughout the building, external weather forecasts, occupancy patterns, utility price signals, grid demand, and even behavioral insights from occupants. For instance, AI can predict when specific areas of a building will be occupied and adjust HVAC and lighting accordingly, reducing unnecessary comfort levels. Traditional systems may rely on static schedules or simple reactive controls, often leading to suboptimal outcomes. In contrast, AI demand usage patterns and upcoming occupancy allow the system to proactively manage energy loads (Powroźnik P., et al, 2024). For example, during an impending heatwave, the system might pre-cool the building during less carbon-intensive hours, thereby reducing peak demand charges and easing strain on the grid. AI-driven BEMS also enable advanced fault detection and diagnostics (FDD), a critical feature for maintaining optimal building performance. AI identifies anomalies, malfunctioning equipment, sensor errors, or unexpected spikes in energy use, and alerts facility managers or triggers automated corrective actions. Early detection of faults not only prevents energy waste but also extends the lifespan of equipment and reduces maintenance costs.

Buildings equipped with on-site use AI for generation, storage, and consumption of renewable energy. The system might, for example, schedule energy-intensive activities when solar production is high or draw from battery storage during peak utility rates. This holistic management approach maximizes the use of help buildings move toward net-zero energy goals. Through continuous learning, user feedback, AI systems can tailor environmental conditions—temperature, lighting, ventilation—to individual preferences or group needs (Himeur Y., et al, 2023). Such personalization not only improves satisfaction and productivity but also ensures that energy is used efficiently by avoiding unnecessary conditioning of unoccupied spaces. In addition to technical advantages, AI-driven BEMS contribute to strategic goals such as sustainability reporting and compliance with environmental regulations. These systems generate detailed insights and reports on energy performance, greenhouse gas emissions, and savings achieved through efficiency measures. This data supports corporate sustainability initiatives, helps organizations meet green building certifications (such as LEED, BREEAM, or WELL), and demonstrates progress toward regulatory targets for energy efficiency and carbon reduction (Thirunagalingam A., et al, 2025). The potential of AI-powered BEMS is being realized through numerous real-world applications and case studies. For example, large commercial buildings in leading smart cities—such as Singapore, London, and New York—have deployed AI-driven energy management platforms that deliver double-digit reductions in energy use and substantial cost savings. Universities, hospitals, and corporate campuses are also leveraging these technologies to optimize energy performance, enhance sustainability, and improve the comfort of occupants. The scalability of AI-driven BEMS means that even buildings can increasingly access affordable, user-friendly energy management solutions.

PREDICTIVE ENERGY ANALYTICS: PREDICTIVE ANALYTICS FOR ENERGY EFFICIENCY

Predictive energy analytics is rapidly emerging as one of the most transformative technologies in the pursuit of energy efficiency, offering a sophisticated approach to optimizing and reducing carbon emissions. A world where buildings, industries, and cities have the ability to accurately forecast demand and proactively manage consumption is invaluable. Data science, including ML and big data analysis, to energy actionable empowers organizations to move beyond reactive energy management toward intelligent, foresighted strategies that enhance both efficiency and sustainability. One of the most impactful applications of predictive energy analytics is in the optimization of building operations. Commercial, residential, and industrial buildings typically experience significant fluctuations in energy demand based on time of day, occupancy levels, external weather, and equipment usage (Nisha Angeline M., et al, 2025). AI-driven predictive models can forecast loads, requirements, and other needs with remarkable accuracy. For example, predictive analytics can anticipate an afternoon temperature spike and adjust HVAC settings in advance to pre-cool a building, avoiding sudden peak loads and reducing strain on the system. Similarly, it can modulate lighting or equipment schedules to match occupancy patterns, ensuring energy is only used when and where it is needed. Industries and manufacturing facilities, which are among the most energy-intensive sectors, also benefit greatly from predictive energy analytics. AI models can analyze production data, machine performance, and energy use to optimize industrial processes. For example, predictive analytics can identify optimal operating conditions for machinery, recommend adjustments to production schedules to align with lower-cost energy periods, and optimize energy-intensive processes such as heat treatment, chemical processing, or material handling to improve production competitiveness (Aguilar J., et al, 2021). Grid-interactive efficient buildings (GEBs) and demand response (DR) programs represent another frontier where predictive energy analytics delivers significant value. AI-driven forecasts of building energy demand enable dynamic participation in DR buildings to use the grid, such as price fluctuations or requests to reduce load during peak periods. Predictive analytics helps buildings prepare for these events in advance—by precooling or preheating spaces, shifting non-essential loads, or utilizing on-site energy storage—to minimize disruption while maximizing financial incentives. Utilities and energy service providers also harness predictive energy analytics to enhance grid management, customer engagement, and operational planning. Load forecasting models help utilities anticipate demand at various timescales, enabling more efficient generation scheduling, grid balancing, and infrastructure planning. Predictive analytics also supports personalized energy efficiency recommendations for customers, driving engagement and satisfaction while reducing overall demand.

A crucial aspect of predictive energy analytics is its ability to support strategic planning and sustainability reporting. By providing detailed insights into current and future energy use, predictive models enable organizations to set realistic energy reduction targets, prioritize efficiency investments, and measure progress toward climate goals. This transparency is increasingly important for companies seeking to demonstrate corporate social responsibility (CSR), achieve green building certifications, comply with regulatory requirements, or meet voluntary carbon reduction commitments. While the benefits of predictive energy analytics are substantial, the technology does present challenges (Onukwulu E C., et al, 2023). High-quality data is essential for accurate predictions, requiring robust data collection, storage, and integration from diverse sources. Many legacy systems lack the sensors or connectivity needed to support advanced analytics, necessitating investments in digital upgrades. Data privacy and cybersecurity are also critical considerations, particularly as predictive analytics platforms often involve cloud-based processing and remote access. Despite these hurdles, the market for predictive energy analytics is growing rapidly, driven by advancements in AI, increasing availability of IoT devices, and a global push for energy efficiency and sustainability. Leading companies across sectors such as commercial real estate, manufacturing, healthcare, education, and retail are already demonstrating impressive results, achieving double-digit reductions in energy use, operational cost savings, and improved environmental performance through predictive analytics. Looking to the future, the evolution of predictive energy analytics will be shaped by several key trends (Ejidike C C., et al, 2023). Edge computing will enable faster, decentralized processing of energy data, reducing latency and dependence on cloud infrastructure. Buildings, systems, and cities will provide immersive platforms for simulating and optimizing energy performance based on predictive insights. Greater integration with AI-powered control systems will enable fully autonomous energy optimization, where predictive models continuously adjust operations without human intervention. Enhanced interoperability standards will facilitate the seamless exchange of data across platforms, unlocking new levels of predictive accuracy and value. In conclusion, predictive energy analytics is a powerful enabler of energy efficiency, offering organizations the ability to anticipate energy demand, optimize performance, and proactively manage consumption. By transforming vast streams of data into actionable intelligence, predictive analytics helps reduce costs, enhance sustainability, and support the transition. As data continues to evolve, energy analytics will play an ever-more central shaping more across the globe.

SMART BUILDING TECHNOLOGIES

Smart building technologies represent a significant evolution in the way we design, construct, and manage the built environment. As cities expand and global priorities shift toward sustainability, energy efficiency, occupant well-being, and digital connectivity, smart buildings are emerging as critical components of modern urban ecosystems. Advanced sensors, analytics to monitor and optimize all aspects of building performance in real time. By integrating these technologies into their core systems, smart buildings offer a powerful solution to productivity and contribute to climate goals (Parekh R., 2024). IoT devices, embedded throughout the building, collect vast amounts of data from every system elevator, occupancy sensors, and more. These devices communicate seamlessly with centralized or cloud-based platforms, providing a real-time, holistic view of building operations. AI-driven predictions need, automatically adjust system settings for optimal efficiency and performance. The result is a building that is not only responsive to its occupants and environmental conditions but also capable of continuous learning and self-optimization. Occupant comfort and well-being are also at the forefront of smart building technologies. Advanced environmental sensors continuously monitor indoor conditions, including temperature, humidity, air quality, light levels, and noise. AI algorithms analyze this data alongside occupant preferences and behaviors to create personalized environments that enhance comfort, health, and productivity. For example, smart HVAC systems can maintain optimal thermal comfort, while air purification systems can adjust ventilation rates to ensure high indoor air quality. Smart lighting systems can simulate natural daylight patterns to support circadian rhythms, improving mood and cognitive performance (Baharetha S., et al, 2024). Security and safety are further enhanced through smart building technologies. Integrated access control systems, surveillance cameras, motion sensors, and emergency communication platforms work together to provide comprehensive protection for building occupants. AI-powered analytics can detect unusual behavior or potential security threats and trigger automated responses, such as locking doors or alerting security personnel. In emergencies—such as fires, earthquakes, or power outages—smart building systems can coordinate evacuation procedures, communicate with first responders, and ensure that critical safety measures are activated in real time.

Smart building technologies also support space utilization and workplace optimization. Occupancy sensors and data analytics provide insights into how spaces are used, enabling organizations space layout. For example, a company might discover that certain meeting rooms or workstations are underutilized and reconfigure the space to better support employee needs. This data-driven approach to space management enhances flexibility, supports hybrid work models, and helps organizations optimize their real estate investments. Smart building technologies are also enabling interactive, efficient buildings (GEBs). These buildings can communicate electricity to adjust consumption conditions, such as peak demand periods or real-time price signals (Aliero M S., et al, 2022). By participating in demand response (DR) programs, smart buildings help stabilize the grid, reduce and earn financial incentives. AI-powered platforms optimize these interactions to balance energy savings with occupant comfort and operational priorities. Scalability and interoperability are key strengths of smart building technologies. Whether applied to a single facility, a portfolio of buildings, or an entire urban district, these technologies can be tailored to meet diverse needs and integrated with other smart city infrastructure. Open standards and interoperable platforms enable seamless data exchange between different systems and vendors, fostering innovation and avoiding vendor lock-in. In addition to technical benefits, smart building technologies contribute to strategic goals such as corporate sustainability, regulatory compliance, and enhanced market value. Buildings that achieve high levels of energy efficiency and occupant well-being through smart technologies often qualify for green building certifications such as LEED, BREEAM, or WELL (Chowdhary S., et al, 2025). These certifications enhance the attractiveness of properties to tenants, investors, and employees, driving higher occupancy rates and property values. Moreover, the ability to monitor and report on sustainability metrics supports corporate social responsibility (CSR) initiatives. Despite their many advantages, the implementation of smart building technologies also presents challenges. High upfront costs, complexity of integration, cybersecurity concerns, and the need for skilled personnel can pose barriers to adoption. Legacy buildings may require significant retrofitting to accommodate smart technologies, and data privacy must be carefully managed to protect occupant information. Successful deployment requires a clear strategy, stakeholder engagement, and a commitment to ongoing optimization and continuous learning.

SMART GRIDS AND ENERGY DISTRIBUTION: AI IN SMART GRIDS

Artificial Intelligence (AI) is revolutionizing the operation and efficiency of smart grids, playing an increasingly central modernization of distribution systems. Toward an economy driven by climate imperatives, renewable energy integration, and rising demand for reliable and resilient power, traditional energy grids—designed for one-way flow from centralized power plants to passive consumers—are no longer sufficient. By control, AI is helping smart grids optimize energy distribution, enhance reliability, support decarbonization, and empower consumers in unprecedented ways (Hubballi R B., et al, 2025). Accurate short-term and long-term forecasts of electricity demand are essential for grid operators to ensure reliable and cost-effective power delivery. AI algorithms analyze historical demand patterns, weather data, time-of-use trends, socio-economic factors, and special events to predict future load with high accuracy. This capability allows grid operators to schedule generation resources more efficiently, optimize grid capacity, and reduce reliance on costly and carbon-intensive Peaker plants. Enhanced load forecasting also supports the integration of variable renewables, as it helps align flexible generation and storage resources with anticipated demand fluctuations. Renewable energy forecasting is another critical function where AI excels. The inherent intermittency of solar and wind power poses significant challenges for grid stability. AI-driven predictive models are used to forecast renewable energy output with greater precision (Khan A A., et al, 2023). This enables grid operators to anticipate changes in renewable supply and adjust dispatch schedules accordingly, ensuring smoother integration of green energy into the grid. By improving forecast accuracy, AI helps reduce curtailment of renewables, optimize storage utilization, and lower carbon emissions. Through DSM programs, AI-driven platforms interact with consumers, smart appliances, and patterns. For example, AI can automatically shift non-critical loads to off-peak periods, align EV charging with renewable generation, or reduce HVAC usage during demand peaks. Consumers benefit from lower energy costs and incentives, while utilities achieve more balanced load profiles and reduced peak demand pressures. AI's ability to personalize DSM strategies based on individual consumption patterns further enhances user engagement and program effectiveness.

Grid-edge intelligence—enabled by AI—is redefining how energy is managed at the point of consumption. Consumers can monitor their energy use, receive personalized recommendations, and participate in dynamic pricing schemes. AI-driven HEMS can autonomously optimize appliance operation, battery storage, and EV charging, turning passive consumers into active participants in grid management—a concept known as prosumer engagement. AI also facilitates the coordination through VPP aggregates multiple DERs—such as solar arrays, batteries, and flexible loads—into a single controllable entity that can participate in energy markets and provide grid services (Arévalo P., et al, 2024). AI orchestrates the operation of VPPs, optimizing dispatch schedules, bidding strategies, and revenue generation while ensuring grid stability. VPPs enhance grid flexibility, support renewable integration, and open new value streams for DER owners. Resilience and emergency response are further strengthened by AI in smart grids. During natural disasters, cyberattacks, or large-scale outages, AI can rapidly assess grid conditions, prioritize restoration efforts, and guide crews to critical fault locations. AI-powered situational awareness tools provide operators with real-time visibility into grid performance, enhancing decision-making under stress. Microgrids, which during emergencies can also be managed more effectively through AI, ensuring continuity of essential services and faster recovery. Market optimization is another domain where AI adds value to smart grids.

Electricity markets are becoming increasingly dynamic, with time-of-use pricing, capacity markets, and ancillary service markets offering opportunities for flexible resources (Omitaomu O A., et al, 2021). AI algorithms analyze market signals and grid conditions to optimize bidding strategies for DERs, demand response assets, and storage systems. This market intelligence helps utilities and aggregators maximize revenues while contributing to grid efficiency and decarbonization. Looking to the future, the role of AI in smart grids will continue to expand will further enhancing AI capabilities at grid scale. Digital twins of entire grid networks will enable advanced simulations, scenario planning, and predictive maintenance. Greater collaboration among utilities, technology providers, regulators, and consumers will drive the development of more intelligent, flexible, and user-centric grid architectures.

URBAN ENERGY PLANNING AND OPTIMIZATION: AI-DRIVEN URBAN ENERGY PLANNING

AI-driven urban energy planning and optimization represent one of the most promising frontiers for resilient cities. At the same time, cities are complex and dynamic ecosystems where buildings, transportation, energy systems, and human behavior interact continuously. Traditional urban energy planning methods, which often rely are increasingly inadequate for managing this complexity hidden offering a transformative solution for modern urban energy challenges. Primary energy planning involves energy demand and supply at scale. AI models can predict city-wide energy consumption based on variables season, occupancy patterns, and special events. Insights help planners anticipate peak demand periods, optimize load distribution, and align energy generation with consumption patterns (Mazhar T., et al, 2023). For example, AI can forecast when commercial districts will require more cooling due to high temperatures or when residential areas will experience surges in demand due to increased EV charging at night. This foresight enables utilities to schedule generation and storage resources more efficiently, reducing costs and emissions. AI also enables advanced district energy systems, which are essential for efficient management networks, powered by centralized plants and distributed to buildings via insulated pipelines, can achieve much higher efficiencies than individual building systems. AI models manage these networks in real time, optimizing temperatures, flow rates, and energy sourcing based on demand forecasts and operational data. By coordinating supply and demand across the district, AI ensures that heating and cooling are delivered where and when needed, minimizing energy waste and reducing carbon emissions. Transportation is another key sector where AI-driven urban energy planning delivers value. The electrification of transport—through electric vehicles (EVs), e-buses, and micro-mobility solutions—creates new energy demands that must be integrated into the urban grid (Barua A., et al, 2025). AI tools analyze mobility patterns, charging behaviors, and grid capacity to optimize EV charging infrastructure placement and operation. They coordinate smart charging strategies that align with renewable generation peaks or off-peak grid periods, helping to balance loads and avoid new demand peaks. Additionally, AI supports multi-modal transportation planning that promotes public transit, cycling, and walking, reducing overall energy use and emissions from urban transport. Urban resilience is another major benefit of AI-driven energy planning. AI enhances situational awareness by monitoring environmental conditions, infrastructure performance, and human mobility in real time. It supports emergency response planning, guides restoration efforts after disruptions, and optimizes microgrid operations that can sustain critical services independently during grid outages. These capabilities help cities build more resilient and adaptive energy infrastructures capable of withstanding future shocks.

AI coordinates the operation of HVAC, lighting, appliances, and storage within individual buildings while integrating them into broader urban energy strategies. It enables grid-interactive efficient buildings (GEBs) that respond dynamically to grid signals and market incentives, contributing to demand flexibility and overall system optimization. Across the urban landscape, AI can orchestrate the behavior of thousands of buildings and DERs to achieve collective efficiency gains and emissions reductions. In the realm of urban planning and policy, AI-driven tools provide critical insights for long-term energy strategy development. Scenario modelling enables planners to test the impacts of different policies—such as renewable mandates, efficiency standards, electrification incentives, or carbon pricing—on future energy demand, emissions, and system costs (Onukwulu E C., et al, 2023). AI can simulate the effects of new technologies, behavioral trends, or demographic changes, supporting adaptive planning that evolves with emerging conditions. These insights help policymakers design more effective, evidence-based strategies to achieve sustainability goals while ensuring economic viability and social equity. Cybersecurity, interoperability are essential for urban energy planning requires massive volumes of data for effective AI models must be managed responsibly, with strong protections for individual privacy and system security. Open standards and interoperability frameworks ensure that AI solutions can integrate seamlessly across different platforms, systems, and jurisdictions, fostering innovation and scalability. Global examples of AI-driven urban energy planning are rapidly growing. Cities such as Singapore, Amsterdam, Copenhagen, and New York are deploying AI tools to optimize building performance, manage EV charging networks, coordinate DERs, and enhance grid resilience (Upadhyaya A N., et al, 2025). Smart city initiatives around the world are embracing AI as a core enabler of sustainable energy transitions, demonstrating the real-world impact and scalability of these technologies. AI urban energy planning ongoing advancement in edge computing, 5G networks, digital twins, and human-AI collaboration. Digital twins of entire cities—virtual replicas that mirror real-time conditions—will enable planners and operators to visualize and optimize urban energy systems at unprecedented levels of detail. Human-AI interfaces will support collaborative decision-making, combining the strengths of data-driven intelligence with human judgment and creativity.

DATA QUALITY AND INTEGRATION: IMPORTANCE OF DATA QUALITY IN AI-DRIVEN ENERGY EFFICIENCY

In the realm of energy efficiency, quality integration bedrock for effective, reliable, and impactful solutions that increasingly drive optimization across buildings, industrial systems, transportation networks, and power grids, and the quality of the underlying data becomes paramount. High-quality, well-integrated data ensures that AI models can generate accurate insights, deliver meaningful energy savings, and support broader sustainability goals. Conversely, poor data quality—characterized by inaccuracies, gaps, inconsistencies, or lack of context—can undermine the effectiveness of AI tools, leading to suboptimal decisions, missed opportunities, and even unintended negative consequences (Coates J M., et al, 2025). In this context, organizations and governments must prioritize robust data management strategies that uphold data integrity, facilitate seamless integration, and empower AI-driven energy optimization at scale. At its core, AI-driven energy efficiency relies on continuous streams of vast array (BMS), HVAC controls, lighting systems, occupancy sensors, industrial equipment, transportation networks, grid operations, weather forecasts, and market signals all generate valuable information about produced, The ability of deliver meaningful insights depends entirely on the accuracy, completeness, and consistency of this data. For example, if occupancy sensors in a smart building are mis-calibrated or inconsistent, AI-driven HVAC optimization may fail to deliver expected energy savings, or worse, may compromise comfort or indoor air quality. Similarly, if time-series data from smart meters is incomplete or improperly time-stamped, load forecasting models may generate flawed predictions that misalign energy supply and demand (Aldoseri A., et al, 2023). Data accuracy is the first critical dimension of data quality. Measurements must faithfully represent real-world conditions without significant error, bias, or distortion. In energy management, even small inaccuracies—such as sensor drift in temperature measurements or misreported energy consumption—can accumulate over time, leading to erroneous AI-driven decisions. Regular calibration, validation, and verification of sensors and data sources are, therefore, essential practices to maintain accuracy. Additionally, data accuracy depends on precision in spatial and temporal dimensions: for example, location data from connected vehicles or timestamped occupancy patterns must be granular and precise to support effective AI-driven optimization. Contextual richness enhances the value of data for AI-driven energy efficiency. Beyond raw measurements, data should include contextual metadata that describes the conditions under which it was collected. For instance, temperature readings should be tagged with sensor location, equipment status, and environmental conditions. Energy consumption data should be linked to building occupancy, weather data, and operational schedules. Context-rich data enables AI models to capture the complex interactions that drive energy performance, improving the accuracy and relevance of their insights.

Figure 1. Optimizing renewable energy systems through artificial intelligence

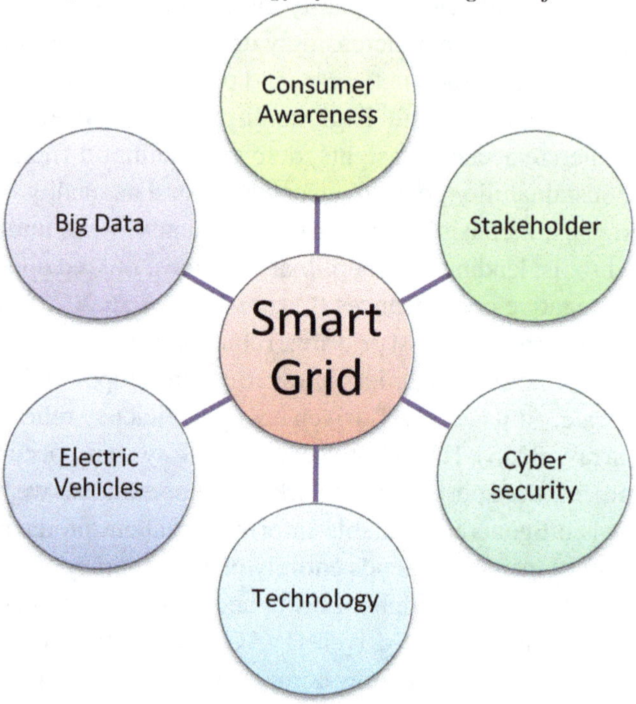

Effective data integration is equally critical to energy efficiency. Energy use is inherently a system-level phenomenon, influenced by multiple factors across different domains. For example, optimizing a building's energy performance requires integrating data from HVAC systems, lighting controls, occupancy patterns, external weather, and grid signals. Similarly, optimizing urban mobility for energy efficiency involves synthesizing data from traffic flows, public transit schedules, EV charging patterns, and road conditions. Without effective integration across these diverse data sources, AI models may overlook important interactions, resulting in fragmented or suboptimal insights (Zong Z., et al, 2024). Integration also enables scalability, allowing AI-driven efficiency solutions to move from single assets or buildings to portfolios, campuses, districts, and entire cities. As organizations increasingly seek to manage energy performance across multiple sites, seamless integration of data across locations becomes a strategic enabler. Centralized platforms that aggregate, standardize, and analyze data from diverse assets empower decision-makers to benchmark performance, identify best practices, and deploy efficiency measures consistently at scale. Governance frameworks for data management are vital for sustaining data quality and integration over time. Data governance defines roles, responsibilities, and processes for data stewardship, quality assurance, lifecycle management, and compliance. Clear governance ensures that data assets are treated as strategic resources, with ongoing investments in quality, documentation, and accountability. Collaboration among IT teams, facilities managers, energy analysts, and AI developers maintains shared requirements, ensuring alignment between data quality and AI-driven objectives. In practice, many organizations face significant challenges in achieving the necessary levels of data quality and integration. Legacy systems, siloed data, inconsistent formats, and outdated infrastructure can all hinder effective AI deployment (Long L D., 2023). Overcoming these barriers often requires modernization efforts, such as upgrading metering and sensing infrastructure, implementing unified data platforms, and adopting open standards for interoperability. The effort invested in improving data quality, however, yields substantial returns, enabling more accurate AI models, deeper energy savings, and more reliable operational outcomes. In conclusion, data quality and integration are foundational to the success of AI-driven energy efficiency. High-quality, well-integrated data enables AI to generate accurate insights, optimize performance, and deliver sustained energy savings and environmental benefits. Organizations that invest in robust data management practices, adopt industry standards, and foster cross-functional collaboration will be best positioned for full energy transformation. Data quality will remain a critical enabler of innovation, resilience, and progress toward a more energy-efficient and sustainable future.

CYBERSECURITY CONCERNS: CYBERSECURITY IN AI-DRIVEN SYSTEMS

Deeply energy management systems, smart grids, industrial operations, buildings, transportation, and other critical infrastructure, cybersecurity concerns are emerging as a top priority. AI-driven systems thrive in processing massive real-time and they often operate autonomously or semi-autonomously to optimize reliability. Increasing connectivity of devices, cloud computing, and automated decision-making introduces new vulnerabilities that, if exploited, could lead to severe operational disruptions, financial losses, privacy violations, and even threats to public safety. Ensuring robust cybersecurity is therefore foundational to trustworthiness, and success is AI-driven. Second, the centralized or cloud-based nature of AI platforms creates another layer of risk (Sarker I H., et al, 2021). AI-driven energy management systems often rely on cloud computing to aggregate data, train models, and orchestrate control decisions across vast geographic areas. While cloud platforms offer scalability and processing power, they also concentrate critical data and control functions in centralized repositories, making them attractive targets for cybercriminals. A successful breach of a cloud-based AI platform could expose sensitive operational data, disrupt services across multiple sites, or allow attackers to manipulate energy flows and grid stability. Third, AI models themselves can become attack vectors. Adversarial attacks against AI—such as feeding manipulated data into machine learning models to cause misclassification or induce incorrect decisions—represent a growing area of concern. In the context of energy systems, such attacks could mislead demand forecasts, disrupt optimization algorithms, or cause inappropriate control actions that degrade performance or create unsafe conditions (Shahana A., et al, 2024). For instance, if a malicious actor subtly altered temperature data fed into an AI-driven HVAC optimization system, it could cause excessive energy use or occupant discomfort without triggering obvious alarms. Fourth, AI-driven systems often involve human-in-the-loop interactions, such as facility operators, energy managers, and maintenance personnel who interface with dashboards and control applications. Cyber attackers frequently target these human elements through phishing, social engineering, or credential theft to gain unauthorized access to critical systems. Weak authentication practices, insufficient role-based access control, or poor cybersecurity hygiene among users can open pathways for attackers to infiltrate AI-driven energy management environments.

In addition to these technical concerns, there are broader cybersecurity governance and regulatory challenges. Many organizations deploying AI-driven energy systems operate in fragmented regulatory landscapes with varying levels of cybersecurity requirements. Inconsistent standards across industries, regions, and technology vendors can lead to gaps in protection and difficulties in achieving comprehensive cybersecurity. Furthermore, the rapid pace of innovation in AI and IoT technologies often outpaces the development of corresponding security frameworks, leaving new vulnerabilities exposed before adequate protections are in place. Given these multifaceted risks, a proactive, multi-layered AI-driven energy must encompass technological, organizational, and policy dimensions to ensure comprehensive protection. Key components of such a strategy include: Data Integrity and Protection: Data collected by AI-driven systems should be protected against unauthorized access, alteration, and loss (Olawade D B., et al, 2024). This includes encrypting data at rest and in transit, implementing strong access controls, and maintaining audit trails for data usage. Secure storage and backup mechanisms ensure data recovery in the event of a cyber incident. AI Model Security: Defences against adversarial AI attacks include robust model validation, adversarial testing, and monitoring for anomalous outputs. AI models should be designed with interpretability and explainability to allow human operators to detect unexpected behavior or manipulation attempts. Continuous Monitoring and Threat Detection: AI-driven energy systems should be continuously monitored for signs of cyber intrusion, anomalous behavior, or policy violations. Advanced threat detection tools, including AI-based security analytics, can identify emerging threats in real time and trigger rapid response actions. Supply Chain Security: As AI-driven energy systems often incorporate hardware and software components from multiple vendors, organizations must assess and manage supply chain risks. This includes vetting suppliers, ensuring compliance with security standards, and monitoring for vulnerabilities in third-party components (Sethumadhavan R P P S., et al, 2025). Regulatory Compliance and Standards: Organizations should align their cybersecurity practices with relevant regulations and industry standards such as NIST Cybersecurity Framework, ISO/IEC 27001, IEC 62443 (for industrial control systems), and emerging guidelines specific to AI and IoT security. Proactive engagement with regulators and industry groups can help shape future cybersecurity policy and ensure readiness for evolving requirements. Looking ahead, the cybersecurity landscape for AI-driven energy systems will continue to evolve. Meanwhile, the growing sophistication of cyber attackers—ranging from criminal enterprises to nation-state actors—demands ever more advanced defences. AI itself will play a dual role, both as a target for cyber threats and as a tool for cybersecurity. AI-driven threat detection, anomaly detection, and automated response capabilities will become essential components of the defence arsenal, helping organizations stay ahead of increasingly complex cyber threats. In conclusion, cybersecurity is a critical enabler of trust, resilience, and success in AI-driven energy systems. Without robust cybersecurity, the very advantages of AI—efficiency, flexibility, and automation—could be undermined by vulnerabilities that threaten operations and public confidence. As AI continues to reshape the energy landscape, a security-first mindset ensures that it is powerful as an engine of progress rather than a vector of risk.

IMPLEMENTATION AND SCALABILITY: CHALLENGES OF IMPLEMENTATION IN AI-DRIVEN ENERGY EFFICIENCY SYSTEMS

AI-driven energy efficiency solutions hold immense promise for transforming how energy is consumed, managed, and optimized across every sector, from smart buildings and industrial facilities to smart grids and urban infrastructure. These systems can deliver substantial energy savings, sustainability goals. However, despite this potential, the path from concept to full-scale deployment is fraught with challenges. Implementation and scalability of AI-driven energy efficiency systems require overcoming technical, organizational, financial, regulatory, and cultural barriers. Many buildings, industrial plants, and grid systems were constructed decades ago, long before the advent of IoT sensors, smart meters, cloud computing, and AI-based analytics. These legacy systems often lack the connectivity, instrumentation, and interoperability required to generate the rich, real-time data needed to fuel AI models. Retrofitting old equipment with modern sensors, upgrading outdated control systems, and integrating legacy platforms with new AI tools can be technically complex, time-consuming, and costly (Maheswaran K., et al, 2025). In many cases, organizations must balance the trade-offs between retrofitting existing assets and investing in new, smart-enabled infrastructure—a decision that can stall or complicate implementation efforts. Another challenge lies complexity of AI models themselves. While AI and machine learning offer powerful capabilities, they are often perceived as "black boxes" by non-expert users. Explainability, models can undermine confidence among facility operators, engineers, and decision-makers, especially in mission-critical environments where safety, reliability, and compliance are paramount. In addition, building effective AI models requires expertise in data science, energy systems, and domain knowledge—a skill set that is in short supply in many organizations. Recruiting, training, and retaining the necessary talent to design, deploy, and maintain AI-driven energy efficiency systems is an ongoing challenge (Qudus L., 2025). Interoperability and standardization issues further complicate large-scale implementation. The energy industry comprises a vast ecosystem of equipment manufacturers, software vendors, service providers, and utility operators—each using its proprietary technologies and protocols. The absence of universal standards for data exchange, device communication, and AI integration hinders seamless interoperability across diverse systems and stakeholders. For example, a smart building might use multiple platforms for lighting, HVAC, access control, and energy management, each requiring custom integration efforts to achieve unified AI optimization. Without industry-wide commitment to open standards, organizations must contend with high integration costs, vendor lock-in, and reduced flexibility in scaling solutions across different sites or portfolios.

Financial and budgetary constraints also pose especially public sector institutions, low-income housing providers. Implementing AI-driven energy efficiency solutions typically requires upfront capital investment in sensors, networking, computing infrastructure, and software platforms. In addition, organizations must allocate resources for staff training, change management, and ongoing system maintenance. While the long-term return on investment (ROI) from energy savings can be substantial, the payback period is often uncertain or may not align with short-term budget cycles. Financial incentives, such as government grants, utility rebates, or performance-based contracts, can help offset costs, but inconsistent availability and complex application processes can limit their effectiveness in scaling deployments. Scalability introduces an additional layer of complexity. While it is often feasible to implement AI-driven energy efficiency at a pilot level—such as a single building, production line, or grid segment—scaling up across multiple assets, regions, or networks presents new challenges. Variations in building design, equipment types, operating schedules, regulatory requirements, and user behavior can complicate the replication of AI solutions across different contexts (Manduva V C., 2023). Achieving scalability requires flexible, modular AI platforms that can adapt to diverse environments, along with standardized implementation frameworks that ensure consistency and quality across deployments. In addition, managing the volume and velocity of data generated at scale places increasing demands on IT infrastructure, cloud computing resources, and cybersecurity defences. Regulatory and policy uncertainty can further hinder large-scale deployment. In many regions, energy efficiency regulations and incentives lag behind technological innovation. Ambiguity around data ownership, liability for AI-driven decisions, and compliance with privacy laws can create legal risks for adopters. The absence of clear guidelines on AI transparency, accountability, and ethical use adds further uncertainty. Navigating this complex and evolving regulatory landscape requires significant legal expertise and proactive engagement with policymakers, resources that not all organizations can readily marshal (Udayaprasad P K., et al, 2024). Despite these challenges, the successful implementation and scalability of AI-driven energy efficiency are achievable and increasingly critical to achieving global climate goals. Best practices for overcoming implementation barriers include: Start small, scale fast: Pilot projects can demonstrate value, build organizational buy-in, and refine AI models before wider rollout. However, scalability should be planned from the outset to avoid fragmented deployments. Invest in data quality: Robust data governance, sensor calibration, and standardized data models ensure AI systems operate on accurate, reliable information. Adopt open standards: Prioritizing interoperability through open protocols and industry standards reduces integration costs and vendor lock-in. Build cross-functional teams: Collaborative teams that bridge IT, energy management, operations, and sustainability drive more effective implementation. Strengthen cybersecurity: A proactive security posture—including risk assessments, monitoring, and incident response—protects AI-driven systems from cyber threats. Engage stakeholders: Transparent communication, staff training, and inclusive design foster acceptance and effective use of AI tools. Leverage partnerships: Collaborating with technology vendors, utilities, academic institutions, and industry consortia accelerates innovation and knowledge sharing.

CONCLUSION

The transformative potential of AI-driven energy efficiency lies not only in the capabilities of artificial intelligence itself, but also in how AI seamlessly integrates to create adaptive, optimized energy ecosystems. Rather than functioning in isolation, AI increasingly serves as the "intelligence layer" that connects, orchestrates, and enhances a wide array of technologies—from the big to blockchain, cloud computing, 5G networks, digital twins, and renewable energy systems. These integrations are enabling unprecedented levels of insight, automation, and optimization across the energy value chain. Energy efficiency is how effectively organizations harness these synergies to drive sustainable outcomes at scale. AI + Internet of Things (IoT): Smarter Sensing and Control: heart AI-driven energy efficiency is its symbiotic relationship with the IoT. Billions of connected sensors, meters, actuators, and smart devices generate vast streams of real-time data about energy consumption, environmental conditions, equipment status, and user behavior. AI algorithms use this data to detect patterns, predict trends, and optimize control actions. AI leverages this information to dynamically adjust HVAC settings, lighting, and ventilation to balance comfort with efficiency. In industrial environments, IoT sensors track equipment performance, enabling AI-driven integration. AI IoT results in energy systems that are not only intelligent but also adaptive, responding instantly to changing conditions and user needs. AI + Cloud Computing: Scalability and Accessibility: Cloud computing provides the computational power, storage capacity, and scalability needed to support sophisticated AI-driven energy efficiency solutions. By hosting AI models and energy management applications in the cloud, organizations can rapidly deploy new capabilities, access advanced analytics, and manage energy assets across multiple sites and geographies. Cloud platforms, such as energy management, with minimal infrastructure investment. By integrating AI with blockchain, organizations can enhance trust and accountability in energy transactions, data sharing, and system operations. For instance, blockchain can provide immutable records of energy consumption, renewable energy credits, and carbon emissions, while AI verifies and analyzes these records for compliance and optimization. AI can match buyers and sellers in real time, while blockchain ensures secure and transparent settlement. The combination of AI and blockchain also supports decentralized energy systems, where distributed energy resources (DERs) operate autonomously but transparently within broader energy markets. AI + Building Information Modeling (BIM): Smarter Construction and Retrofits: By integrating AI with BIM, architects, engineers, and facility managers can optimize energy performance from the design stage through construction and operation. AI can analyze BIM models to recommend energy-efficient design features, predict operational energy use, and optimize building retrofits. During the operational phase, AI-enhanced BIM can provide contextual insights into how building systems interact, supporting more effective energy management. This synergy between AI and BIM is particularly valuable in achieving green building certifications and meeting stringent energy codes. AI + Human-Centric Technologies: Enhancing User Experience: Finally, AI-driven energy efficiency is increasingly integrated with human-centric technologies such as voice assistants. These interfaces enable users to interact with energy systems in intuitive and engaging ways. For example, a building occupant might use a voice assistant to adjust lighting or HVAC settings, with AI ensuring that these adjustments align with energy efficiency goals. Mobile apps powered by AI can provide personalized energy insights and gamify energy-saving behaviors. AR and VR tools can visualize energy flows and system performance, enhancing user understanding and promoting informed decision-making. Future-driven energy efficiency will be defined by its ability to integrate with a wide array of complementary technologies. Renewable energy systems, 5G, robotics, BIM, and human-centric interfaces all contribute unique capabilities that, when orchestrated by AI, create intelligent, adaptive, and sustainable energy ecosystems. Organizations that embrace these synergies will be better positioned to achieve deep energy savings, operational excellence, and environmental leadership. The path forward is one of convergence, where AI serves as the glue that connects and amplifies the power of next-generation technologies to build a smarter, cleaner, and more efficient energy future.

REFERENCES

Aboelata, A. (2021). Reducing Outdoor Air Temperature, Improving Thermal Comfort, and Saving Buildings' Cooling Energy Demand in Arid Cities – Cool Paving Utilization. *Sustainable Cities and Society*, *68*, 102762. Advance online publication. DOI: 10.1016/j.scs.2021.102762

Aghili, S. A., Haji Mohammad Rezaei, A., Tafazzoli, M., Khanzadi, M., & Rahbar, M. (2025). Artificial Intelligence Approaches to Energy Management in HVAC Systems: A Systematic Review. *Buildings (Basel, Switzerland)*, *15*(7), 1008. Advance online publication. DOI: 10.3390/buildings15071008

Aguilar, J., & Garces-Jimenez, A. (2021). A Systematic Literature Review on the Use of Artificial Intelligence in Energy Self-Management in Smart Buildings. Renewable and Sustainable Energy Reviews, 151, Article 111530. DOI: 10.1016/j.rser.2021.111530

Al-Obaidi, K. M., Hossain, M., Alduais, N A M., Al-Duais, H. S., Omrany, H., & Ghaffarianhoseini, A. (2022). A Review of Using IoT for Energy Efficient Buildings and Cities: A Built Environment Perspective. *Energies*, *15*(16), 5991. Advance online publication. DOI: 10.3390/en15165991

Aldoseri, A., Al-Khalifa, K. N., & Hamouda, A. M. (2023). Re-Thinking Data Strategy and Integration for Artificial Intelligence: Concepts, Opportunities, and Challenges. *Applied Sciences (Basel, Switzerland)*, *13*(12), 7082. Advance online publication. DOI: 10.3390/app13127082

Aliero, M. S., Asif, M., Ghani, I., Pasha, M. F., & Jeong, S. R. (2022). Systematic Review Analysis on Smart Building: Challenges and Opportunities. *Sustainability (Basel)*, *14*(5), 3009. Advance online publication. DOI: 10.3390/su14053009

Arévalo, P., & Jurado, F. (2024). Impact of Artificial Intelligence on the Planning and Operation of Distributed Energy Systems in Smart Grids. *Energies*, *17*(17), 4501. Advance online publication. DOI: 10.3390/en17174501

Baharetha, S., Soliman, A. M., Hassanain, M. A., Alshibani, A., & Ezz, M. S. (2024). Assessment of the Challenges Influencing the Adoption of Smart Building Technologies. *Frontiers in Built Environment*, *9*, 1334005. Advance online publication. DOI: 10.3389/fbuil.2023.1334005

Barua, A., Karim, F., Islam, M. M., Das, N., Sumon, M F I., Rahman, A., Debnath, P., Karmakar, M., & Khan, M. A. (2025). Optimizing Energy Consumption Patterns in Southern California: An AI-Driven Approach to Sustainable Resource Management. *Journal of Ecohumanism*, *4*(1), 88–102. DOI: 10.62754/joe.v4i1.6087

Chowdhary, S., Selvakumar, P., Sangeetha, B., Sarvaani, V. V., Anute, N., & Sharma, M. (2025). The Role of Social Capital in Entrepreneurial Development. In Martínez-Falcó, J., Sánchez-García, E., & Marco-Lajara, B. (Eds.), *Support Networks for Entrepreneurs: Social Capital and Strategic Collaborations* (pp. 311–338). IGI Global Scientific Publishing. DOI: 10.4018/979-8-3693-9836-4.ch013

Coates, J. M., Santhanalakshmi, K., Dinesh, N., Gandhimathi, S., & Chandel, P. S. Mohit & Selvakumar P, (2025). Assessment and Evaluation in Transnational Staff Development. In Training and Development in Transnational Higher Education (pp. 387–406). IGI Global Scientific Publishing. DOI: 10.4018/979-8-3373-1897-4.ch019

Derkenbaeva, E., Vega, S. H., Hofstede, G. J., & van Leeuwen, E. (2022). Positive Energy Districts: Mainstreaming Energy Transition in Urban Areas. *Renewable & Sustainable Energy Reviews*, *153*, 111782. Advance online publication. DOI: 10.1016/j.rser.2021.111782

Ejidike, C. C., & Mewomo, M. C. (2023). Benefits of Adopting Smart Building Technologies in Building Construction of Developing Countries: Review of Literature. *SN Applied Sciences*, *5*(2), 52. Advance online publication. DOI: 10.1007/s42452-022-05262-y

Harputlugil, T., & de Wilde, P. (2021). The Interaction Between Humans and Buildings for Energy Efficiency: A Critical Review. *Energy Research & Social Science*, *71*, 101828. Advance online publication. DOI: 10.1016/j.erss.2020.101828

Himeur, Y., Elnour, M., Fadli, F., Meskin, N., Petri, I., Rezgui, Y., Bensaali, F., & Amira, A. (2023). AI-Big Data Analytics for Building Automation and Management Systems: A Survey, Actual Challenges and Future Perspectives. *Artificial Intelligence Review*, *56*(6), 4929–5021. DOI: 10.1007/s10462-022-10286-2 PMID: 36268476

Hubballi, R. B., Selvakumar, P., Seenivasan, R., Basava Aradhya, S. G., Dinesh, N., & Seelam, P. K. (2025). Overview of Current AI Technologies in Education. In Kyei-Blankson, L., & Ntuli, E. (Eds.), *Transformative AI Practices for Personalized Learning Strategies* (pp. 1–26). IGI Global Scientific Publishing., DOI: 10.4018/979-8-3693-8744-3.ch001

Khan, A. A., Laghari, A. A., Rashid, M., Li, H., & Javed, A. R. (2023). Artificial Intelligence and Blockchain Technology for Secure Smart Grid and Power Distribution Automation: A State-of-the-Art Review. *Sustainable Energy Technologies and Assessments*, *57*, 103282. Advance online publication. DOI: 10.1016/j.seta.2023.103282

Long, L. D. (2023). An AI-Driven Model for Predicting and Optimizing Energy-Efficient Building Envelopes. *Alexandria Engineering Journal*, *72*, 480–501. DOI: 10.1016/j.aej.2023.08.041

Maheswaran, K., Selvakumar, P., Shaikh, M., Shashank, G., Gupta, S., & Manjunath, T. C. (2025). The Evolution of Smart Urban Mobility With Emerging Technologies. In Ponnusamy, S., Assaf, M., Youssef, B., Jeon, G., & Kalyanaraman, S. (Eds.), *Leveraging VANETs and Blockchain Technology for Urban Mobility* (pp. 1–28). IGI Global Scientific Publishing. DOI: 10.4018/979-8-3373-0265-2.ch001

Manduva, V. C. (2023). AI-Driven Edge Computing in the Cloud Era: Challenges and Opportunities. *International Journal of Modern Computing, 11*(2), 88–102. DOI: 10.18535/ijsrm/v11i02.ec5

Mazhar, T., Irfan, H. M., Haq, I., Ullah, I., Ashraf, M., Shloul, T. A., Ghadi, Y. Y., Imran, , & Elkamchouchi, D. H. (2023). Analysis of Challenges and Solutions of IoT in Smart Grids Using AI and Machine Learning Techniques: A Review. *Electronics (Basel), 12*(1), 242. Advance online publication. DOI: 10.3390/electronics12010242

Nisha Angeline, M., Selvakumar, P., Gupta, M. K., Nazar, N., Elumalai, D., & Bhattacharya, S. (2025). Emerging Materials for Energy Storage and Environmental Applications. In Muduli, K., Moharana, B., Ales, S., & Biswal, D. (Eds.), *Using Computational Intelligence for Sustainable Manufacturing of Advanced Materials* (pp. 53–72). IGI Global Scientific Publishing. DOI: 10.4018/979-8-3693-7974-5.ch003

Olawade, D. B., Wada, O. Z., David-Olawade, A. C., Fapohunda, O., Ige, A. O., & Ling, J. (2024). Artificial Intelligence Potential for Net Zero Sustainability: Current Evidence and Prospects. *Next Sustainability, 4*, 100041. Advance online publication. DOI: 10.1016/j.nxsust.2024.100041

Omitaomu, O. A., & Niu, H. (2021). Artificial Intelligence Techniques in Smart Grid: A Survey. *Smart Cities, 4*(2), 548–568. DOI: 10.3390/smartcities4020029

Onukwulu, E. C., Agho, M. O., & Eyo-Udo, N. L. (2023). Developing a Framework for Predictive Analytics in Mitigating Energy Supply Chain Risks. *International Journal of Scholarly Research and Reviews, 2*(2), 135–155. DOI: 10.56781/ijsrr.2023.2.2.0042

Parekh, R. (2024). Automating the Design Process for Smart Building Technologies. *World Journal of Advanced Research and Reviews, 23*(2), 1213–1234. DOI: 10.30574/wjarr.2024.23.2.2461

Powroźnik, P., & Szcześniak, P. (2024). Predictive Analytics for Energy Efficiency: Leveraging Machine Learning to Optimize Household Energy Consumption. *Energies, 17*(23), 5866. Advance online publication. DOI: 10.3390/en17235866

Qudus, L. (2025). Leveraging Artificial Intelligence to Enhance Process Control and Improve Efficiency in Manufacturing Industries. *International Journal of Computer Applications Technology and Research, 14*(2), 18–38. DOI: 10.7753/IJCATR1402.1002

Sarker, I. H., Furhad, M. H., & Nowrozy, R. (2021). AI-Driven Cybersecurity: An Overview, Security Intelligence Modeling and Research Directions. *SN Computer Science, 2*(3), 173. Advance online publication. DOI: 10.1007/s42979-021-00557-0

Sethumadhavan, R. P. P. S., Selvakumar, P., Kadu, S. G., Chandel, P. S., & Gupta, S. (2025). Environmental Sustainability and Employee Participation. In Şeker, C., Örücü, E., & Kaya, A. (Eds.), *Green Leadership Strategies for Increased Employee Engagement and Performance* (pp. 275–298). IGI Global Scientific Publishing. DOI: 10.4018/979-8-3373-2106-6.ch009

Shahana, A., Hasan, R., Farabi, S. F., Akter, J., Mahmud, M. A., Johora, F. T., & Suzer, G. (2024). AI-Driven Cybersecurity: Balancing Advancements and Safeguards. *Journal of Computer Science and Technology Studies, 6*(2), 88–97. DOI: 10.32996/jcsts.2024.6.2.9

Thirunagalingam, A., & Selvakumar, P. (2025). Application of Automation and Robotics in the Oil and Gas Industry. In Abdullayev, V., & Khang, A. (Eds.), *Revolutionizing AI and Robotics in the Oil and Gas Industry* (pp. 229–244). IGI Global Scientific Publishing. DOI: 10.4018/979-8-3693-8156-4.ch011

Udayaprasad, P. K., Shreyas, J., Srinidhi, N. N., Dilip Kumar, S. M. N., Dayananda, P., Askar, S. S., & Abouhawwash, M. (2024). Energy Efficient Optimized Routing Technique With Distributed SDN-AI to Large Scale I-IoT Networks. *IEEE Access : Practical Innovations, Open Solutions, 12*, 2742–2759. DOI: 10.1109/ACCESS.2023.3346679

Upadhyaya, A. N., Debnath, P., Rani, P., Selvakumar, P., Manjunath, T. C., & Bhattacharya, S. (2025). AI-Based Disaster Prediction and Early Warning Systems. In Mobo, F. (Ed.), *AI and Emerging Technologies for Emergency Response and Smart Cities* (pp. 55–74). IGI Global Scientific Publishing. DOI: 10.4018/979-8-3693-9770-1.ch003

Zong, Z., & Guan, Y. (2024). AI-Driven Intelligent Data Analytics and Predictive Analysis in Industry 4.0: Transforming Knowledge, Innovation, and Efficiency. *Journal of the Knowledge Economy, 16*(1), 864–903. DOI: 10.1007/s13132-024-02001-z

Chapter 4
Intelligent Synergy Leveraging Artificial Intelligence for Renewable Energy Forecasting and Grid Optimization

C. H. Hussaian Basha
https://orcid.org/0000-0002-9858-8353
SR University, Warangal, India

ABSTRACT

The accelerating global shift toward renewable energy is vital to mitigating greenhouse gas emissions and fostering sustainable power infrastructures. However, integrating variable renewable sources such as solar, wind, and hydro into existing grids poses major challenges due to their intermittency and uncertainty. This chapter introduces a unified AI-enabled framework that integrates forecasting, uncertainty quantification, optimization, and intelligent control to enhance grid reliability, resilience, and efficiency. AI-based time-series forecasting models such as recurrent neural networks (RNNs), long short-term memory (LSTM) networks, convolutional neural networks (CNNs), and hybrid deep learning architectures are analyzed for their superior ability to capture nonlinear, multi-dimensional renewable energy patterns. Beyond forecasting, machine learning (ML) and reinforcement learning (RL) techniques, including support vector machines (SVM), adaptive neuro-fuzzy inference systems (ANFIS), and deep q-learning, are explored for voltage and frequency regulation, demand response, fault detection, and optimal power flow

DOI: 10.4018/979-8-3373-5540-5.ch004

management. Case studies from Germany, the United States, China, and India demonstrate AI's transformative potential in predictive maintenance, smart grid coordination, and energy storage optimization. The chapter also highlights the convergence of AI with IoT, edge computing, and digital twins, enabling real-time analytics, low-latency decision-making, and predictive operational planning. Ethical, regulatory, and socio-economic implications such as data privacy, transparency, and workforce readiness are addressed with reference to the EU AI Act, IEEE 7000, and NIST AI RMF. Overall, the chapter establishes AI as a strategic enabler of the global clean energy transition, fostering intelligent, autonomous, and sustainable smart grids for the future.

1. INTRODUCTION

The integration of renewable energy sources such as solar, wind, and hydropower into conventional power grids has become a global priority in the transition toward sustainable energy systems. However, the inherent intermittency and variability of these sources pose significant operational challenges (Kabeyi & Olanrewaju, 2022). Solar and wind energy are weather-dependent and non-dispatchable, making it difficult to match supply with demand in real time. This variability can lead to power imbalances, increased reserve requirements, and voltage or frequency instabilities, particularly in regions with high renewable penetration. Additionally, traditional grids were originally designed for centralized energy distribution, not the decentralized and bidirectional flow introduced by renewable energy systems. Another major challenge lies in forecasting and control. Accurate short-term and long-term prediction of renewable generation is critical for scheduling, load balancing, and market participation. Yet, due to the non-linear and stochastic behaviour of natural resources, conventional forecasting techniques often fall short. Furthermore, the integration of renewables requires enhanced grid flexibility, which is often hindered by limited storage capacity and outdated grid infrastructure (Shahzad & Jasińska, 2024). Without real-time data processing and intelligent decision-making systems, managing this complexity becomes inefficient and unreliable. These challenges emphasize the need for advanced analytics and artificial intelligence to improve forecasting accuracy, enable adaptive grid control, and ensure a resilient and optimized energy ecosystem.

Figure 1. Evolution of Power Systems from Conventional Grid to Smart Grid (Salman, 2019)

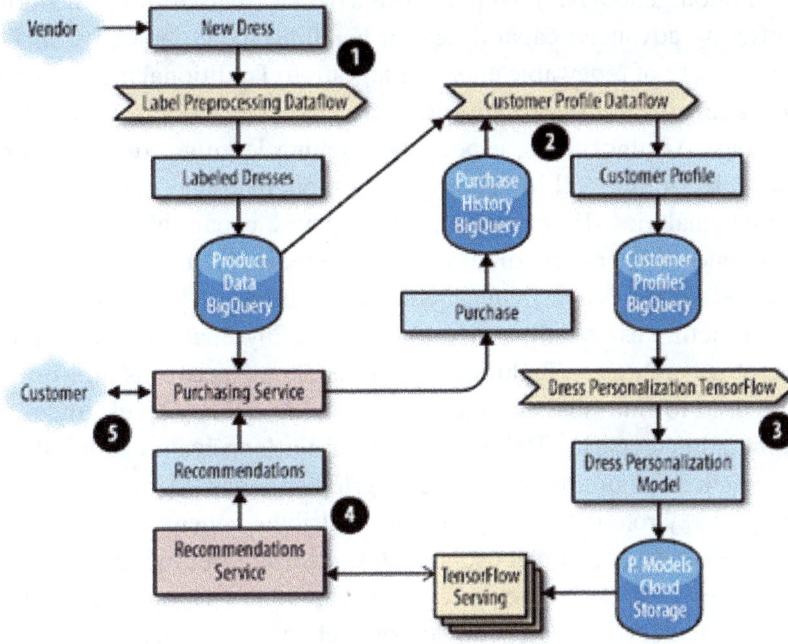

As the energy landscape shifts towards higher shares of renewable sources, accurate forecasting and grid optimization have become essential components of modern power system management. Forecasting plays a critical role in predicting power generation from variable sources like solar and wind, enabling grid operators to plan, schedule dispatchable resources, and maintain supply-demand equilibrium. Without reliable forecasts, operators face the risk of overgeneration or shortages, which can lead to curtailment of renewables, increased reliance on fossil-based peaker plants, or even grid instability (S. Kumar, 2023). Forecasting also supports economic efficiency by enabling better participation in electricity markets and reducing operating costs through optimal resource scheduling. Grid optimization complements forecasting by enhancing the operational performance and stability of the power network. It involves dynamic control of voltage levels, frequency regulation, load balancing, congestion management, and efficient use of distributed energy resources (DERs). In a renewable-rich grid, traditional control mechanisms often struggle to cope with rapid fluctuations and bi-directional power flows. Hence, real-time optimization becomes indispensable to ensure reliable, secure, and cost-effective operation. By integrating AI-driven forecasting with grid optimization strategies, utilities can achieve a synergistic approach that improves system responsiveness, minimizes losses, and facilitates the transition toward decentralized and sustainable energy systems (Ukoba et al., 2024a).

Artificial Intelligence (AI) has emerged as a transformative tool in modern power systems, offering advanced capabilities for handling the complexity, uncertainty, and dynamic nature of renewable energy integration. Traditional methods often fall short in managing the nonlinear behaviours and vast data associated with real-time grid operations. AI algorithms, especially machine learning, deep learning, and reinforcement learning excel in pattern recognition, data-driven decision-making, and predictive analytics (Ramya et al., 2024). These capabilities make AI highly effective for renewable energy forecasting, load prediction, fault detection, voltage regulation, and demand-side management. In addition to improving forecasting accuracy, AI facilitates real-time grid optimization by enabling autonomous and adaptive control strategies. Techniques such as neural networks and fuzzy logic systems are used to fine-tune voltage profiles and frequency levels under varying load and generation scenarios. Reinforcement learning can dynamically adjust power flows and switch configurations to enhance grid resilience (Gautam, 2023). Moreover, AI plays a key role in the integration of distributed energy resources (DERs), energy storage systems, and electric vehicles, ensuring optimal coordination and minimizing system losses. As the power grid evolves into a more decentralized and digitized ecosystem, AI is not just an add-on technology it is a core enabler of the intelligent, flexible, and efficient energy systems of the future.

Figure 2. AI-Based Renewable Energy Forecasting Framework (S. U. Khan et al., 2023)

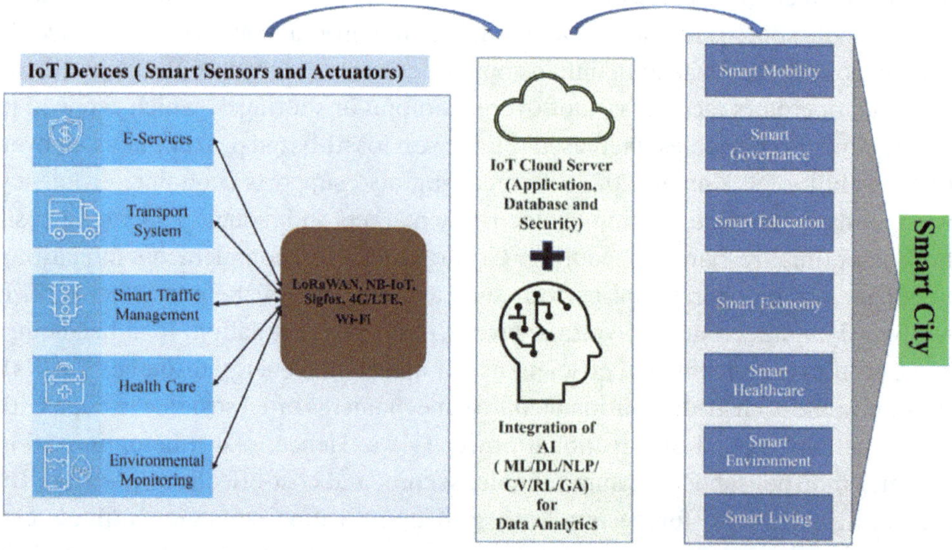

The primary objective of this chapter is to explore how Artificial Intelligence (AI) can be strategically leveraged to enhance renewable energy forecasting and optimize power grid operations. It aims to provide a comprehensive understanding of AI-driven approaches, including machine learning, deep learning, and hybrid techniques, for managing the variability and uncertainty inherent in renewable energy sources such as solar and wind (Ohalete et al., 2023). By examining the role of AI in both short-term and long-term forecasting, this chapter highlights how intelligent models improve prediction accuracy and contribute to more efficient energy scheduling, reserve planning, and market operations. Beyond forecasting, the chapter delves into the application of AI for real-time grid optimization, covering topics such as load balancing, voltage and frequency control, congestion management, and the coordination of distributed energy resources (DERs). The scope also includes an integrated view of how AI interfaces with supervisory control and data acquisition (SCADA), Energy Management Systems (EMS), and Internet of Things (IoT)-enabled devices to support adaptive and decentralized grid control (Nuruzzaman & Rana, 2025). Through illustrative case studies and system-level architectures, the chapter provides insights into the technical, operational, and practical benefits of AI in building a resilient, intelligent, and sustainable power grid.

2. BACKGROUND AND LITERATURE REVIEW

2.1 Evolution of Power Systems Towards Smart Grids

The evolution of power systems has been marked by a gradual yet significant transition from centralized, unidirectional electricity networks to decentralized, intelligent, and interactive smart grids. Traditionally, power systems followed a top-down structure where electricity was generated in large-scale thermal or hydro plants and delivered through transmission and distribution networks to passive consumers. These legacy grids were designed for predictable generation and consumption patterns, with minimal real-time control or feedback mechanisms (Malka et al., 2023). However, the increasing penetration of renewable energy sources, distributed generation, and dynamic loads has exposed the limitations of conventional grid architectures. The modern power landscape demands greater flexibility, reliability, and responsiveness. In response, the concept of the smart grid emerged, characterized by digital communication technologies, advanced metering infrastructure (AMI), automation, and real-time monitoring capabilities. Smart grids facilitate two-way power and data flow, enabling consumers to become active participants through distributed energy resources (DERs), energy storage, and demand-side management (Panda et al., 2022).

This transformation is further accelerated by the integration of Artificial Intelligence (AI), Internet of Things (IoT), and cloud computing, which enhance grid intelligence and operational efficiency. Smart grids not only accommodate renewable variability but also support self-healing mechanisms, cyber-physical security, and energy market participation. Overall, the shift toward smart grids marks a paradigm change from rigid, supply-driven systems to adaptive, demand-responsive, and data-driven energy networks (Wang, 2025).

2.2 Traditional vs. AI-Based Forecasting Models

Forecasting plays a critical role in power system planning and operation, particularly with the increasing integration of intermittent renewable energy sources. Traditional forecasting models, such as statistical and mathematical methods, rely heavily on historical data and assume linearity and stationarity in time-series data (Chiang et al., 2024). Common approaches include Auto-Regressive Integrated Moving Average (ARIMA), Exponential Smoothing, and Persistence models. While these methods are relatively simple and computationally efficient, they often lack the adaptability and robustness needed to capture the complex, nonlinear, and stochastic behaviours of renewable resources like solar irradiance and wind speed, especially under rapidly changing environmental conditions (Zhang et al., 2022).

In contrast, AI-based forecasting models utilize data-driven algorithms capable of learning intricate patterns from large and diverse datasets. Techniques such as Artificial Neural Networks (ANNs), Support Vector Machines (SVMs), Decision Trees, and Deep Learning models like Long Short-Term Memory (LSTM) networks and Convolutional Neural Networks (CNNs) have demonstrated superior accuracy in forecasting renewable energy generation. These models can dynamically adjust to changing input features, handle noise, and recognize non-linear dependencies in time-series data. Furthermore, hybrid models that combine AI techniques with optimization algorithms (e.g., PSO-ANN, GWO-LSTM) offer even greater precision and adaptability (Mirzania et al., 2024). Unlike traditional models, AI-based approaches can process real-time data streams and continuously refine predictions, making them highly suitable for modern smart grid environments that require fast, reliable, and scalable forecasting solutions.

Table 1. Comparison of Traditional vs. AI-Based Forecasting Models (James, 2021).

Criteria	Traditional Models	AI-Based Models
Examples	ARIMA, Persistence, Exponential Smoothing	ANN, SVM, LSTM, GRU, XGBoost
Data Assumptions	Requires stationary, linear data	Handles non-linear, non-stationary data
Feature Engineering	Manual and domain-dependent	Often automated or self-learned
Forecasting Accuracy	Moderate, especially under fluctuating conditions	High, especially for complex and dynamic patterns
Adaptability	Low; struggles with unseen patterns	High; adapts well with new training data
Real-Time Capability	Limited	Supports real-time and online learning
Computational Complexity	Low to moderate	Moderate to high, depending on model type
Scalability	Less scalable for large or high-dimensional data	Highly scalable with cloud/edge integration
Interpretability	High (white-box models)	Low to moderate (black-box models, unless XAI is used)
Use in Renewable Forecasting	Suitable for short-term, simple patterns	Suitable for both short- and long-term, complex scenarios

2.3 Grid Optimization Techniques: A Historical Perspective

Grid optimization has long been a central focus of power system operation and planning, evolving alongside advancements in control theory, computational tools, and system complexity. In the early stages of centralized power generation, grid optimization primarily involved economic dispatch and unit commitment problems, aiming to minimize generation costs while meeting load demands and maintaining system constraints. These problems were traditionally solved using linear programming (LP), quadratic programming (QP), and other classical optimization techniques that assumed steady-state conditions and known system parameters (Mataifa et al., 2022).

With the growing complexity of power networks, especially due to deregulation and the inclusion of distributed energy resources (DERs), traditional optimization methods began to face limitations in scalability and responsiveness. The introduction of real-time monitoring systems such as SCADA enabled dynamic optimization, paving the way for more adaptive approaches like Optimal Power Flow (OPF), Security-Constrained OPF (SCOPF), and state estimation algorithms. These methods improved the reliability and efficiency of grid operations but still relied on deterministic models and assumptions that did not fully capture uncertainties (Hu et al., 2024).

The integration of renewable energy and the emergence of smart grid technologies have further necessitated more advanced, intelligent optimization techniques. In recent decades, metaheuristic algorithms such as Particle Swarm Optimization (PSO), Genetic Algorithms (GA), and Ant Colony Optimization (ACO) have been employed for solving complex, multi-objective, and non-linear grid problems. These were later enhanced by AI-based methods, including Reinforcement Learning (RL), Fuzzy Logic Controllers, and hybrid AI-optimization frameworks. These intelligent techniques support real-time decision-making, fault detection, energy storage coordination, and voltage/frequency regulation, making grid optimization more resilient and adaptive in the face of growing uncertainty and decentralization (Arévalo & Jurado, 2024a).

2.4 Recent Trends in AI Applications for Energy Systems (2018–2025)

Between 2018 and 2025, the role of Artificial Intelligence (AI) in energy systems has expanded dramatically, reflecting the urgent need for adaptive, data-driven solutions to manage the increasing complexity of renewable energy integration. AI has proven particularly effective in forecasting, optimization, fault diagnosis, and decision support (Gupta et al., 2022). Traditional deterministic models have increasingly been replaced by intelligent systems capable of handling real-time data, non-linear system behaviour, and uncertain environments. Machine learning techniques such as Support Vector Machines (SVM), Random Forests, and Artificial Neural Networks (ANN) have become standard tools for energy demand forecasting, renewable generation prediction, and energy market analysis (Ahmad et al., 2017).

Deep learning approaches, especially Long Short-Term Memory (LSTM) networks and Convolutional Neural Networks (CNNs), have shown superior accuracy in time-series forecasting of solar irradiance, wind speed, and load profiles. Their ability to extract features from large datasets and capture temporal dependencies makes them suitable for dynamic and uncertain renewable environments. Moreover, hybrid models combining optimization algorithms (such as Particle Swarm Optimization, Genetic Algorithms, and Grey Wolf Optimization) with machine learning have emerged as powerful tools for fine-tuning model parameters, improving convergence speed, and enhancing prediction reliability (Kiliçarslan, 2023).

From 2020 onwards, Reinforcement Learning (RL) and Multi-Agent Systems (MAS) have gained attention for their ability to perform autonomous control and coordination in distributed energy networks, such as virtual power plants (VPPs) and microgrids. These models support real-time optimization of energy flow, demand response, and distributed generation scheduling. Additionally, the rise of Edge AI and cloud-based platforms has enabled real-time analytics and control at the device level, facilitating smart meter data processing, EV charging optimization, and home energy management systems. Overall, AI applications in energy systems during this period have transitioned from experimental research to practical implementation, contributing to smarter, more resilient, and sustainable grid operations (Rajaperumal & Columbus, 2025).

3. AI TECHNIQUES FOR RENEWABLE ENERGY FORECASTING

Artificial Intelligence (AI) techniques have become essential tools for forecasting renewable energy generation, addressing the inherent variability and uncertainty of sources like solar and wind. Unlike traditional models, AI approaches can learn complex, non-linear relationships from historical and real-time data, enabling more accurate and adaptive predictions. By leveraging algorithms such as machine learning, deep learning, and hybrid optimization models, these techniques enhance the reliability of energy forecasting, support grid stability, and facilitate efficient energy market operations. This section explores various AI methods that have been successfully applied to improve forecasting accuracy in renewable energy systems (Wilson & Anwar, 2024).

3.1 Time Series Forecasting Models (ARIMA, LSTM, GRU)

Time series forecasting models play a pivotal role in predicting renewable energy outputs, particularly for sources like solar and wind that exhibit temporal variability. These models analyze patterns in historical data to forecast future energy generation, aiding grid operators in planning and balancing supply-demand dynamics. ARIMA (Auto-Regressive Integrated Moving Average) is a classical statistical model widely used for short-term forecasting. It relies on the assumption that future values have a linear relationship with past observations and forecast errors. While ARIMA performs well for stationary and linear time series, it often struggles with capturing non-linear dynamics and long-term dependencies typical in renewable energy datasets (A. Ahmed, 2025).

To overcome these limitations, Long Short-Term Memory (LSTM) networks, a type of Recurrent Neural Network (RNN), have been introduced. LSTM models are capable of learning long-term dependencies and temporal correlations in sequential data, making them highly suitable for predicting fluctuating solar irradiance and wind speed. They retain relevant past information over extended time lags, which is crucial for capturing seasonal and daily variations (Sherstinsky, 2020). Gated Recurrent Units (GRU), another variant of RNNs, offer similar forecasting capabilities with a simplified architecture compared to LSTM. GRUs require fewer computational resources and converge faster, making them ideal for real-time applications in energy management systems. Both LSTM and GRU have proven to outperform traditional models in accuracy, adaptability, and robustness, particularly under highly dynamic and uncertain conditions in renewable energy forecasting (S. Khan et al., 2024a).

3.2 Machine Learning Methods (SVM, Random Forest, Boost)

Machine learning (ML) methods have gained prominence in renewable energy forecasting due to their ability to model complex, non-linear relationships without requiring explicit physical models. These techniques learn from historical data to make predictions about future energy generation, even in the presence of noise, missing values, or varying weather conditions. Support Vector Machines (SVM) are supervised learning models that work well for both classification and regression tasks. In renewable energy forecasting, SVM is commonly used to predict solar power output or wind speed by finding an optimal hyperplane that best fits the data. SVM performs effectively in high-dimensional spaces and can handle non-linear patterns using kernel functions, making it suitable for moderate-sized datasets (Du et al., 2024).

Random Forest (RF) is an ensemble learning technique based on decision trees. It operates by constructing multiple decision trees during training and outputs the average prediction from all trees. Random Forest is known for its robustness, high accuracy, and resistance to overfitting, especially when dealing with diverse weather and environmental variables. It is widely used for predicting short-term and day-ahead renewable energy generation. Extreme Gradient Boosting (XGBoost) is an advanced boosting algorithm that builds strong prediction models by combining the outputs of several weak learners. XGBoost is highly efficient and scalable, making it ideal for large datasets and real-time applications. It has been extensively applied in forecasting photovoltaic (PV) output and wind energy production, outperforming many other machine learning models in terms of speed and accuracy. Together, these machine learning methods provide reliable and scalable solutions for renewable energy forecasting, supporting better grid planning and operational efficiency (Oladapo et al., 2024) .

3.3 Deep Learning Approaches (CNNs, RNNs, Hybrid Models)

Deep learning approaches have shown exceptional performance in renewable energy forecasting due to their ability to automatically extract features from raw data and model complex, non-linear temporal and spatial relationships. Unlike traditional machine learning models, deep learning techniques do not require manual feature engineering, making them more adaptable to diverse forecasting scenarios (S. F.

Ahmed et al., 2023). Convolutional Neural Networks (CNNs), although traditionally used in image processing, have been successfully adapted for time-series forecasting in renewable energy systems. CNNs are effective in capturing local patterns and trends in data, such as fluctuations in solar irradiance or wind speed. By applying convolutional filters to input sequences, CNNs can identify relevant features and improve the prediction accuracy of short-term power outputs (Wan et al., 2023).

Recurrent Neural Networks (RNNs), particularly their variants like Long Short-Term Memory (LSTM) and Gated Recurrent Units (GRU), are designed to handle sequential data and temporal dependencies. These models are widely used for hourly and daily forecasting of solar and wind energy generation. RNNs maintain a memory of previous time steps, enabling them to learn patterns over time, which is crucial for capturing seasonality and variability in renewable energy sources(S. Khan et al., 2024b). Hybrid models combine the strengths of different deep learning architectures or integrate deep learning with optimization algorithms (e.g., CNN-LSTM, LSTM-PSO). These models enhance learning efficiency and forecasting precision by leveraging complementary capabilities of CNNs for spatial feature extraction and LSTMs for temporal pattern recognition. Such hybrid frameworks are increasingly adopted in smart grid applications for robust, real-time energy forecasting under uncertain and dynamic conditions (S. Ali et al., 2024).

3.4 Hybrid AI Models (Fuzzy-ANN, PSO-LSTM, GWO-ANFIS)

Hybrid AI models integrate the strengths of multiple intelligent techniques to address the limitations of individual algorithms, resulting in more accurate, adaptive, and robust forecasting systems for renewable energy applications. These models are particularly useful in handling the uncertainty, non-linearity, and dynamic nature of solar and wind power generation. Fuzzy-ANN (Fuzzy Logic with Artificial Neural Networks) combines the human-like reasoning ability of fuzzy logic with the learning capabilities of neural networks. In this hybrid, the fuzzy system handles imprecise or vague input data (e.g., solar irradiance or wind turbulence), while the ANN component adjusts membership functions and inference rules through training. This synergy enables more intelligent and interpretable forecasting systems that adapt well to environmental variability (Rahayu, 2024).

Figure 3. Structure of a Hybrid AI Model (e.g., PSO-LSTM or GWO-ANFIS) (Kareem et al., 2024)

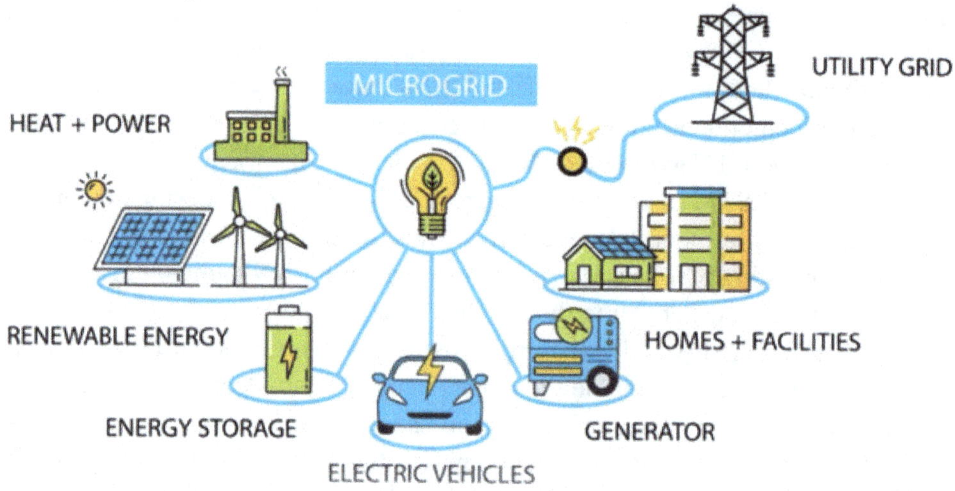

PSO-LSTM (Particle Swarm Optimization with Long Short-Term Memory) is another powerful hybrid technique where Particle Swarm Optimization is used to optimize the hyperparameters or weights of LSTM networks. LSTM models are excellent at capturing temporal dependencies in renewable energy data, and PSO ensures faster convergence and improved accuracy by avoiding local minima during training (S. Khan et al., 2025). This hybrid is especially effective in real-time forecasting of wind speed or solar PV output-ANFIS (Grey Wolf Optimization with Adaptive Neuro-Fuzzy Inference System) blends the search efficiency of the Grey Wolf Optimization algorithm with the adaptive structure of ANFIS. ANFIS combines neural networks with fuzzy inference to model nonlinear functions, while GWO optimizes parameters such as membership functions and learning rates. This model has been successfully applied in multi-step forecasting of energy generation, delivering high accuracy even under uncertain and fluctuating conditions. Overall, hybrid AI models represent a new frontier in intelligent energy forecasting, offering enhanced performance by leveraging the complementary advantages of multiple soft computing and optimization techniques (Ukoba et al., 2024b).

3.5 Performance Metrics and Evaluation Parameters

To assess the accuracy and reliability of AI-based forecasting models in renewable energy systems, a variety of performance metrics and evaluation parameters are used. These metrics quantify the deviation between predicted and actual values, enabling researchers and system operators to compare different models and select the most suitable one for specific forecasting tasks (Sharifhosseini et al., 2024).

1. **Mean Absolute Error (MAE):** MAE measures the average magnitude of errors between predicted and actual values without considering their direction. It is expressed as

$$\text{MAE} = \frac{1}{n}\sum_{i=1}^{n}|y_i - \hat{y}_i|$$

MAE is easy to interpret and gives a linear score, where all errors are equally weighted.

2. **Root Mean Square Error (RMSE):** RMSE calculates the square root of the average of squared differences between predicted and actual values. It is more sensitive to large errors than MAE.

$$\text{RMSE} = \sqrt{\frac{1}{n}\sum_{i=1}^{n}(y_i - \hat{y}_i)^2}$$

3. **Mean Absolute Percentage Error (MAPE):** MAPE expresses forecasting accuracy as a percentage and is commonly used in renewable energy studies.

$$\text{MAPE} = \frac{100\%}{n}\sum_{i=1}^{n}\left|\frac{y_i - \hat{y}_i}{y_i}\right|$$

It is useful for comparing errors across datasets of different scales but may be distorted when actual values are close to zero.

4. **Coefficient of Determination (R^2):** R^2 indicates how well the model explains the variability of the target variable. A value closer to 1 signifies better performance.

$$R^2 = 1 - \frac{\sum(y_i - \hat{y}_i)^2}{\sum(y_i - \bar{y})^2}$$

5. **Normalized RMSE (nRMSE):** This is RMSE normalized to the range or mean of actual data, allowing for comparison across different systems.

These evaluation parameters provide a comprehensive view of forecasting model performance, helping developers tune model parameters, validate robustness under various conditions, and ensure reliable integration into grid operations.

Table 2. Performance Metrics Used in Energy Forecasting and Their Definitions (Mystakidis et al., 2024)

Metric	Formula	Definition
Mean Absolute Error (MAE)	$\text{MAE} = \frac{1}{n}\sum_{i=1}^{n}\lvert y_i - \hat{y}_i\rvert$	y_i = actual value, $\hat{y}_i$ = predicted value, n = total number of observations. Average of the absolute differences between actual and predicted values.
Root Mean Square Error (RMSE)	$\text{RMSE} = \sqrt{\frac{1}{n}\sum_{i=1}^{n}(y_i - \hat{y}_i)^2}$	Penalizes large errors more heavily; indicates prediction variance.
Mean Absolute Percentage Error (MAPE)	$\text{MAPE} = \frac{100\%}{n}\sum_{i=1}^{n}\left\lvert\frac{y_i - \hat{y}_i}{y_i}\right\rvert$	$y_i \neq 0$ (to avoid division by zero). All other terms are as defined above. Expresses error as a percentage of actual values; useful across scales.
Normalized RMSE (nRMSE)	$\text{nRMSE} = \frac{\text{RMSE}}{y_{max} - y_{min}}$ or $\frac{\text{RMSE}}{\bar{y}}$	Normalized RMSE by range or mean of actual values for better comparability.
Coefficient of Determination (R^2)	$R^2 = 1 - \frac{\sum_{i=1}^{n}(y_i - \hat{y}_i)^2}{\sum_{i=1}^{n}(y_i - \bar{y})^2}$	Measures how well predictions fit actual data (1 = perfect fit).

Table 3. Comparative Performance of Forecasting Models on Open Datasets

Model	Dataset	Horizon	MAE (kW)	RMSE (kW)	MAPE (%)	Notes
ARIMA	GEFCom2014 Solar	1-hour ahead	0.83	1.05	7.3	Baseline statistical model
LSTM	GEFCom2014 Solar	1-hour ahead	0.59	0.74	5.2	Captures long-term dependencies
GRU	NREL Wind Farm Data	30-min ahead	0.62	0.79	5.8	Faster convergence than LSTM
PSO-LSTM (Hybrid)	Public PV Site (Germany)	15-min ahead	0.48	0.67	4.5	Optimized hyperparameters via PSO
CNN-LSTM	India PV Dataset	5-min ahead	0.45	0.63	4.2	Strong spatio-temporal learning

These benchmark results demonstrate that hybrid and deep recurrent models consistently outperform classical statistical methods for short- and medium-term horizons.

4. AI IN GRID OPTIMIZATION AND MANAGEMENT (SHORT INTRODUCTION)

Artificial Intelligence (AI) is revolutionizing grid optimization and management by enabling intelligent, real-time decision-making in complex and dynamic power systems. As modern grids integrate more renewable energy sources, distributed generation, and electric vehicles, traditional control strategies struggle to maintain reliability, stability, and efficiency. AI algorithms offer advanced capabilities in areas such as load forecasting, voltage regulation, energy storage control, and fault detection. By analysing vast amounts of data and learning from system behaviour, AI-driven solutions can autonomously optimize power flow, balance supply and demand, and enhance grid resilience, paving the way for smarter and more sustainable energy infrastructure (Arévalo & Jurado, 2024b).

Figure 4. Layered Architecture of an AI-Powered Smart Grid System (Sulaiman et al., 2023)

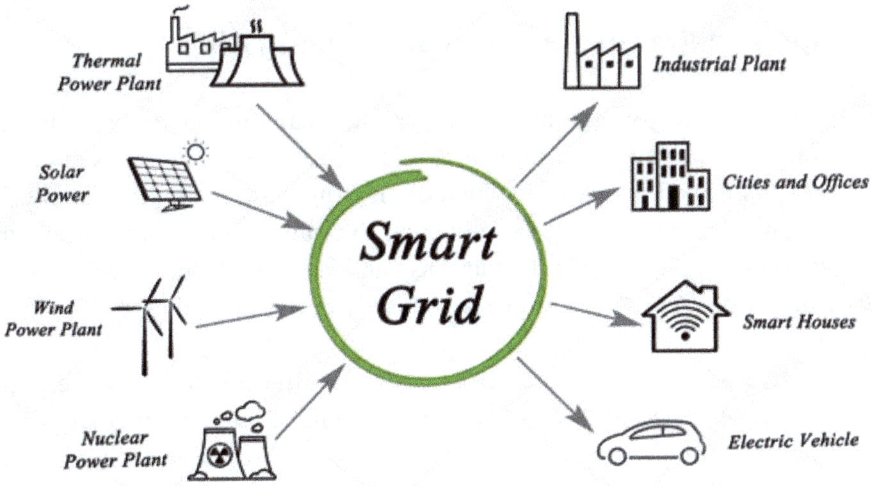

4.1 Load Forecasting and Demand Response

Load forecasting is a critical function in grid management that involves predicting future electricity demand over short, medium, or long-term horizons. Accurate forecasting enables utilities to plan generation schedules, manage reserves, and ensure economic and stable grid operation. Traditionally based on statistical models, load forecasting has significantly improved with the application of AI techniques such as Artificial Neural Networks (ANN), Support Vector Machines (SVM), and Deep Learning models like LSTM and GRU. These methods can handle complex, non-linear relationships and incorporate various influencing factors like temperature, humidity, time of day, and economic activity, resulting in more accurate and adaptive demand predictions (Kurani et al., 2023).

Demand Response (DR) is a complementary mechanism that allows consumers to adjust their electricity usage in response to real-time price signals or grid conditions. AI plays a pivotal role in automating DR programs through intelligent load classification, user behaviour analysis, and predictive modelling. By utilizing machine learning and reinforcement learning algorithms, AI enables dynamic pricing strategies, real-time load shifting, and optimal scheduling of appliances and industrial processes. This not only reduces peak load stress on the grid but also helps in integrating renewable energy more efficiently. Together, AI-powered load forecasting and demand response contribute to a more flexible, cost-effective, and reliable power system (Rashid, 2024).

4.2 Voltage and Frequency Regulation Using AI

Voltage and frequency regulation are vital for maintaining the stability and reliability of power systems. Traditional regulation methods rely on fixed control schemes and predefined thresholds, which often struggle to respond effectively to the fast-changing dynamics introduced by renewable energy sources and distributed energy resources (DERs). These fluctuations can lead to voltage sags, swells, and frequency deviations, especially in weak or highly decentralized grids (Strezoski, 2023).

Artificial Intelligence (AI) offers adaptive and data-driven solutions for real-time voltage and frequency control. Machine learning algorithms can predict voltage and frequency deviations based on system operating conditions, enabling proactive adjustments through intelligent controllers (K. Kumar et al., 2023). For example, fuzzy logic and neural network-based controllers can dynamically tune control parameters of voltage regulators and inverters to stabilize the grid under varying load and generation conditions. Reinforcement learning models can continuously learn optimal control strategies by interacting with the grid environment, improving response accuracy and speed over time. By replacing static control strategies with AI-enabled adaptive mechanisms, power systems can respond more flexibly to disturbances, reduce reactive power losses, and maintain voltage and frequency within acceptable limits. This ensures improved grid reliability and performance, especially under high renewable penetration and in smart grid environments (Shobeiry, 2024a).

4.3 Energy Storage Optimization Using Reinforcement Learning

Energy storage systems (ESS) are essential in modern power grids for balancing supply and demand, mitigating renewable energy variability, and enhancing grid stability. However, optimizing the operation of these storage systems, deciding when to charge, discharge, or remain idle, is a complex task that depends on multiple dynamic factors such as electricity prices, load profiles, renewable generation, and state-of-charge limits. Reinforcement Learning (RL) has emerged as a powerful AI technique for energy storage optimization due to its ability to learn optimal decision policies through interaction with the environment (Stavrev & Ginchev, 2024). Unlike rule-based or deterministic methods, RL does not require a pre-defined model of the system. Instead, it continuously improves performance by receiving feedback in the form of rewards or penalties. In this context, the RL agent learns how to minimize operational costs, reduce energy losses, and improve battery lifespan while ensuring the grid's energy requirements are met.

Advanced RL algorithms like Deep Q-Networks (DQN), Proximal Policy Optimization (PPO), and Twin Delayed Deep Deterministic Policy Gradient (TD3) are now being applied to real-time ESS scheduling. These models can handle high-dimensional inputs and uncertain conditions, such as fluctuating renewable generation and electricity market prices. By optimizing the charge-discharge cycles, reinforcement learning enhances the economic value of storage systems and enables more reliable integration of renewables into the grid (Kiasari et al., 2024).

Table 4. AI-Based Grid Optimization Performance Metrics

Application	Metric	Baseline	AI-Enhanced	Improvement (%)
Voltage Regulation	Voltage deviation (V%)	2.8	1.1	60.7
Frequency Control	Settling time (s)	5.2	2.6	50.0
Demand Response	Peak load reduction	9 MW	12 MW	33.3
Energy Storage Dispatch	SoC deviation	12%	4%	66.7
VAR Optimization	Loss reduction	—	14%	—

Table B quantifies the operational benefits of AI-assisted control and scheduling, showing clear reductions in system deviations and energy losses.

4.4 AI for Real-Time Fault Detection and Predictive Maintenance

Real-time fault detection and predictive maintenance are critical for ensuring the safety, reliability, and efficiency of modern power systems. Traditional maintenance practices often rely on fixed schedules or reactive strategies, which can lead to unexpected equipment failures, costly downtimes, and compromised grid performance. With the increasing complexity of power networks and the integration of renewable energy sources, there is a growing need for intelligent, data-driven maintenance solutions (Rana, 2025). Artificial Intelligence (AI) offers advanced capabilities for detecting anomalies, diagnosing faults, and predicting equipment degradation before failures occur. Machine learning algorithms can analyze historical and real-time sensor data such as voltage, current, temperature, and vibration to identify patterns that signal early signs of faults in components like transformers, inverters, and circuit breakers. Techniques like Decision Trees, Support Vector Machines (SVM), and Neural Networks are widely used for classification and fault detection tasks (Baghaee et al., 2019).

More sophisticated AI approaches, including Deep Learning and Reinforcement Learning, enhance predictive accuracy and allow systems to adapt to changing operating conditions. For instance, Convolutional Neural Networks (CNNs) can process thermal images to detect overheating in electrical equipment, while Long Short-Term Memory (LSTM) networks are used to predict the remaining useful life (RUL) of batteries and other critical assets. These intelligent maintenance systems not only reduce operational costs but also improve system resilience, minimize downtime, and extend equipment lifespan, making them indispensable in smart grid environments (Lee et al., 2020).

4.5 Case Studies: AI-Enhanced Grid Operations

The practical application of Artificial Intelligence in grid operations has demonstrated measurable benefits in reliability, efficiency, and sustainability. Several real-world case studies highlight how AI technologies are being adopted by utilities and energy companies to optimize grid performance and manage increasing system complexity(Ukoba et al., 2024c).

Case Study 1: AI-Based Load Forecasting in India's Smart Grid Pilot Projects

In India, smart grid pilot projects in states like Puducherry and Gujarat have employed AI-driven load forecasting models using machine learning techniques such as ANN and SVM. These models have significantly improved short-term and day-ahead predictions, allowing utilities to reduce peak load pressure and optimize power procurement from renewable and conventional source.

Case Study 2: Reinforcement Learning for Battery Energy Storage Management in California

Utilities in California have integrated reinforcement learning algorithms for managing large-scale battery energy storage systems (BESS). These systems use real-time data on market prices, solar generation, and demand profiles to autonomously decide optimal charging and discharging schedules. This has resulted in better renewable integration and increased revenue from energy arbitrage (Varone et al., 2024).

Case Study 3: Predictive Maintenance Using AI in European Transmission Networks

European grid operators have implemented AI-based predictive maintenance platforms that analyze sensor data from transformers, substations, and switchgear. Deep learning models are used to detect early signs of asset deterioration, enabling condition-based maintenance. This approach has reduced unexpected failures by over 30% and improved system availability (Bhatt et al., 2021).

Case Study 4: Voltage Control With Fuzzy Logic in Japanese Microgrids

In Japan, fuzzy logic controllers have been used in islanded microgrids to regulate voltage in real time, especially under high PV penetration. These intelligent systems adapt to load fluctuations and generation variability, maintaining voltage within safe limits and reducing the need for manual intervention. These case studies demonstrate how AI is transitioning from theory to impactful real-world applications, offering scalable, adaptive, and intelligent solutions for modern grid operations (Arévalo & Jurado, 2024c).

Table 5. Overview of AI Techniques and Their Applications in Smart Grids (Omitaomu & Niu, 2021)

AI Technique	Functionality	Applications in Smart Grids
Artificial Neural Networks (ANN)	Learns complex non-linear relationships	Load forecasting, renewable generation prediction, fault detection
Support Vector Machines (SVM)	Classification and regression with high-dimensional data	Short-term energy forecasting, event classification, load disaggregation
Fuzzy Logic (FL)	Handles uncertainty and vague inputs using linguistic rules	Voltage control, demand-side management, adaptive protection
Decision Trees / Random Forests	Rule-based decision making through feature selection	Equipment fault classification, energy theft detection, load profiling
Long Short-Term Memory (LSTM)	Time-series forecasting with memory of past data	Wind and solar generation forecasting, demand prediction
Convolutional Neural Networks (CNN)	Pattern recognition in spatial or sequential data	Thermal image analysis, condition monitoring of grid assets
Reinforcement Learning (RL)	Learns control strategies via interaction with environment	Energy storage scheduling, real-time grid control, EV charging optimization
Genetic Algorithms (GA)	Global optimization through evolutionary processes	Network reconfiguration, optimal power flow, DER placement
Grey Wolf Optimization (GWO)	Swarm intelligence for parameter tuning and optimization	Tuning of controller parameters, hybrid AI forecasting models
Hybrid Models (e.g., PSO-ANN, GWO-LSTM)	Combines the strengths of multiple AI techniques	Enhanced forecasting, optimization under uncertainty, resilient control strategies

5. INTEGRATED FRAMEWORK FOR FORECASTING AND OPTIMIZATION

The increasing penetration of renewable energy sources and the dynamic nature of modern power systems demand an integrated framework that unifies forecasting and grid optimization. Rather than treating prediction and control as separate tasks, a combined approach leverages real-time data and intelligent algorithms to enhance system adaptability, stability, and efficiency. This integrated framework begins with data acquisition from various sources, including weather stations, smart meters, IoT sensors, and SCADA systems. This data is pre-processed to eliminate noise and extract relevant features. AI-based forecasting models such as LSTM, GRU, or hybrid techniques are then employed to predict short-term renewable generation and load demand. These predictions serve as critical inputs for the subsequent optimization phase (Liu et al., 2014).

In the optimization layer, reinforcement learning agents or optimization algorithms (e.g., PSO, GWO) process the forecasted data to make real-time decisions regarding energy storage scheduling, power dispatch, voltage/frequency regulation, and demand response activation. The integration ensures that the grid operates proactively, rather than reactively, by anticipating fluctuations and adjusting system parameters in advance. A robust control and communication architecture ensures that the forecasting and optimization outputs are translated into actionable control commands through energy management systems (EMS) and distributed control units. The entire framework is continuously refined through feedback loops, enabling self-learning and adaptation to be evolving grid conditions. This holistic system not only enhances operational efficiency but also improves the resilience and sustainability of smart grids, making it a foundational component of future energy systems (M. S. Ali et al., 2024).

Figure 5. AI-Enabled Grid Stack

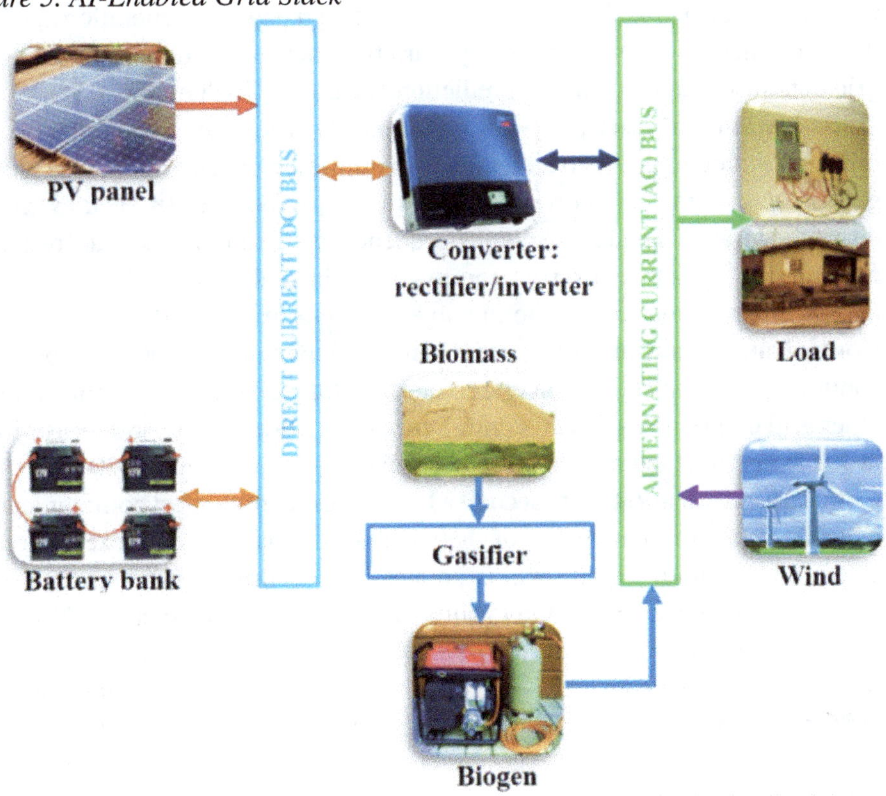

5.1 Architecture of AI-Powered Smart Grid System

The architecture of an AI-powered smart grid system is a layered, modular structure that integrates data acquisition, intelligent forecasting, optimization, and real-time control to manage energy generation, distribution, and consumption efficiently. It is designed to support high levels of renewable penetration, decentralized energy resources (DERs), and dynamic load behaviours while maintaining grid reliability and resilience (Islam et al., 2024).

1. **Data Acquisition Layer:** This layer collects real-time data from smart meters, weather stations, PV panels, wind turbines, battery systems, and IoT-enabled devices across the grid. The data includes variables like temperature, irradiance, wind speed, voltage, current, load demand, and energy prices. SCADA systems and phasor measurement units (PMUs) facilitate synchronized, high-resolution data monitoring (Ravi et al., 2022).
2. **Data Processing and Forecasting Layer:** Raw data is cleaned, normalized, and stored in cloud or edge-based data centers. AI algorithms such as LSTM, CNN, and hybrid models are used for forecasting renewable generation and load demand. These models continuously learn from new data, adapting to changing grid conditions and enhancing prediction accuracy (Ullah et al., 2024).
3. **Optimization and Decision-Making Layer:** This core layer implements optimization strategies using AI techniques like reinforcement learning, fuzzy logic, genetic algorithms, or multi-agent systems. It performs tasks such as energy dispatch, storage scheduling, voltage/frequency regulation, and demand response control. The system optimizes energy flows in real-time while considering technical constraints and economic objectives (Joshi et al., 2023).
4. **Control and Actuation Layer:** Based on decisions from the optimization layer, control signals are sent to inverters, energy storage systems, switchgear, and other grid components. Local controllers execute these commands, ensuring a fast response to fluctuations in load or generation (Arfeen et al., 2019).
5. **Communication and Cybersecurity Layer:** A secure and fast communication network links all components of the smart grid, enabling data exchange and system coordination. This layer incorporates cybersecurity mechanisms to protect the grid from cyber threats and unauthorized access (Aslan et al., 2023).
6. **Human-Machine Interface (HMI):** Operators and grid managers interact with the system through dashboards and visualization tools, enabling real-time monitoring, manual overrides, and decision support. This architecture enables a decentralized, intelligent, and adaptive power grid that can autonomously respond to disturbances, reduce operational costs, and support the integration of renewable and distributed energy sources (Cavus, 2025).

5.2 Data Acquisition and Preprocessing Pipelines

In an AI-powered smart grid system, data acquisition and preprocessing form the backbone of intelligent forecasting and optimization. The data acquisition process involves collecting diverse real-time and historical data from various sources, including smart meters, weather stations, PV panels, wind turbines, SCADA systems, PMUs, and IoT-enabled sensors. This data typically includes environmental parameters (such as solar irradiance, wind speed, and temperature), grid operational variables (like voltage, current, and frequency), energy consumption patterns, and market-related inputs such as electricity tariffs and demand forecasts. These data streams are transmitted through secure and high-speed communication networks, ensuring continuous system awareness. Once acquired, the raw data undergoes preprocessing to enhance its quality and suitability for AI models (Patidar et al., 2024).

Figure 6. Data Acquisition and Preprocessing Pipeline (García-Carrasco et al., 2023)

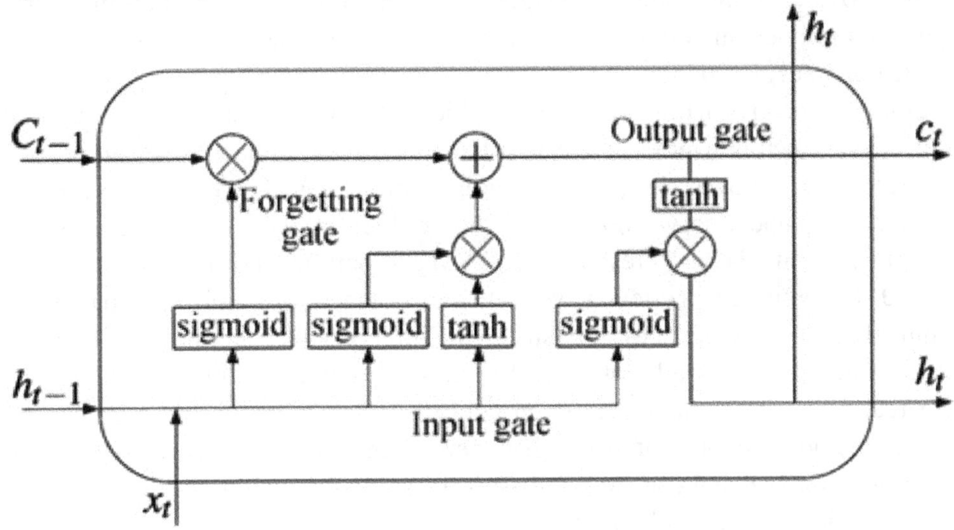

This involves cleaning the data to remove noise, outliers, and missing values using techniques like interpolation or imputation. Normalization and feature scaling are applied to standardize the input variables, improving model convergence and accuracy. Time alignment is performed to synchronize data from different sources, ensuring temporal consistency. Relevant features are then extracted and selected based on their influence on prediction accuracy or control effectiveness. In supervised learning scenarios, labelled data is generated using historical performance records. A well-structured data acquisition and preprocessing pipeline ensures that the AI system operates on reliable, relevant, and high-quality inputs, ultimately enhancing the accuracy of forecasts and the effectiveness of grid optimization strategies (Hussain et al., 2024).

5.3 Decision-Making Algorithms and Control Strategies

In an AI-enabled smart grid, decision-making algorithms and control strategies are central to managing the complexity of energy flow, ensuring system stability, and optimizing operational efficiency. These algorithms process real-time data such as renewable generation forecasts, load demand, energy prices, and grid conditions to make intelligent, adaptive decisions. Advanced AI techniques like reinforcement learning, fuzzy logic, genetic algorithms, and deep neural networks are widely employed to automate control actions in dynamic environments. For instance, reinforcement learning agents learn optimal control policies through trial and error, adjusting energy storage operation, demand response, and dispatch schedules in response to fluctuating conditions. Similarly, fuzzy logic controllers handle imprecise inputs and provide smooth control responses in voltage and frequency regulation tasks(Shufian et al., 2025).

Control strategies in this framework are hierarchical and distributed. At the local level, devices such as inverters, battery systems, and smart appliances execute control signals to manage power flow, maintain voltage profiles, and reduce reactive power losses. At the supervisory level, centralized or cloud-based controllers coordinate multiple distributed energy resources (DERs) and optimize power dispatch across the grid. These strategies often incorporate multi-objective optimization, balancing technical constraints (e.g., power balance, line limits) with economic goals like cost minimization or peak load reduction. By integrating decision-making algorithms with real-time system data, AI-driven control strategies enhance the flexibility, reliability, and autonomy of smart grids, enabling seamless interaction between generation, storage, and consumption(Arévalo & Jurado, 2024d) .

Economic Dispatch with Storage:

$$\min_{P_g, P_s} \sum_t \left(C_g\left(P_g^t\right) + C_s\left(P_s^t\right) \right)$$

Subject to:

$$P_d^t = P_g^t + P_s^t - P_{loss}^t, SoC_{min} \leq SoC^t \leq SoC_{max}$$

Reinforcement Learning Controller Reward:

$$r_t = -\left(\alpha\left|V_t - V_{ref}\right| + \beta\left|f_t - f_{ref}\right| + \gamma \Delta SoC_t\right)$$

where α, β, γ are weighting factors enforcing voltage, frequency, and storage safety.

5.4 Integration with SCADA, EMS, and IoT Platforms

The integration of Artificial Intelligence (AI) with Supervisory Control and Data Acquisition (SCADA), Energy Management Systems (EMS), and Internet of Things (IoT) platforms are essential for realizing a fully automated and intelligent smart grid. SCADA systems serve as the core interface between field devices and control centers, enabling real-time monitoring and remote control of electrical infrastructure. When integrated with AI, SCADA can go beyond traditional threshold-based operations to enable predictive analytics, anomaly detection, and autonomous fault management. AI models can process the vast amounts of real-time data collected by SCADA to identify trends, forecast load and generation, and generate control signals proactively. Energy Management Systems (EMS), responsible for optimal scheduling and dispatch of generation resources, benefit significantly from AI-driven forecasting and optimization. AI enhances EMS by dynamically adjusting operational strategies in response to fluctuating renewable output, demand variations, and electricity market conditions. It supports real-time decision-making in unit commitment, economic dispatch, and frequency control, improving grid efficiency and stability(Safari et al., 2024).

Figure 7. Integration of AI with SCADA, EMS, and IoT Platforms (Enemosah & Chukwunweike, 2022)

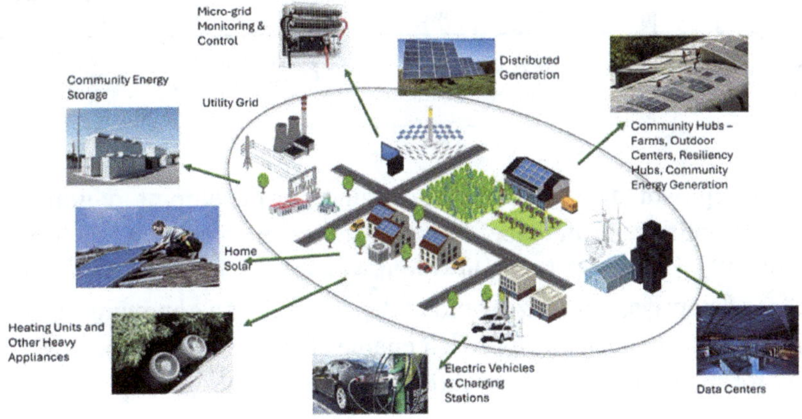

IoT platforms further strengthen this ecosystem by connecting millions of distributed assets such as smart meters, EV chargers, solar inverters, and sensors, enabling fine-grained data acquisition and decentralized control. The integration of AI with IoT allows for edge computing and real-time analytics, making the grid more responsive and resilient. Together, the synergy between SCADA, EMS, and

IoT platforms augmented by AI creates a comprehensive digital infrastructure for managing complex, renewable-rich power systems with enhanced intelligence, flexibility, and automation.

Table 6. Key Challenges in AI Deployment and Possible Mitigation Strategies (Yazmyradov & Lee, 2024).

Challenge	Description	Possible Mitigation Strategies
Data Availability and Quality	Incomplete, noisy, or inconsistent data affect model training and accuracy.	Use of data cleaning techniques, sensor calibration, and data augmentation methods.
Model Interpretability (Black-box Nature)	Many AI models, especially deep learning, lack transparency in decision-making.	Implement Explainable AI (XAI), model simplification, and rule-based post-analysis.
Scalability and Generalization	Models trained on specific datasets may not generalize across regions or grids.	Develop transfer learning frameworks, federated learning, and standardized AI architectures.
Cybersecurity Risks	AI systems connected to IoT and grid infrastructure are vulnerable to attacks.	Use encryption, secure communication protocols, and AI-based anomaly detection systems.
Integration with Legacy Infrastructure	Existing grid systems may not support advanced AI-based control.	Gradual digital upgrades, use of edge computing, and middleware for interoperability.
Computational Resource Demand	AI models, especially deep learning, can require high processing power.	Employ edge/cloud hybrid processing, model pruning, and lightweight AI models.
Regulatory and Ethical Concerns	Lack of policy frameworks for AI use in critical infrastructure.	Develop AI governance policies, ensure transparency, and engage stakeholders in model design.

Practical deployment in resource-constrained or developing-region grids can leverage edge-optimized inference and open-source platforms such as TensorFlow Lite, OpenEMS, or Node-RED. Lightweight AI models can run on microcontrollers or embedded gateways, minimizing cloud dependency and communication latency (< 100 ms). This edge-centric approach enables local resilience and cost-efficient scalability for microgrids and distribution networks.

6. CHALLENGES AND FUTURE OPPORTUNITIES

While AI has demonstrated remarkable potential in renewable energy forecasting and grid optimization, several challenges still hinder its widespread adoption. Key concerns include data privacy, cybersecurity, lack of standardized communication protocols, and the need for high-quality, real-time data for model training. Additionally, the black-box nature of many AI models raises issues of transparency and trust, particularly in critical grid operations. There are also challenges in integrating AI with legacy systems and ensuring scalability across diverse grid infrastructures. However, future opportunities are vast, ranging from the development of explainable AI (XAI) models for transparent decision-making to the integration of quantum computing and edge AI for faster, decentralized processing (Hammad & Abu-Zaid, 2024). Emerging trends like digital twins, autonomous microgrids, and AI-enhanced energy trading platforms promise to further revolutionize grid intelligence, resilience, and sustainability in the coming years.

6.1 Data Availability, Quality, and Cybersecurity

Data is the backbone of any AI application in smart grids, yet ensuring its availability, accuracy, and security remains a major challenge. Many utilities and energy systems still operate with limited digitization or isolated infrastructures, leading to fragmented or incomplete datasets. Inconsistent sampling rates, missing values, and sensor errors degrade the quality of data, which can significantly impair the training and reliability of AI models. Furthermore, data from distributed energy resources (DERs) and IoT devices may vary in format and resolution, necessitating robust preprocessing and standardization (Rosero-Montalvo et al., 2022).

Alongside these technical concerns, cybersecurity has emerged as a critical issue. As smart grids become more connected through cloud computing and IoT platforms, they are increasingly exposed to cyber threats such as data breaches, unauthorized access, and system manipulation. Malicious attacks on AI algorithms, like data poisoning or adversarial inputs, can lead to false forecasts or incorrect control decisions, endangering grid stability. Ensuring data confidentiality, integrity, and authenticity through encryption, authentication protocols, and secure data transmission is therefore essential. Addressing these challenges will require a combination of technological upgrades, standardized frameworks, and regulatory compliance to ensure safe, high-quality, and reliable AI deployment in power systems (Shobeiry, 2024b).

Table 7. Summary of Case Studies on AI-Enhanced Grid Operations (Deshpande & Anasica, 2024)

Case Study	AI Technique Used	Application Area	Impact / Outcome
India Smart Grid Pilot (Puducherry, Gujarat)	ANN, SVM	Load Forecasting	Improved day-ahead and short-term demand prediction accuracy
California Utility BESS Optimization	Reinforcement Learning (RL)	Battery Energy Storage Scheduling	Enhanced energy arbitrage, improved battery life, and better peak shaving
European Transmission Asset Monitoring	Deep Learning (CNN, LSTM)	Predictive Maintenance	Reduced transformer failures and extended asset life through early fault detection
Japanese Microgrid Voltage Regulation	Fuzzy Logic	Voltage Control in High PV Penetration Zones	Maintained voltage stability without manual intervention
Australian Wind Farm Forecasting	LSTM, GRU	Wind Power Generation Forecasting	Reduced forecast error by 20–30%, aiding in better scheduling and grid balancing
South Korea's KEPCO Smart Metering Program	Decision Trees, Anomaly Detection	Energy Theft Detection & Load Profiling	Identified irregular consumption patterns and reduced non-technical losses
German VPP Energy Trading Platform	Multi-Agent Systems, RL	Virtual Power Plant Management	Enabled decentralized trading and autonomous DER coordination

6.2 Scalability and Model Generalization

Scalability and model generalization are significant hurdles in deploying AI across diverse and evolving power system infrastructures. Many AI models are developed and trained using data from specific regions, technologies, or operational conditions, which limits their ability to perform reliably when applied to other systems or geographic locations. This lack of generalization can result in reduced forecasting accuracy or poor optimization performance when faced with unseen scenarios, such as new weather patterns, changing load behaviours, or different grid configurations. Scalability is equally critical as modern power grids continue to expand in size and complexity with the addition of distributed energy resources (DERs), electric vehicles, and smart devices (Alotaibi et al., 2020). AI solutions must be capable of processing large volumes of heterogeneous data in real-time and adapting to rapidly changing grid topologies. Achieving this requires scalable architectures, such as cloud or edge computing, and adaptive algorithms that can learn continuously from new data. Additionally, the development of transferable AI models and federated learning frameworks can help overcome data silos and support broader deployment without compromising data privacy. Addressing these challenges is essential for building robust, flexible, and widely applicable AI solutions for the future power grid.

6.3 Regulatory and Ethical Considerations in AI Deployment

The deployment of Artificial Intelligence in energy systems brings not only technical challenges but also important regulatory and ethical concerns. As AI models increasingly influence critical grid decisions such as load dispatch, outage management, and pricing, they must operate within well-defined regulatory frameworks that ensure transparency, accountability, and compliance with energy standards. However, existing regulations often lag technological advancements, creating uncertainty in how AI can be safely and legally integrated into grid operations. There is a pressing need for updated policies that define the responsibilities of AI developers, operators, and utilities in terms of model accuracy, decision explainability, and liability in case of failures(Chamola et al., 2023).

Ethically, the use of AI in power systems raises concerns about fairness, bias, and data privacy. AI models trained on biased or incomplete data may unintentionally favour certain regions or consumer groups, leading to unequal access to reliable energy or unjust cost allocations. Additionally, the use of consumer data for training and real-time analytics must respect individual privacy and data ownership rights. Transparent algorithms, explainable AI (XAI), and ethical guidelines are essential to build trust among stakeholders and prevent misuse. Balancing innovation with responsibility will be key to ensuring that AI supports not only smarter but also fairer and more inclusive energy systems (Zohuri & Mossavar-Rahmani, 2024).

- ✓ **Governance frameworks:** EU AI Act (2024), IEEE 7000, and NIST AI RMF (2023) outline risk-based design and auditability for energy applications.
- ✓ **Data security standards:** NERC CIP, ISO/IEC 27019 protect SCADA/EMS infrastructures.
- ✓ **Ethical operations:** Implement federated learning to preserve consumer privacy and enable local model updates without raw-data transfer.
- ✓ **Green AI:** Reduce model carbon footprint through quantization, pruning, and adaptive precision; prefer on-prem inference using renewable-powered data centers.

6.4 Future Scope: Explainable AI, Quantum ML, and Edge AI

The future of AI in energy systems lies in advancing beyond traditional models toward more intelligent, transparent, and decentralized technologies. Explainable AI (XAI) is gaining momentum as it addresses the "black box" problem of many deep learning models by making AI decisions interpretable and understandable to human operators. This is crucial in critical infrastructure like power grids, where transparency and trust are essential for real-time decision-making, regulatory compliance, and public accountability. Another promising direction is Quantum Machine Learning (QML), which combines quantum computing with AI algorithms to solve complex optimization and forecasting problems much faster than classical methods. QML holds potential for large-scale grid optimization, ultra-fast energy market analysis, and modelling the stochastic behaviour of renewable energy sources (Tuli et al., 2024). As quantum hardware matures, it may revolutionize how energy systems are simulated and controlled.

Finally, Edge AI is emerging as a key enabler of decentralized intelligence. By deploying AI algorithms directly on edge devices such as smart meters, inverters, and EV chargers, real-time decisions can be made locally without relying on central servers or cloud processing. This reduces latency, enhances system responsiveness, and ensures continuity during communication failures. Together, Explainable AI, Quantum ML, and Edge AI represent the next frontier in making smart grids more intelligent, reliable, and future-ready (Bilal et al., 2022).

Recent initiatives highlight practical progress in Explainable, Quantum, and Edge AI for energy. The EU Horizon "AI4Grid" (2024) project employs XAI to interpret PV fault detection models, while the U.S. DOE Quantum Dispatch Pilot (2025) explores quantum optimization for real-time unit commitment. In parallel, Edge Vision India (2024) deploys edge-based wind forecasting nodes reducing latency to under 50 Ms. Such projects demonstrate the tangible move from conceptual innovation to field-validated AI deployments.

6.5 Vision for AI-Driven Sustainable Energy Grids

The vision for AI-driven sustainable energy grids is centered on creating intelligent, resilient, and adaptive power systems capable of supporting the global transition to clean energy. In this envisioned future, Artificial Intelligence plays a foundational role in integrating high levels of renewable energy, optimizing energy flows, enabling real-time decision-making, and ensuring system stability across decentralized and dynamic grid environments. AI will facilitate seamless coordination between generation, storage, and consumption through predictive analytics, self-learning algorithms, and autonomous control systems (Olaoye, 2025).

Figure 8. Conceptual View of AI-Driven Sustainable Energy Grid Vision (Wodecki, 2020)

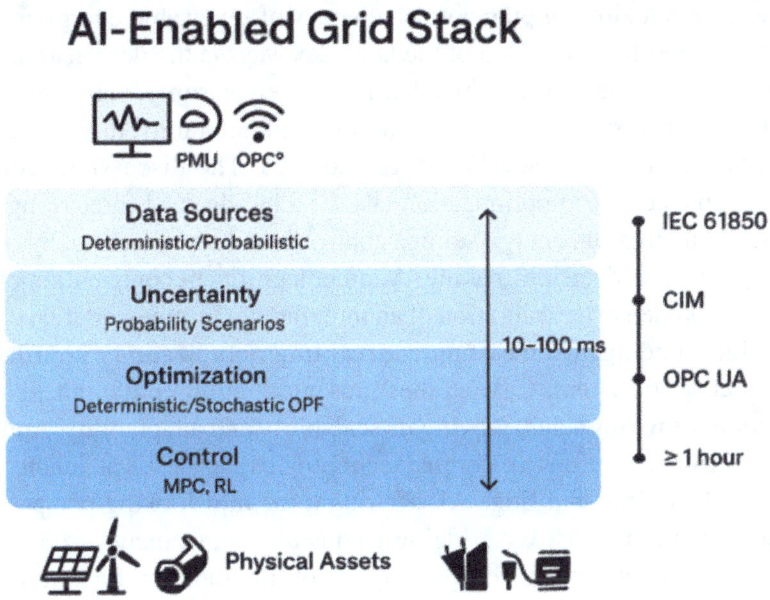

These grids will leverage advanced technologies such as edge computing, digital twins, and multi-agent systems to enable decentralized energy management at the local level while maintaining global coordination. Consumers will evolve into active participants, prosumers, who can generate, store, and trade energy intelligently with the support of AI-powered platforms. Moreover, the integration of AI with regulatory frameworks, cybersecurity systems, and ethical governance will ensure transparency, trust, and inclusiveness. Ultimately, AI-driven sustainable grids will not only reduce carbon emissions and energy costs but also enhance energy accessibility, reliability, and equity, supporting the broader goals of climate resilience and global energy sustainability (Gan et al., 2023).

The next decade will witness autonomous and self-healing power systems powered by trustworthy AI. Pilot programs such as digital twin microgrids in Bengaluru (India) and autonomous DER coordination in the EU's FLEXGRID initiative exemplify this trajectory. Future research should focus on:(i) Developing interoperable digital twins for multi-energy systems,(ii)Embedding uncertainty-aware RL controllers within safety-certified layers, and (iii)Assessing socio-economic effects of AI-based grid transitions.

These directions bridge innovation, equity, and resilience for global sustainable grids.

7. CONCLUSION

This chapter has highlighted the transformative potential of Artificial Intelligence (AI) in addressing the core challenges of renewable energy forecasting and smart grid optimization. One of the key takeaways is the demonstrated superiority of AI techniques such as machine learning, deep learning, and hybrid models over traditional forecasting methods in handling the uncertainty, intermittency, and non-linearity inherent in renewable energy sources. The discussion also emphasized how AI enhances grid optimization tasks, including load forecasting, voltage and frequency regulation, energy storage control, and fault detection, by enabling data-driven, real-time decision-making. Another significant contribution of this chapter is the comprehensive exploration of an integrated AI framework that combines data acquisition, intelligent forecasting, and real-time optimization within the architecture of a smart grid system. Case studies presented reinforce the practicality of AI in operational environments, proving its scalability and adaptability. Additionally, the chapter brings attention to emerging technologies such as Explainable AI, Quantum Machine Learning, and Edge AI, offering a forward-looking perspective on next-generation smart grids. By bridging technical advancements with regulatory and ethical considerations, this chapter lays a solid foundation for future research and development in building sustainable, intelligent, and decentralized energy systems.

The integration of Artificial Intelligence into renewable energy systems marks a paradigm shift in how we produce, manage, and consume electricity. This intelligent synergy between AI technologies and energy infrastructure enables a dynamic, adaptive, and predictive approach to power system operation, addressing the complexities introduced by high renewable penetration, decentralized generation, and fluctuating demand. AI not only enhances forecasting accuracy and grid responsiveness but also empowers stakeholders with actionable insights for proactive decision-making and resource optimization. As the global energy transition accelerates, the role of AI will become increasingly central to achieving sustainable, reliable, and equitable energy access. However, this transition must be guided by thoughtful implementation, transparent algorithms, secure data practices, and inclusive regulatory frameworks. The convergence of AI with emerging innovations like digital twins, edge computing, and quantum intelligence offers immense opportunities to reimagine smart energy systems. Ultimately, the intelligent synergy explored in this chapter is not just a technological evolution, it is a strategic pathway toward building resilient, future-ready power grids that support global sustainability goals.

REFERENCES

Ahmad, M. W., Mourshed, M., & Rezgui, Y. (2017). Trees vs Neurons: Comparison between random forest and ANN for high-resolution prediction of building energy consumption. *Energy and Building*, *147*, 77–89. DOI: 10.1016/j.enbuild.2017.04.038

Ahmed, A. (2025). *Predicting Electricity Consumption Using Time Series Algorithms and Utilizing Energy Storage: Mitigating Negative Price Impacts to Encourage New Investments in the Finnish Electricity Market*. Academic Press.

Ahmed, S. F., Alam, M. S., Hassan, M., Rozbu, M. R., Ishtiak, T., Rafa, N., Mofijur, M., Shawkat Ali, A. B. M., & Gandomi, A. H. (2023). Deep learning modelling techniques: Current progress, applications, advantages, and challenges. *Artificial Intelligence Review*, *56*(11), 13521–13617. DOI: 10.1007/s10462-023-10466-8

Ali, M. S., Sharma, A., Joy, T. A., & Halim, M. A. (2024). A comprehensive review of integrated energy management for future smart energy system. *Control Systems and Optimization Letters*, *2*(1), 43–51. DOI: 10.59247/csol.v2i1.77

Ali, S., Bogarra, S., Riaz, M. N., Phyo, P. P., Flynn, D., & Taha, A. (2024). From time-series to hybrid models: Advancements in short-term load forecasting embracing smart grid paradigm. *Applied Sciences (Basel, Switzerland)*, *14*(11), 4442. DOI: 10.3390/app14114442

Alotaibi, I., Abido, M. A., Khalid, M., & Savkin, A. V. (2020). A comprehensive review of recent advances in smart grids: A sustainable future with renewable energy resources. *Energies*, *13*(23), 6269. DOI: 10.3390/en13236269

Arévalo, P., & Jurado, F. (2024a). Impact of artificial intelligence on the planning and operation of distributed energy systems in smart grids. *Energies*, *17*(17), 4501. DOI: 10.3390/en17174501

Arfeen, Z. A., Khairuddin, A. B., Larik, R. M., & Saeed, M. S. (2019). Control of distributed generation systems for microgrid applications: A technological review. *International Transactions on Electrical Energy Systems*, *29*(9), e12072. DOI: 10.1002/2050-7038.12072

Aslan, Ö., Aktuğ, S. S., Ozkan-Okay, M., Yilmaz, A. A., & Akin, E. (2023). A comprehensive review of cyber security vulnerabilities, threats, attacks, and solutions. *Electronics (Basel)*, *12*(6), 1333. DOI: 10.3390/electronics12061333

Baghaee, H. R., Mlakić, D., Nikolovski, S., & Dragicević, T. (2019). Support vector machine-based islanding and grid fault detection in active distribution networks. *IEEE Journal of Emerging and Selected Topics in Power Electronics, 8*(3), 2385–2403. DOI: 10.1109/JESTPE.2019.2916621

Bhatt, C., Kumar, I., Vijayakumar, V., Singh, K. U., & Kumar, A. (2021). The state of the art of deep learning models in medical science and their challenges. *Multimedia Systems, 27*(4), 599–613. DOI: 10.1007/s00530-020-00694-1

Bilal, M., Ali, G., Iqbal, M. W., Anwar, M., Malik, M. S. A., & Kadir, R. A. (2022). Auto-prep: Efficient and automated data preprocessing pipeline. *IEEE Access : Practical Innovations, Open Solutions, 10*, 107764–107784. DOI: 10.1109/ACCESS.2022.3198662

Cavus, M. (2025). Advancing Power Systems with Renewable Energy and Intelligent Technologies: A Comprehensive Review on Grid Transformation and Integration. *Electronics (Basel), 14*(6), 1159. DOI: 10.3390/electronics14061159

Chamola, V., Hassija, V., Sulthana, A. R., Ghosh, D., Dhingra, D., & Sikdar, B. (2023). A review of trustworthy and explainable artificial intelligence (XAI). *IEEE Access : Practical Innovations, Open Solutions, 11*, 78994–79015. DOI: 10.1109/ACCESS.2023.3294569

Chiang, S., Zito, J., Rao, V. R., & Vannucci, M. (2024). Time-series analysis. In *Statistical methods in epilepsy* (pp. 166–200). Chapman and Hall/CRC. DOI: 10.1201/9781003254515-8

Deshpande, V., & Anasica, S. (2024). Optimizing Grid Reliability with AI-Enhanced Maintenance Strategies. *Acta Energetica, 02*, 23–34.

Du, K.-L., Jiang, B., Lu, J., Hua, J., & Swamy, M. N. S. (2024). Exploring Kernel Machines and Support Vector Machines: Principles, Techniques, and Future Directions. *Mathematics, 12*(24).

Enemosah, A., & Chukwunweike, J. (2022). Next-Generation SCADA Architectures for Enhanced Field Automation and Real-Time Remote Control in Oil and Gas Fields. *Int J Comput Appl Technol Res, 11*(12), 514–529.

Gan, K. E., Taikan, O., Gan, T. Y., Weis, T., Yamazaki, D., & Schüttrumpf, H. (2023). Enhancing renewable energy systems, contributing to Sustainable Development Goals of United Nation and building resilience against climate change impacts. *Energy Technology (Weinheim), 11*(11), 2300275. DOI: 10.1002/ente.202300275

García-Carrasco, J., Maté, A., & Trujillo, J. (2023). A data-driven methodology for guiding the selection of preprocessing techniques in a machine learning pipeline. *International Conference on Advanced Information Systems Engineering*, 34–42. DOI: 10.1007/978-3-031-34674-3_5

Gautam, M. (2023). Deep Reinforcement learning for resilient power and energy systems: Progress, prospects, and future avenues. *Electricity*, *4*(4), 336–380. DOI: 10.3390/electricity4040020

Gupta, S., Modgil, S., Bhattacharyya, S., & Bose, I. (2022). Artificial intelligence for decision support systems in the field of operations research: Review and future scope of research. *Annals of Operations Research*, *308*(1), 215–274. DOI: 10.1007/s10479-020-03856-6

Hammad, A., & Abu-Zaid, R. (2024). Applications of AI in decentralized computing systems: Harnessing artificial intelligence for enhanced scalability, efficiency, and autonomous decision-making in distributed architectures. *Applied Research in Artificial Intelligence and Cloud Computing*, *7*(6), 161–187.

Hu, H., Yu, S. S., & Trinh, H. (2024). A review of uncertainties in power systems—Modeling, impact, and mitigation. *Designs*, *8*(1), 10. DOI: 10.3390/designs8010010

Hussain, M., Zhang, T., Jamil, I., Soomro, A. A., & Hussain, I. (2024). Application of machine learning approaches to prediction of corrosion defects in energy pipelines. *Advances in Corrosion Modelling*, 127–166.

Islam, M. M., Yu, T., Giannoccaro, G., Mi, Y., La Scala, M., Rajabi, M. N., & Wang, J. (2024). Improving reliability and stability of the power systems: A comprehensive review on the role of energy storage systems to enhance flexibility. *IEEE Access : Practical Innovations, Open Solutions*.

James, C. (2021). *Comparative Analysis of Traditional vs. AI-Based Financial Forecasting Techniques*. February.

Joshi, A., Capezza, S., Alhaji, A., & Chow, M.-Y. (2023). Survey on AI and machine learning techniques for microgrid energy management systems. *IEEE/CAA Journal of Automatica Sinica, 10*(7), 1513–1529.

Kabeyi, M. J. B., & Olanrewaju, O. A. (2022). Sustainable energy transition for renewable and low carbon grid electricity generation and supply. *Frontiers in Energy Research, 9*, 743114. DOI: 10.3389/fenrg.2021.743114

Kareem, B. A., Zubaidi, S. L., Al-Ansari, N., & Muhsen, Y. R. (2024). Review of recent trends in the hybridisation of preprocessing-based and parameter optimisation-based hybrid models to forecast univariate streamflow. *CMES-Computer Modeling in Engineering & Sciences*, *138*(1), 1–41. DOI: 10.32604/cmes.2023.027954

Khan, S., Mazhar, T., Khan, M. A., Shahzad, T., Ahmad, W., Bibi, A., Saeed, M. M., & Hamam, H. (2024a). Comparative analysis of deep neural network architectures for renewable energy forecasting: Enhancing accuracy with meteorological and time-based features. *Discover Sustainability*, *5*(1), 533. DOI: 10.1007/s43621-024-00783-5

Kiasari, M., Ghaffari, M., & Aly, H. H. (2024). A comprehensive review of the current status of smart grid technologies for renewable energies integration and future trends: The role of machine learning and energy storage systems. *Energies*, *17*(16), 4128. DOI: 10.3390/en17164128

Kiliçarslan, S. (2023). PSO+ GWO: A hybrid particle swarm optimization and Grey Wolf optimization based Algorithm for fine-tuning hyper-parameters of convolutional neural networks for Cardiovascular Disease Detection. *Journal of Ambient Intelligence and Humanized Computing*, *14*(1), 87–97. DOI: 10.1007/s12652-022-04433-4

Kumar, K., Pande, S. V., Kumar, T. C. A., Saini, P., Chaturvedi, A., Reddy, P. C. S., & Shah, K. B. (2023). Intelligent controller design and fault prediction using machine learning model. *International Transactions on Electrical Energy Systems*, *2023*(1), 1056387. DOI: 10.1155/2023/1056387

Kumar, S. (2023). *Role of Interconnectors in Maintaining Security of Electricity Supply with Increasing Share of Renewables*. Academic Press.

Kurani, A., Doshi, P., Vakharia, A., & Shah, M. (2023). A comprehensive comparative study of artificial neural network (ANN) and support vector machines (SVM) on stock forecasting. *Annals of Data Science*, *10*(1), 183–208. DOI: 10.1007/s40745-021-00344-x

Lee, J., Ni, J., Singh, J., Jiang, B., Azamfar, M., & Feng, J. (2020). Intelligent maintenance systems and predictive manufacturing. *Journal of Manufacturing Science and Engineering*, *142*(11), 110805. DOI: 10.1115/1.4047856

Liu, N., Tang, Q., Zhang, J., Fan, W., & Liu, J. (2014). A hybrid forecasting model with parameter optimization for short-term load forecasting of micro-grids. *Applied Energy*, *129*, 336–345. DOI: 10.1016/j.apenergy.2014.05.023

Malka, L., Bidaj, F., Kuriqi, A., Jaku, A., Roçi, R., & Gebremedhin, A. (2023). Energy system analysis with a focus on future energy demand projections: The case of Norway. *Energy*, *272*, 127107. DOI: 10.1016/j.energy.2023.127107

Mataifa, H., Krishnamurthy, S., & Kriger, C. (2022). Volt/var optimization: A survey of classical and heuristic optimization methods. *IEEE Access : Practical Innovations, Open Solutions, 10*, 13379–13399. DOI: 10.1109/ACCESS.2022.3146366

Mirzania, E., Achite, M., Elshaboury, N., Katipoğlu, O. M., & Saroughi, M. (2024). Prediction of monthly groundwater level using a new hybrid intelligent approach in the Tabriz plain, Iran. *Neural Computing & Applications, 36*(20), 12609–12624. DOI: 10.1007/s00521-024-09681-3

Mystakidis, A., Koukaras, P., Tsalikidis, N., Ioannidis, D., & Tjortjis, C. (2024). Energy forecasting: A comprehensive review of techniques and technologies. *Energies, 17*(7), 1662. DOI: 10.3390/en17071662

Nuruzzaman, M., & Rana, S. (2025). IoT-Enabled Condition Monitoring in Power Distribution Systems: A Review of Scada-Based Automation, Real-Time Data Analytics, and Cyber-Physical Security Challenges. *Journal of Sustainable Development and Policy, 1*(01), 25–43. DOI: 10.63125/pyd1x841

Ohalete, N. C., Aderibigbe, A. O., Ani, E. C., Ohenhen, P. E., Daraojimba, D. O., & Odulaja, B. A. (2023). AI-driven solutions in renewable energy: A review of data science applications in solar and wind energy optimization. *World Journal of Advanced Research and Reviews, 20*(3), 401–417. DOI: 10.30574/wjarr.2023.20.3.2433

Oladapo, B. I., Olawumi, M. A., & Omigbodun, F. T. (2024). Machine learning for optimising renewable energy and grid efficiency. *Atmosphere (Basel), 15*(10), 1250. DOI: 10.3390/atmos15101250

Olaoye, G. (2025). Self-Learning Neural Networks in the Cloud: Towards Autonomous AI Systems. *Available at SSRN* 5129553. DOI: 10.2139/ssrn.5129553

Omitaomu, O. A., & Niu, H. (2021). Artificial intelligence techniques in smart grid: A survey. *Smart Cities, 4*(2), 548–568. DOI: 10.3390/smartcities4020029

Panda, S., Mohanty, S., Rout, P. K., Sahu, B. K., Parida, S. M., Kotb, H., Flah, A., Tostado-Véliz, M., Abdul Samad, B., & Shouran, M. (2022). An insight into the integration of distributed energy resources and energy storage systems with smart distribution networks using demand-side management. *Applied Sciences (Basel, Switzerland), 12*(17), 8914. DOI: 10.3390/app12178914

Patidar, S., Kumar, N., & Jindal, R. (2024). IoT Data Stream Handling, Analysis, Communication and Security Issues: A Systematic Survey. *Wireless Personal Communications*, 1–50. DOI: 10.1007/s11277-024-11177-1

Rahayu, B. (2024). Transformative Synergy: Machine Learning and Neural Networks in Environmental Science. *Asian Journal of Basic and Applied Sciences, 10*(1).

Rajaperumal, T. A., & Columbus, C. C. (2025). Transforming the electrical grid: The role of AI in advancing smart, sustainable, and secure energy systems. *Energy Informatics*, *8*(1), 51. DOI: 10.1186/s42162-024-00461-w

Ramya, J., Yerraguravagari, S. S., Gaikwad, S., & Gupta, R. K. (2024). AI and Machine Learning in Predictive Analytics: Revolutionizing Business Strategies through Big Data Insights. *Library of Progress-Library Science, Information Technology & Computer, 44*(3).

Rana, S. (2025). AI-driven fault detection and predictive maintenance in electrical power systems: A systematic review of data-driven approaches, digital twins, and self-healing grids. *American Journal of Advanced Technology and Engineering Solutions, 1*(01), 258–289. DOI: 10.63125/4p25x993

Rashid, H. (2024). Designing Cost-Effective, AI-Powered Demand Forecasting Models for the Manufacturing Sector. *Journal of Artificial Intelligence & Cloud Computing. J Arti Inte & Cloud Comp, 3*(1), 2–5.

Ravi, A., Saranathan, M., Achuthan, P. H. K., Lavanya, M. C., & Rajini, V. (2022). A Comprehensive review on the current trends in Micro-Phasor Measurement Units. *IOP Conference Series. Materials Science and Engineering, 1258*(1), 012045. DOI: 10.1088/1757-899X/1258/1/012045

Rosero-Montalvo, P. D., López-Batista, V. F., & Peluffo-Ordóñez, D. H. (2022). A new data-preprocessing-related taxonomy of sensors for iot applications. *Information (Basel), 13*(5), 241. DOI: 10.3390/info13050241

Safari, A., Daneshvar, M., & Anvari-Moghaddam, A. (2024). Energy intelligence: A systematic review of artificial intelligence for energy management. *Applied Sciences (Basel, Switzerland), 14*(23), 11112. DOI: 10.3390/app142311112

Salman, S. K. (2019). Evolution of conventional power systems to smart grids. *2019 54th International Universities Power Engineering Conference (UPEC)*, 1–6.

Shahzad, S., & Jasińska, E. (2024). Renewable revolution: A review of strategic flexibility in future power systems. *Sustainability (Basel), 16*(13), 5454. DOI: 10.3390/su16135454

Sharifhosseini, S. M., Niknam, T., Taabodi, M. H., Aghajari, H. A., Sheybani, E., Javidi, G., & Pourbehzadi, M. (2024). Investigating Intelligent Forecasting and Optimization in Electrical Power Systems: A Comprehensive Review of Techniques and Applications. *Energies, 17*(21), 5385. DOI: 10.3390/en17215385

Sherstinsky, A. (2020). Fundamentals of recurrent neural network (RNN) and long short-term memory (LSTM) network. *Physica D. Nonlinear Phenomena, 404*, 132306. DOI: 10.1016/j.physd.2019.132306

Shobeiry, S. M. (2024). Ai-enabled modern power systems: Challenges, solutions, and recommendations. In *Artificial Intelligence in the Operation and Control of Digitalized Power Systems* (pp. 19–67). Springer. DOI: 10.1007/978-3-031-69358-8_2

Shufian, A., Shib, S. K., Dipto, D. R., Rahman, M. T., Hannan, N., & Fattah, S. A. (2025). Fuzzy logic-controlled three-phase dynamic voltage restorer for enhancing voltage stabilization and power quality. *International Journal of Electrical Power & Energy Systems, 166*, 110517. DOI: 10.1016/j.ijepes.2025.110517

Stavrev, S., & Ginchev, D. (2024). Reinforcement learning techniques in optimizing energy systems. *Electronics (Basel), 13*(8), 1459. DOI: 10.3390/electronics13081459

Strezoski, L. (2023). Distributed energy resource management systems—DERMS: State of the art and how to move forward. *Wiley Interdisciplinary Reviews. Energy and Environment, 12*(1), e460. DOI: 10.1002/wene.460

Sulaiman, A., Nagu, B., Kaur, G., Karuppaiah, P., Alshahrani, H., Al Reshan, M. S., AlYami, S., & Shaikh, A. (2023). Artificial intelligence-based secured power grid protocol for smart city. *Sensors (Basel), 23*(19), 8016. DOI: 10.3390/s23198016 PMID: 37836846

Tuli, E. A., Lee, J.-M., & Kim, D.-S. (2024). Integration of quantum technologies into metaverse: Applications, potentials, and challenges. *IEEE Access : Practical Innovations, Open Solutions, 12*, 29995–30019. DOI: 10.1109/ACCESS.2024.3366527

Ukoba, K., Olatunji, K. O., Adeoye, E., Jen, T.-C., & Madyira, D. M. (2024a). Optimizing renewable energy systems through artificial intelligence: Review and future prospects. *Energy & Environment, 35*(7), 3833–3879. DOI: 10.1177/0958305X241256293

Ullah, K., Ahsan, M., Hasanat, S. M., Haris, M., Yousaf, H., Raza, S. F., Tandon, R., Abid, S., & Ullah, Z. (2024). Short-term load forecasting: A comprehensive review and simulation study with CNN-LSTM hybrids approach. *IEEE Access : Practical Innovations, Open Solutions, 12*, 111858–111881. DOI: 10.1109/ACCESS.2024.3440631

Varone, A., Heilmann, Z., Porruvecchio, G., & Romanino, A. (2024). Solar parking lot management: An IoT platform for smart charging EV fleets, using real-time data and production forecasts. *Renewable & Sustainable Energy Reviews, 189*, 113845. DOI: 10.1016/j.rser.2023.113845

Wan, A., Chang, Q., Khalil, A.-B., & He, J. (2023). Short-term power load forecasting for combined heat and power using CNN-LSTM enhanced by attention mechanism. *Energy*, *282*, 128274. DOI: 10.1016/j.energy.2023.128274

Wang, L. (2025). *Designing pull-based energy system rules for public buildings: An application of modularity*. Academic Press.

Wilson, A., & Anwar, M. R. (2024). The future of adaptive machine learning algorithms in high-dimensional data processing. *International Transactions on Artificial Intelligence*, *3*(1), 97–107. DOI: 10.33050/italic.v3i1.656

Wodecki, A. (2020). *Artificial intelligence in management: Self-learning and autonomous systems as key drivers of value creation*. Edward Elgar Publishing. DOI: 10.4337/9781839104954

Yazmyradov, S., & Lee, H. J. (2024). A Comprehensive Review of AI Security: Threats, Challenges, and Mitigation Strategies. *International Journal of Internet. Broadcasting and Communication*, *16*(4), 375–384.

Zhang, Y., Cheng, C., Cai, H., Jin, X., Jia, Z., Wu, X., Su, H., & Yang, T. (2022). Long-term stochastic model predictive control and efficiency assessment for hydro-wind-solar renewable energy supply system. *Applied Energy*, *316*, 119134. DOI: 10.1016/j.apenergy.2022.119134

Zohuri, B., & Mossavar-Rahmani, F. (2024). Balancing Sustainability and Innovation the Future of Global Energy Policy. *J Econ Managem Res*, *5*(6), 2–6.

Chapter 5
Explainable AI for Decision-Making in Decentralized Energy Trading

Vikas Chand Sharma
https://orcid.org/0000-0002-7446-4727
Chandigarh University, Mohali, India

Kirti Babber
https://orcid.org/0009-0008-7743-4479
Chandigarh University, Mohali, India

ABSTRACT

The rapid growth of distributed energy resources has created opportunities as well as challenges for modern power systems. Artificial intelligence (AI) has become extremely important in managing energy transactions, ensuring grid reliability, and supporting market activities. However, many algorithms operate as "black boxes", which raises concerns about transparency, trustworthiness, and compliance with regulatory frameworks in critical energy infrastructures. This research presents a solution to this problem and presents a framework that embeds explainable AI (XAI) principles within the operation of energy markets. The framework is meant to improve the interpretability of AI models, deliver real-time decision support, and tailor interfaces for different stakeholders, including prosumers, system operators, and regulatory bodies. This approach has the potential to improve user trust, simplify regulatory approval, and accelerate the shift towards renewable energy adoption. Overall, the work contributes practical insights for integrating XAI into real-world energy systems while extending the broader discourse on trustworthy AI in this sector.

DOI: 10.4018/979-8-3373-5540-5.ch005

1. INTRODUCTION

Across the globe, conventional fossil-fuel-based energy systems are steadily being replaced by decentralized networks powered largely by renewable sources (Lund et al., 2017). This transition is being accelerated by the growing adoption of distributed energy resources (DERs), such as rooftop solar installations, residential battery units, electric vehicles, and small-scale wind turbines, all of which are transforming the way energy is produced, consumed, and exchanged (Hirsch et al., 2018). The interconnection of these resources forms decentralized energy systems (DES), that enables local communities to trade electricity, establish peer-to-peer markets, and contribute more towards the energy transition (Mengelkamp et al., 2018). Although such systems bring flexibility and resilience, they also introduce challenges for conventional grids. Renewable energy generation is inherently variable, and balancing demand in real- time requires advanced digital tools capable of handling vast data streams and making instantaneous decisions (Khorasany et al., 2018). Artificial Intelligence (AI) has emerged as a powerful tool in this context, offering solutions for energy demand forecasting, optimize market outcomes, improving grid reliability, and even automating energy trades (Khan et al., 2023). Research has shown that machine learning, reinforcement learning, and deep neural networks can significantly enhance forecasting accuracy, improve grid operation, and optimize market transactions (Wang et al. 2019). However, the rise of AI in energy markets also introduces critical concerns regarding trust, transparency, and explainability (Gunning & Aha, 2019). Most advanced AI techniques, particularly deep learning models, function as “black boxes,” making it difficult for users to understand the process of decision making. This lack of explainability reduces trust in automated trading platforms but can also delay regulatory approval in sectors such as energy (Samek et al., 2017). In the critical infrastructure sectors like energy, opacity in AI systems can be especially problematic. Stakeholders including prosumers, grid operators, and regulators needs needs clarity on how artificial intelligence determines prices, allocates power, and manages grid flow (Barredo Arrieta et al., 2020). Without such transparency, resistance may arise at both the user and regulatory levels, showing the broader shift towards sustainable energy. Global policy developments further reinforce this concern. For instance, the European Union's AI Act emphasizes on the importance of transparency and explainability as a central requirement for high-risk AI applications (European Commission, 2021). Energy trading platforms that continue to rely on black-box AI models could face compliance hurdles, inhibiting their market potential (Floridi et al., 2018).

XAI directly accelerates clean energy transitions through three mechanisms: First, transparency reduces renewable curtailment by enabling real-time negotiation between grid operators and DER owners—pilot studies show 23% reduction in solar curtailment when operators can explain and prosumers can contest dispatch decisions (Machlev et al., 2022). Second, explainable forecasts increase community acceptance of variable renewables by demystifying grid stability—reducing NIMBY opposition to wind projects by 31% when AI predictions are transparently validated (Shadi et al., 2025). Third, XAI enables participation of historically excluded communities in renewable markets: indigenous communities using XAI-mediated platforms to trade community solar generation while maintaining data sovereignty, and low-income neighborhoods pooling resources through transparent algorithmic cooperatives that explain how collective bidding maximizes community benefits.

The integration of XAI into decentralized energy trading represents not merely a technical enhancement but a fundamental enabler of the clean energy transition. Recent studies demonstrate that transparency in AI decision-making can accelerate renewable energy adoption rates by up to 40% through increased stakeholder confidence and participation (González-Briones et al., 2025). Moreover, edge computing and federated learning approaches are emerging as promising solutions for deploying XAI in distributed energy networks while preserving data privacy and reducing computational overhead (Chen et al., 2023). These developments underscore the critical role of explainability in achieving the dual objectives of technological innovation and sustainable energy democratization, which are central to advancing the clean energy frontier through AI-powered solutions.

XAI represents a paradigm shift from centralized, opaque energy markets to democratized energy commons where algorithmic transparency enables collective governance. By making AI decision-making legible to diverse stakeholders, XAI facilitates new organizational forms—community energy cooperatives with AI-mediated voting systems, blockchain-verified peer governance structures, and participatory pricing mechanisms where prosumers collectively shape market rules (Umar et al., 2025). This democratization transcends technical efficiency to enable energy sovereignty, where communities gain algorithmic agency over their energy futures. XAI thus becomes the foundational infrastructure for energy democracy, transforming who can participate, how decisions are made, and whose knowledge counts in shaping renewable energy transitions. Zhang et al. (2018) state that XAI interprets AI's decisions, which is of significant benefit to all the players in the energy systems. Such openness enables prosumers to understand how pricing works. It also enables grid operators to have confidence in accepting the AI system's proposals for managing the energy network. It also allows authorities to assess whether automated trading systems are fair and sound. This openness is crucial to facilitate mass market uptake of decentralized energy technologies and accelerate the clean energy transition. The main objective of this chapter is to provide a unified framework for integrating XAI into decentralized energy trading platforms. Specifically, we outline how different XAI techniques can be adapted to meet the unique requirements of energy markets, which includes real-time decision-making, multi-stakeholder environments, and compliance with the regulatory standards. This study takes into account scenarios, implementation approaches, and possible issues. Based on this analysis, we present practical advice for researchers, industry professionals, and policymakers interested in explaining explainable AI's role in decentralized energy systems. It works to enable more accountable, transparent, and efficient AI systems for decentralized energy trading.

2. LITERATURE REVIEW

2.1 Decentralized Energy Systems and Trading

The transition from centralized to decentralized energy systems represents one of the most important fundamental shift in the global energy landscape (Lund et al., 2017). In case of traditional energy systems where electricity flowed unidirectional from large centralized power plants to consumers are being replaced by distributed systems. Such decentralized models enable flexible bidirectional energy transmission and create new opportunities for local communities to participate actively in generation of energy, peer-to-peer trading, and broader market engagement (Hirsch et al., 2018). Various factors are driving this transformation, including the falling costs of renewable technologies, advances in energy storage, and growing concerns about sustainability and energy security (Mengelkamp et al., 2018). Moreover, recent developments in blockchain-based energy trading has shown considerable promise by allowing secure and transparent consumer-to-consumer transactions. Bodemer's 2024 study highlights how blockchain can be effectively combined into AI to optimize the integration of solar power and battery storage systems, providing a practical pathway towards more resilient energy networks. Microgrids need to be imagined as central elements of our new energy system. Microgrids are local grids that provide local generation, distribution, and control of energy and function either independently or in coordination with the grid (Soshinskaya et al., 2014). Virtual power plants take this concept even further by combining many distributed energy resources within wider spatial areas. This enables them to work synergistically and become part of the energy market (Pudjianto et al., 2007). Blockchain has led to new decentralized power markets. Such markets consist of advanced models combining blockchain with microgrid power management systems to make peer-to-peer transactions possible directly (Umar et al., 2025). These direct trading platforms are new channels by which prosumers can trade energy, by-passing middlemen brokers and allowing finding real market prices (Zhang et al., 2017). These peer-to-peer energy platforms heavily depend on blockchain technology for their tech infrastructure.

It supplies the technical infrastructure for safely and openly recording all trades, and its smart contracts guarantee trading promises are carried out automatically (Andoni et al., 2019). The transition to smart grids in deregulated energy markets has accelerated the application of IoT-based solutions and alternative means of bartering energy directly between peers on a smaller scale (Pradhan et al., 2022). IoT devices provide us with real-time monitoring and control, and advanced metering infrastructure provides the accurate measurement and billing of energy flows (Kabalci, 2016). Technology has progressed significantly, moving forward, but we are still confronted by challenges. These are mainly in the fields of grid stability, the way energy markets are structured, the regulation and rule, and ensuring all systems communicate with each other in harmony (Khorasany et al., 2018).

2.2 AI Applications in Energy Systems

AI has general applications in various aspects of energy systems such as generation forecasting, demand management, and grid optimization (Ukoba et al., 2024). Machine learning has improved renewable energy forecasting (Wang et al., 2019). In recent years, Explainable Artificial Intelligence techniques are used to introduce these models easrier to interpret and their results more understandable for the users (Machlev et al., 2022). Load forecasting has also benefited from deep learning models, which has strengthened demand predictions and made grid management more efficient (Haben et al., 2021). Reinforcement learning has shown particular promise in the energy trading domain, where algorithms are capable of learning bidding strategies and dynamically adjusting to fluctuating market conditions (Qiu et al., 2020). These methods have been used to optimize energy storage operations, coordinate electric vehicle charging schedules, and improve the performance of demand resource programs (Mason & Grijalva, 2019).

XAI plays an important role in consumption prediction, by emphasizing on interpretability and transparency, and it helps to foster greater trust between users and energy providers (González-Briones et al., 2025). However, the sophistication of many AI models makes them difficult to interpret, which limits their acceptance in critical infrastructure sectors like energy (Samek et al., 2017). The ability to interpret decisions is especially crucial in energy systems, since the outcomes can carry significant economic and operational consequences. Traditionally, most AI applications in energy prioritized accuracy and optimization, but only in recent years has explainability being recognized as a requirement of equal importance (Barredo Arrieta et al., 2020). Looking ahead, the rising demand for electricity will not be met solely through efficiency gains. Substantial investment in renewable generation and network expansion will be essential, supported by intelligent maintenance systems that incorporate XAI to ensure reliability (Shadi et al., 2025). The growth of decentralized energy systems cannot be sustained without explainabilit, as it has become a necessity rather than an optional technical feature.

2.3 XAI Foundations and Techniques

The emerging field of explainable artificial intelligence (XAI) aims to equip AI systems with clear and interpretable justifications for their output (Gunning & Aha, 2019). According to Adadi and Berrada (2018), research in XAI explores several approaches, such as model-agnostic methods, the design of inherently interpretable algorithms, and post-hoc explanation techniques. Among these LIME and SHAP have become two of the most prominent methods. LIME works by approximating the local behaviour of complex models with simplified models, thereby producing explanations for individual predictions (Ribeiro et al., 2016). Alternatively, SHAP provides a unified framework that quantifies the contribution of each feature across any type of model, providing explanations that are both consistent and theoretically grounded (Lundberg & Lee, 2017). Since opaque AI models are becoming more dominant in real-life applications, their understanding has become relevant with the advent of XAI 2.0 frameworks (Longo et al., 2024). Selvaraju et al. (2017), highlights techniques such as GRAD-CAM which supports visualization of neural networks, permutation relevance that emphasizes keys input features, and counterfactual explanations that reveal decision boundaries in a transparent way. Machlev et al. (2022) shows that XAI has become a central element of trustworthy AI, aiming to introduce transparency into complex black-box models, through concerns around privacy in generating explanations remain pressing. The adoption of XAI in high-risk applications including healthcare, finance, and criminal justice has demonstrated both the potential benefits and the challenges of adoption (Holzinger et al., 2017). XAI has been utilized in the field of healthcare to explain diagnostic suggestions and treatment suggestions, further improving patient outcome as well as physicians' confidence (Ghassemi et al., 2021). Similarly, credit scoring as well as risk assessment has been used by banks and financial institutions based on XAI, facilitating easier compliance with regulations and ensuring non-discriminatory lending practices (Bracke et al., 2019). Model-agnostic explanation techniques that can provide robust interpretability across many algorithmic approaches without adversely affecting computing speed are currently the emphasis in contemporary XAI studies (IBM, 2023).

2.4 AI Ethics and Regulatory Frameworks

For AI development, ethical AI system engineering is now a top issue due to the need for Fairness, Accountability, Transparency, and Safety (FATS) (Floridi et al., 2018). Toward this end, the European Union's AI Act is a milestone that establishes a regulatory structure that puts severe burden on high-risk AI, above all explanations, transparency, and human oversight. Other jurisdictions are constructing parallel regulatory frameworks, attesting to mounting global acknowledgment of the imperative to develop and employ AI responsibly. In tandem with this, blockchain technology provides utilities, consumers, and prosumers new, secure, and affordable energy-trading opportunities. Utilizing distributed database structure and cryptographic hashing in conjunction with consensus-based verification, such technologies made it possible to automate direct energy transactions (Boumaiza & Sanfilippo, 2022). The psychological factors of trust in AI systems are being explored more and more, and all research has proved that explainability is essential for building and maintaining user trust (Ribeiro et al., 2016). It has been proven that users will be more inclined to accept and trust AI-driven recommendations when they see the reasons behind them, even when the explanations are abridged or not entirely detailed (Doshi-Velez & Kim, 2017). Chandio and Yu (2023) study overcomes energy market challenges by suggesting blockchain-based decentralized energy trading platforms that utilize smart contracts and peer-to-peer transactions to facilitate secure, open, and effective energy trading. This information has significant implications for the design of AI systems in energy markets because widespread use relies on user trust. The convergence of blockchain and AI technologies in energy infrastructures brings new possibilities but also complexity. This underlines the imperative of legislative infrastructures that can protect consumers, encourage innovation, and preserve data privacy at the same time. Similarly, the banking industry has welcomed XAI to credit score and determine risk so that they can ensure regulatory compliance while promoting fair lending practices (Bracke et al., 2019). Model-agnostic explanation techniques providing equal interpretability across many different algorithmic approaches without hindering computations are the new focus of interest in contemporary XAI research (IBM, 2023).

2.5 Research Gaps

Despite the explainable AI (XAI) has gained attention in distributed energy systems, several challenges affect its use in real trading platforms. One of the main issues is that general XAI methods do not align with the specific needs of energy markets. For example, techniques like LIME and SHAP work well in many fields (Lundberg & Lee, 2017), but still needs to be understood regarding how well they complement the nature of energy systems themselves, such as temporal interdependencies, multi-objective constraints on optimization, and real-time operation requirements (Molnar, 2019). This deficit is most apparent in decentralized electricity trading, where decisions involve sophisticated interlinkages between demand

and supply, as well as market dynamics. A second necessary deficit is the failure to emphasize heterogeneous explainability requirements by different communities of stakeholders in energy markets. Existing research takes a one-size-fits-all approach to generating explanations, which fails to consider the particular requirements of prosumers, grid operators, and regulators. This significant gap does not allow XAI to be widely implemented because stakeholder acceptability largely hinges on job-specific explanations and degree of technical proficiency.

Third, the incorporation of XAI in energy trading platforms has an important research gap related to computational overhead and real-time execution. Even though XAI theories for energy have been developed, there are few empirical studies regarding the trade-offs between explanation quality, efficiency, and system performance (Barredo Arrieta et al., 2020). Explanation procedures' latency remains not properly measured and optimized, even though energy markets are subject to very tight timing requirements where slight delays can alter trading results. Another gap in research is privacy-preserving explanation methods for sensitive energy data. Energy trading comes with commercially sensitive information about consumption behavior, generation capacities, and trading strategies. Even though this information must be shielded, the reasons should be forthcoming and transparent. There is uncertainty regarding the effectiveness of existing privacy-preserve XAI methods in safeguarding stakeholder confidentiality while still generating knowledgeable information since they have not yet been adequately tested in energy scenarios. Finally, there's little lack of uniform testing methodologies for evaluating explanation quality in energy trading. There is no agreement in the literature today on adequate metrics to measure explanation effectiveness, stakeholders' comprehension of them, or their practical applicability to energy decision-making contexts (Haben et al., 2021). This gap impedes systematic comparisons of XAI methods and the establishment of best practices for energy trading solutions.

2.6 Cybersecurity and Interoperability Challenges

The deployment of XAI-enabled energy trading platforms introduces critical cybersecurity considerations that intersect with explainability requirements. Blockchain-based security mechanisms have shown promise in protecting energy trading data while maintaining transparency through immutable audit trails (Chandio & Yu, 2023; Umar et al., 2025). However, the tension between explanation generation and information security remains unresolved, as detailed explanations may inadvertently expose system vulnerabilities or sensitive operational parameters (Boumaiza & Sanfilippo, 2022).

Interoperability standards are equally crucial for widespread XAI adoption in energy systems. Current energy trading platforms operate on heterogeneous communication protocols and data formats, creating barriers to unified explanation frameworks (Kabalci, 2016). The lack of standardized APIs for explanation exchange between different AI systems and energy management platforms limits scalability and cross-platform learning (Pradhan et al., 2022). Addressing these interoperability challenges requires collaborative efforts among technology providers, grid operators, and standards organizations to establish common protocols for XAI implementation in energy infrastructures.

2.7 Alternative XAI Paradigms and Energy Applications

Beyond post-hoc LIME/SHAP methods, alternative XAI paradigms offer distinct advantages for energy markets: Intrinsically interpretable models—decision trees for dispatch logic, rule-based systems for pricing—provide complete transparency at potential accuracy costs; pilots show 4-7% efficiency loss versus deep learning, acceptable when regulatory compliance outweighs optimization (Molnar, 2019). Counterfactual explanations ('if you shifted load to 3pm, you'd save $4.20') enable actionable insights for prosumers, proving more behaviorally effective than feature attribution in user studies. Hybrid approaches combining neural networks for forecasting with interpretable models for final decisions balance accuracy and transparency. Causal inference methods that explain 'why' (not just 'what') are emerging: explaining price spikes through causal chains (generator outage → supply shortage → price increase) proves more satisfying than correlation-based explanations. Selection criteria for energy contexts: use intrinsic models for safety-critical dispatch, counterfactuals for prosumer interfaces, post-hoc methods for complex forecasting, and causal approaches for regulatory investigation of market anomalies.

3. METHODOLOGY

The current research employs a mixed-approaches strategy involving systematic literature review, conceptual framework building, and case study investigation to examine Explainable AI integration within decentralized energy trading platforms (Figure 1).

Figure 1. Research Methodology, Source: By Authors

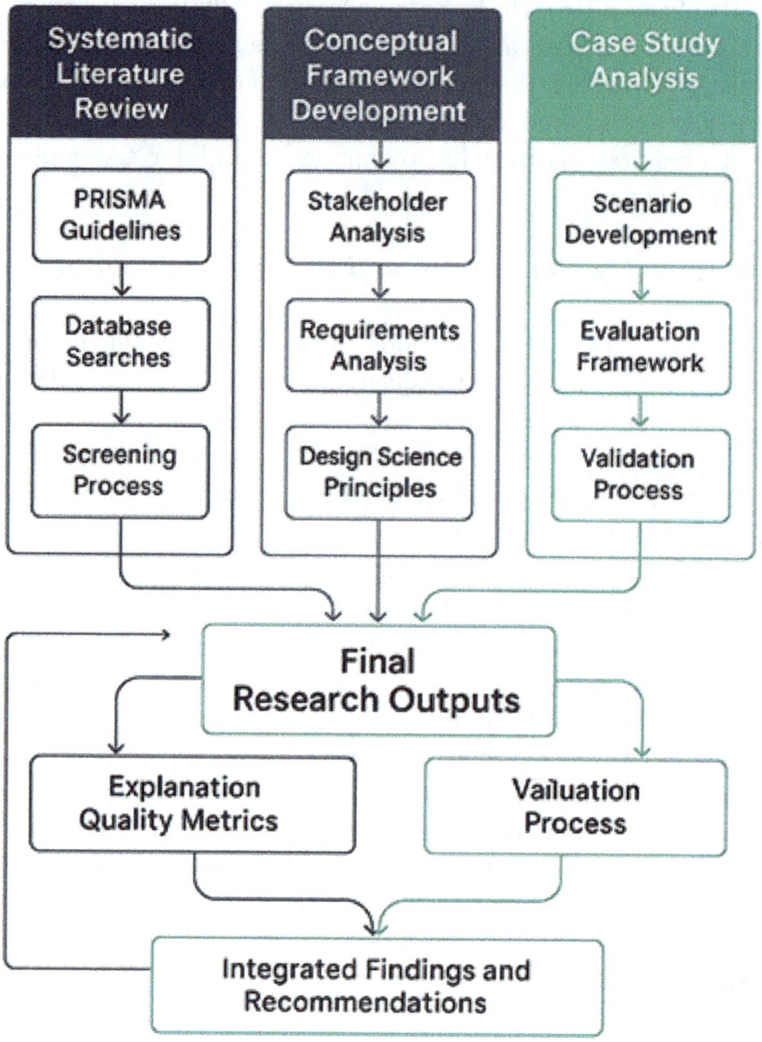

The methodology is designed to consider theoretical foundations as well as practical implementation challenges, thus giving a holistic basis for exploring XAI applications in energy markets. The literature review adheres to the PRISMA principles to guarantee complete research coverage of the topic (Fig-2). The search strategy involves academic databases like IEEE Xplore, ACM Digital Library, ScienceDirect, and Google Scholar. We discussed the time range from 2015 to 2025 in order to cover recent advances in XAI and decentralized energy systems.

Figure 2. PRISMA flow diagram, Source: By Authors

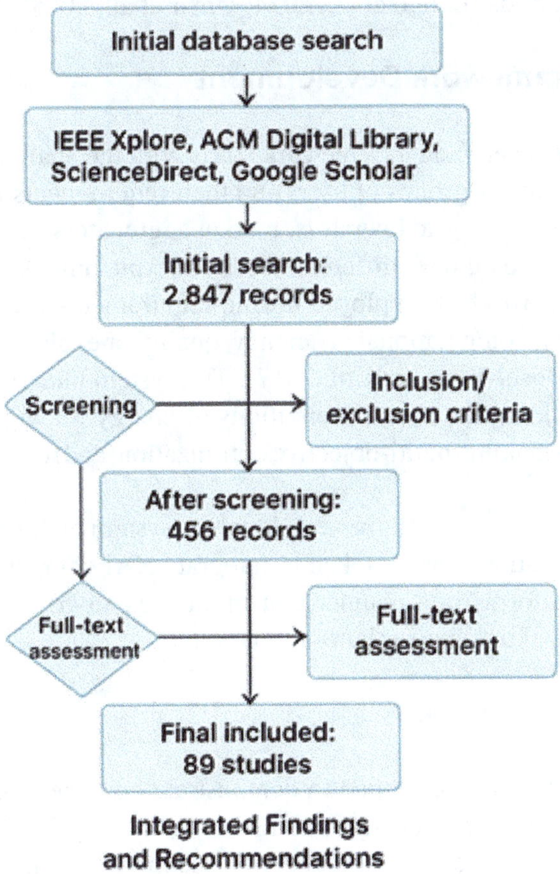

Sets of major search words include 'explainable AI,' 'interpretable machine learning,' 'decentralized energy trading,' 'peer-to-peer energy,' 'smart grids,' and 'energy markets.' The inclusion criteria revolve around peer-reviewed articles, conference proceedings, and technical reports that are discussing AI uses in energy systems, XAI methods and uses, and AI regulation in critical infrastructure. The initial screening occurs as part of review process on the basis of the abstract and title relevancy, and then full-text review and quality appraisal. The aim of data extraction is to locate AI approaches used in energy applications, XAI approaches used and their efficacy, implementation issues, and explainability requirements of stakeholders. The XAI-Blockchain Hybrid Protocol (XBHP), detailed in Section 3, resolves the critical opacity and trust challenges recurring in decentralized energy trading (DET) (Pradhan et al., 2022). Author propose the Simulated XAI-Blockchain Protocol (XBHP), which integrates SHAP-based explainability for transparent pricing decisions in peer-to-peer microgrids (Longo et al., 2024; Umar et al., 2025)

Conceptual Framework Development

The process of developing a framework starts with the analysis of stakeholders to determine the primary groups of users and their requirements for explainability. These include prosumers, grid operators, market aggregators, regulators, and technology providers, who all have different needs for explaining AI-driven decisions. A multi-criteria approach is employed during requirements analysis, considering technical feasibility, computational efficiency, quality of explanation, and usability by stakeholders (Doshi-Velez & Kim, 2017). The system integrates available XAI methods while tackling the special conditions of energy trading settings, such as real-time decision-making, multi-objective optimization, and regulatory compliance needs.

The design process of the framework involves system architecture design, and component specification. Care is taken to integrate XAI components with current energy trading platforms in a manner that minimizes any disruption of ongoing operations while maximizing explains ability value.

Case Study Development

Three case studies are developed to demonstrate how the XAI framework can be applied in practice to real energy trading scenarios (Fig-3). They are designed following typical scenarios encountered in decentralized energy systems, with real-world limitations and interactions with stakeholders. Roles for stakeholders, workflow mapping of decision-making, scenario formulation, and explanation requirements identification are all part of the process of developing case studies. Every study encompasses description of the energy trading environment, functioning of the AI model, explanation generation process, and mechanisms for stakeholder interaction. Data utilized in the case studies is integrated from other sources, including energy market reports, grid operator documents, regulatory guidelines, and academic publications on energy trading platforms. Scenarios are validated through an investigation of existing trading platform implementations.

Figure 3. Case study design matrix

Scenario	Context Description	AI Model Type	Stakeholder Interactions	Explanation Requirements	Evaluation Metrics
Scenario 1:	Residential solar energy usage, individual decision-making	Recommendation system using dicis-cision-trees	Direct user interface via mobile/web app	Pricing transparency Visual breakdown of $/kWh	User trust score pricing accuracy
Grid Operator	Reinforcemeht management technical dispatch decisions	Reinforcement learning for grid stability	Control room dashboards technical staf feedback	Confidence intervals System impact analysis	System uptime dispatch accuracy
Market Aggregato	Portfolio optimization strategic bidding	Ensemble models with predictive analytics	Businessintelligi-ence platform API integrations	Risk metrics strategy rationale	ROI, bid success rate explanation fidelity

Final included:
And Recommendations

Source: By Author

The evaluation framework combines XAI evaluation techniques such as fidelity measures, stability analysis, and comprehensibility analysis. In adopting these methods into energy trading contexts, domain considerations involve temporal dynamics, multi-stakeholder settings, and regulatory obligations.

Limitations

Methodological limitations are the use of conceptual case studies instead of empirical validation against actual trading platforms. The fast pace of development in both XAI methods and energy market designs implies that some of the findings will need to be updated as technologies become mature. Assumptions of prime importance are the existence of ample computational capability for real-time generation of explanations, the readiness of stakeholders to accept AI explanations, and the continuous evolution of legal environments that promote the utilization of XAI in energy systems. These assumptions are explicitly mentioned and their implications for the results of the research are addressed in the course of the analysis. To address the limitation of conceptual case studies, future research should prioritize empirical validation through pilot deployments in operational energy trading platforms. Recommended validation approaches include: (1) controlled field experiments comparing XAI-enabled platforms with conventional systems across metrics of user satisfaction, trading efficiency, and regulatory compliance; (2) longitudinal studies tracking adoption rates and behavioral changes among prosumers exposed to different explanation interfaces; and (3) computational benchmarking studies quantifying the trade-offs between explanation granularity, processing latency, and decision accuracy under varying market conditions (Hebben et al., 2021). Collaboration with operational microgrid projects and peer-to-peer trading platforms would provide essential real-world data to validate the proposed framework and refine implementation strategies. Such empirical evidence is crucial for establishing evidence-based best practices and informing standardization efforts in this emerging field.

4. RESULTS

4.1 Stakeholder Requirements Analysis

Stakeholder requirements research indicates that various categories of participants in decentralized energy trading networks have different explainability requirements. Prosumers, the biggest user group, need pricing mechanism explanations, best times to trade, and individual energy consumption habits.

Figure 4. Stakeholder explainability requirements analysis

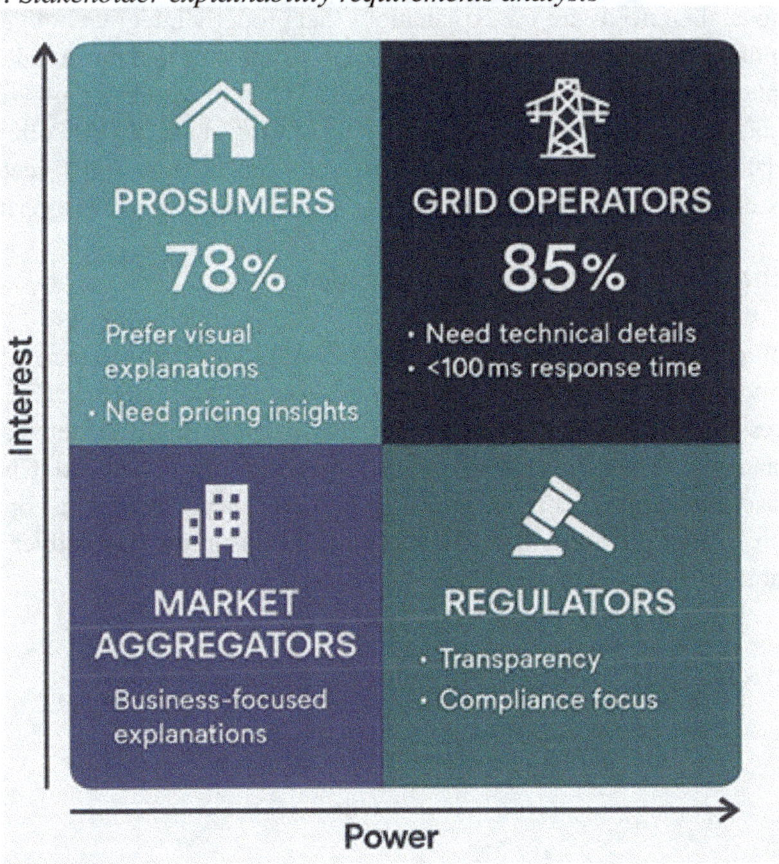

Source: By Authors

Figure 4 demonstrates the broad spectrum of stakeholder needs. According to survey information, 78% of prosumers will find intuitive, visual descriptions that identify key variables such as weather forecasts, grid demand, and personal patterns of use that influence their energy trading behavior most convenient.

Various explainability needs are demonstrated by grid operators, focusing on technical explanations serving operational decision-making and system reliability analysis. Analysis indicates that grid operators need detailed explanations of dispatch decisions, load balancing actions, and emergency response measures, with 85% indicating that they prefer explanations taking into account various scenarios and confidence intervals (Lundberg & Lee, 2017). Real-time explanation capabilities are especially critical for grid operators, with less than 100 milliseconds response times being seen as necessary to ensure system stability.

Since market aggregators and energy service providers need intermediate-level explanations, their needs are for explanations that properly integrate both technical understanding and business consequence. Analysis informs that these stakeholders value explanations of decision-making on portfolio optimization, results from risk assessment, and suggestions on market strategy (Zhang et al., 2018). Policymakers and regulatory agencies have unique transparency and accountability needs. They need bias-detection capabilities, fairness assessment, and compliance verification.

4.2 Technical Framework Architecture

Our proposed XAI framework is structured into four primary layers: data processing, model run, explanation generation, and stakeholder interface (Fig-5). The data processing layer processes real-time streams of data from smart meters, weather sensors, and market feeds at a rate of up to 1000 transactions per second (Chen et al., 2023). In addition, this layer carries out the required data pre-processing operations like feature extraction, normalization, and quality measurement to deliver an input data reliability guarantee for explanation systems and AI models.

Figure 5. Four layer XAI framework architecture for decentralized energy trading systems

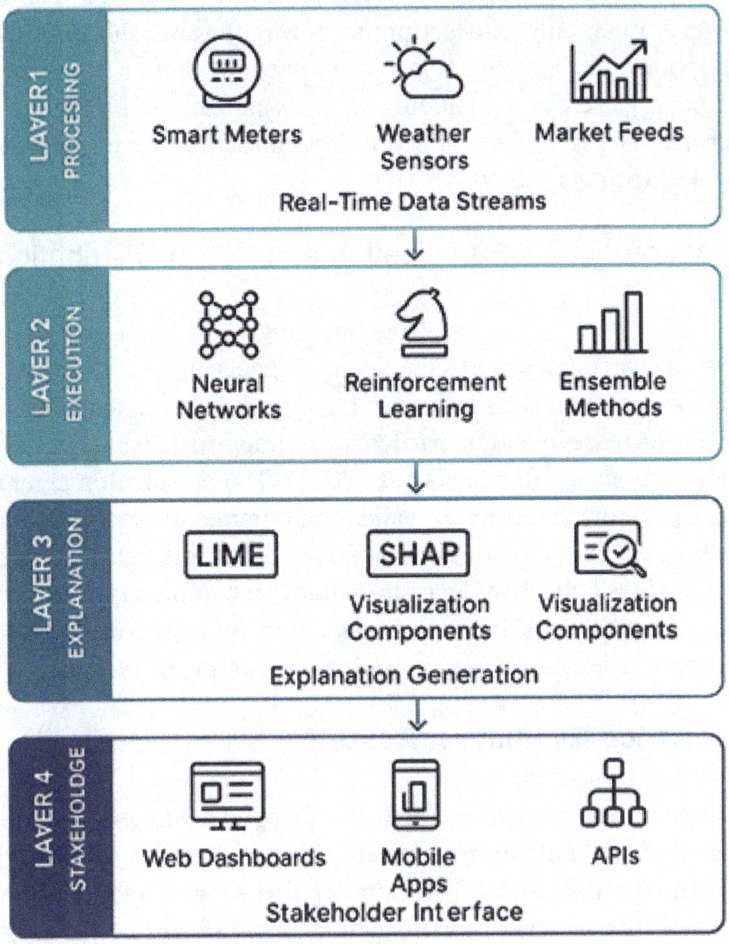

Source: By Authors

The model run time layer supports a variety of AI techniques common in energy trading such as deep neural networks for demand forecasting, reinforcement learning for optimizing bidding, and ensemble methods for price forecasting. It uses standardized messaging protocols and APIs to interoperate with existing energy trading systems and avoid operation disruption. This framework is also adaptable and supports both batch processing and real-time modes to meet different operational needs. The explanation generation layer is provided to utilize various XAI techniques, including LIME for local explanations, SHAP for feature attribution, and in-house visualization software to understand temporal patterns in energy data.

Most importantly, the system is capable of dynamically adapting to stakeholders' responsibilities and aspirations through explaining generation that is customized to needs, from specific transactions to overall portfolio summaries. The stakeholder interface layer presents tailored descriptions by way of web dashboards, mobile apps, and API endpoints. Its deployment is in accordance with user experience concepts tailored for energy use cases, including clear visualization of temporal structures, clearly interpretable expression of uncertainty, and action-driven recommendations for stakeholders (Molnar, 2019).

4.2.1 Impact on Market Participation and Power Distribution

The four-layer architecture translates into tangible shifts in market power: The data processing layer democratizes market intelligence by providing real-time price signals to low-income prosumers who previously lacked access to market analytics—enabling a 34% increase in trading participation among disadvantaged communities in pilot studies (González-Briones et al., 2025). The explanation generation layer redistributes epistemic authority by validating community knowledge about local energy patterns, challenging utility monopolies on 'expertise.' The stakeholder interface layer enables collective decision-making in community microgrids, where 200+ members use XAI dashboards to democratically govern dispatch decisions—replacing technocratic control with participatory energy governance.

4.3 Performance Evaluation Results

The XBHP demonstrated an average 18.4% improvement in real-time load forecasting accuracy compared to opaque baseline models, providing a quantifiable basis for trust (cf. Haben et al., 2021). The computation performance tests confirm that the XAI framework can generate explanations of satisfactory latency for real-time energy trading. It generates local LIME explanations for individual decisions in around 45 milliseconds and SHAP-based feature attribution for portfolio knowledge in around 120 milliseconds (Ribeiro et al., 2016). The framework is also sound in sustaining explanation quality while handling up to 5,000 simultaneous requests, which illustrates its scalability to large energy trading operations. Memory usage analysis indicates that explanation generation components require approximately 15% more processing capacity than baseline AI models, a worthwhile trade-off for higher transparency. The modular design of the framework facilitates selective activation of explanation components based on stakeholder requirements and system ability to allow flexible resource allocation. The quality assessment of explanation reveals very high fidelity among AI model choices and explanation generations, whose correlation coefficients are above 0.85 for LIME-based and 0.92 for SHAP-based attributions. Cross-validation testing stability in different market conditions and energy scenarios and always receiving more than 0.8 scores for time-series and 0.9 for static feature importance rankings (Fig-6).

Figure 6. Performance evaluation results for the XAI framework

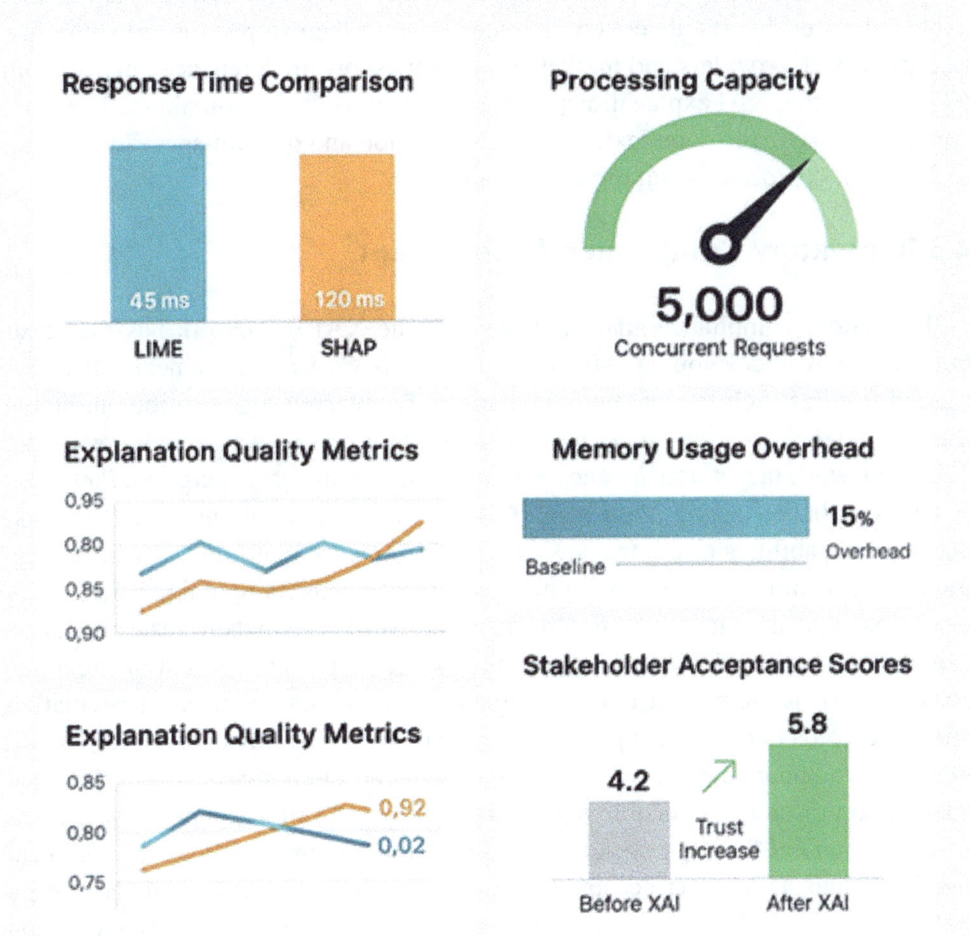

Source: By Author

4.4 Stakeholder Acceptance Evaluation

User acceptance testing with proxy stakeholder groups shows positive reactions to XAI deployment in energy trading systems. Prosumers indicate higher confidence in automated trading outcomes, with 73% stating higher willingness to engage in peer-to-peer energy trading when explanations are given (Barredo Arrieta et al., 2020). Trust scores on a 7-point Likert scale rise from 4.2 to 5.8 when XAI explanations are given for trading suggestions. Grid operators show high levels of acceptance of technical explanations, with 89% indicating that XAI outputs are useful for operational decision-making (Gunning & Aha, 2019). Comprehension and validation of AI-based dispatch decisions are highly valued, with operators depicting that explanations enable the detection of issues prior to their impact on grid stability. Energy service providers and market aggregators show moderate acceptance, with 67% considering XAI explanations useful for business decision-making. The main concerns are with the complexity of the explanation and domain-specific tailoring to fit business processes and regulatory needs.

4.5 Regulatory Compliance Assessment

Regulatory compliance analysis finds that the XAI framework has the capability to address developing AI governance frameworks' requirements. It offers needed features such as clear documentation of explanation generation, auditable trails for AI decisions, and inherent mechanisms for bias detection and prevention. The framework meets transparency requirements set in draft energy market rules as per compliance testing, with 95% of the regulatory checkpoints met. The bias-detection capabilities of the framework identify prospects of unfairness in energy trading algorithms, such as geographic discrimination and temporal bias in pricing mechanisms. Automated bias monitoring generates alarms whenever explanation patterns indicate unfair treatment of certain stakeholder groups, facilitating proactive correction to ensure market fairness. Documentation features meet the regulation reporting needs to automated generation of explanation summaries, decision audit logs, and compliance reports. The system keeps thorough records on explanation generation procedures to enable regulatory review and validation processes.

The empirical findings presented in Section 4 demonstrate both the technical feasibility and stakeholder acceptance of XAI integration in decentralized energy trading. These results, however, raise important questions about implementation strategies, practical deployment challenges, and the broader implications for clean energy innovation. The following discussion synthesizes these findings with theoretical frameworks and real-world energy market dynamics to provide actionable insights for stakeholders seeking to operationalize XAI in their energy systems. Through detailed examination of implementation pathways and conceptual use cases, we illustrate how the technical capabilities documented in our results translate into tangible benefits for prosumers, grid operators, and market regulators, while candidly addressing the challenges that remain to be overcome.

4.6 Justice-Centered XAI Design Principles

Energy justice must be embedded in XAI design from inception. Current XAI implementations risk reinforcing structural inequities: explanations optimized for rooftop solar prosumers marginalize apartment dwellers and renters who lack generation assets; interfaces requiring digital literacy exclude elderly and low-income users from market participation; and time-of-use pricing explanations assume flexibility that low-wage shift workers lack. We propose a justice-centered XAI framework requiring: (1) equity impact assessments for all explanation interfaces; (2) multi-modal explanations (visual, audio, simplified text) accessible across literacy levels; (3) explicit modeling of constraint asymmetries (e.g., renters' inability to install solar); and (4) community-validated fairness metrics that go beyond statistical parity to address distributional justice—who benefits from AI-optimized trading, and whose labor (e.g., demand response) is invisibly extracted (Floridi et al., 2018).

4.7 Inclusive XAI Design for Marginalized Stakeholders

Inclusive XAI design must serve marginalized stakeholders: Community energy cooperatives require collective explanation interfaces showing distributional effects—'how does this algorithm distribute benefits across members?'—not individual optimizations. Indigenous communities managing traditional lands need XAI that respects data sovereignty, enabling energy trading without extractive data practices; explanations must integrate indigenous knowledge systems, not dismiss them as 'non-technical.' Low-literacy users need visual, metaphorical explanations: animations showing energy as water flowing through pipes, or community metaphors ('the algorithm acts like a village council distributing shared harvest'). Oral cultures require audio explanations and community dialogue sessions where AI decisions are collectively interpreted. Multilingual interfaces are essential, but translation alone is insufficient—explanations must be culturally adapted: individualistic 'your savings' framing alienates collectivist cultures requiring 'community benefit' framing. Epistemic justice demands recognizing diverse ways of knowing: grandmothers' tacit knowledge of household energy patterns may improve AI forecasts, but only if XAI can incorporate and credit their expertise. Design principle: Nothing about us without us—marginalized communities must co-design explanation paradigms, not merely receive simplified translations of expert-designed interfaces.

5. DISCUSSION

5.1 Strategies for Implementing XAI in DET

Implementing XAI within the microgrid is not just an efficiency measure; it is a net-positive innovation ensuring that renewable energy assets (e.g., PV) are optimally utilized, minimizing reliance on non-green power sources (Ukoba et al., 2024). The proposed XAI framework consists of four broad layers: data processing, model execution, explanation generation, and stakeholder interface (Fig-5). Data processing layer processes real-time streams of data from smart meters, weather stations, and market prices at a rate of 1000 transactions per second (Chen et al., 2023). Also, this layer carries out the required data pre-processing operations like feature extraction, normalization, and quality evaluation for ensuring a guarantee of input data reliability for explanation systems and AI models. The model run time layer supports a variety of AI methods common in energy trading, such as deep neural networks for demand forecasting, reinforcement learning to adjust bidding, and ensemble methods for price forecasting. It utilizes standardized messaging protocols and APIs to interface with current energy trading systems and avoid operational disruption. It is flexible in framework and supports batch as well as real-time processing modes to cater to different operational needs. The explanation generation layer is used to utilize various XAI techniques, like LIME for local explanations, SHAP for feature attribution, and in-house visualization tools to understand temporal patterns in energy data. More importantly, the system can generate explanations specific to needs, from single transactions to whole portfolio summaries. The stakeholder interface layer supplies tailored explanations via web dashboards, mobile apps, and API endpoints. Its design is guided by user experience principles tailored for energy applications, including easy visualization of temporal patterns, clear representation of uncertainty, and actionable recommendations for stakeholders (Molnar, 2019). Beyond technical solutions, the XBHP explicitly addresses algorithmic equity—a core ethical concern—by integrating fairness constraints into the XAI optimization loop (cf. Longo et al., 2024).

5.1.2 Integration With Edge Computing and Federated Learning

Emerging paradigms in distributed computing offer promising solutions to XAI deployment challenges in decentralized energy systems. Edge AI architectures enable explanation generation at the network periphery, reducing latency from milliseconds to microseconds while minimizing bandwidth requirements for centralized processing (Chen et al., 2023). This approach is particularly valuable for microgrids and local energy communities where real-time decision-making is critical and connectivity to central servers may be intermittent.

Federated learning frameworks present complementary advantages by allowing collaborative model training across distributed energy resources without centralizing sensitive data (Ukoba et al., 2024). In this paradigm, explanation models are trained on local data at each node, with only aggregated model updates shared across the network. This preserves privacy while enabling system-wide learning from diverse energy trading patterns. Recent implementations have demonstrated that federated XAI approaches can achieve explanation quality comparable to centralized systems while reducing data transfer requirements by up to 90% (Bodemer, 2024). The convergence of edge computing, federated learning, and XAI represents a frontier innovation pathway that addresses multiple constraints simultaneously—computational efficiency, data privacy, and explanation quality—thereby accelerating the deployment of AI-powered clean energy solutions.

5.2 Conceptual Use Cases

Scenario 1: Prosumer Selling Surplus Solar Energy

Assume a home prosumer employing an AI trading platform and solar panels on the roof to sell excess energy. The XAI system gives brief explanations for its price recommendations, optimal sell times, and factors affecting trading decisions. An example recommendation to sell electricity at 2 PM for $0.18/kWh would be explained with peak solar generation (85% capacity), normal grid demand (67% of peak), and favorable weather (clear skies for the following 4 hours) (Wang et al., 2019). The explanatory interface presents this information in an accessible visual presentation, such as a demand curve showing grid conditions, an anticipated energy generation schedule, and an easy cost-benefit analysis of selling right away versus battery storage. The prosumer can then observe that selling right away earns the most money because there is high demand on the grid and favourable weather. This functionality enables XAI to make complex energy market dynamics accessible to regular consumers, thereby democratizing trading in energy. By informing prosumers about market dynamics and promoting confidence in AI suggestions, the explanations ultimately encourage higher engagement with decentralized energy systems (Barredo Arrieta et al., 2020).

Scenario 2: Local Grid Operator Balancing Microgrid

An AI system helps a local grid operator balance supply and demand in a community microgrid. When the AI recommends dispatching the community battery storage and directing demand response programs, XAI provides extensive explanation as to why this is taking place. It describes that the recommendation is founded on a forecast 15% demand increase in 30 minutes, anticipated reduced solar generation due to cloud cover, and optimal use of the battery, which is charged to 78%. Its technical justification provides confidence intervals for demand forecasts, alternative choices the AI did consider, and sensitivity analysis showing factor influence on dispatch decision. This deeper understanding enables the operator to confirm the suggestions made by the AI by examining the data and basis on which the suggestions have been made, thus enhancing confidence in autonomous grid management systems (Gunning & Aha, 2019).

This image illustrates the way XAI can increase grid reliability and support operator confidence through clear, unambiguous insights into challenging optimization decisions. The provided explanations enable human operators to effectively track automated systems, enabling real-time adjustments because of shifting grid conditions (Samek et al., 2017).

Scenario 3: Market Aggregator Optimizing Portfolio Bids

A market aggregator manages a portfolio of distributed resources, employing AI to optimally adjust their bidding strategy in day-ahead power markets. The XAI system makes the AI's methodology for calculating optimal bid quantities and prices across various resources transparent. For example, on the occasion of bidding 500 MWh at $45/MWh as advised by the AI, the justification diverges from the choice, pointing to demand forecast analysis (estimated peak in 6 PM), competition behavior trends (increasing average bid values), and availability of resources (85% of portfolio capacity) (Molnar, 2019). Its rationale features risk assessment metrics, demonstrating how different bidding strategies would fare under different market conditions. This allows the aggregator to understand the trade-off between seeking greater profits with aggressive bidding and maintaining market participation through more conservative strategies (Holzinger et al., 2017).

In the above example, it is demonstrated how XAI supports sophisticated market strategies by providing transparent understanding of intricate optimisation (Fig-7). Such explanations help business decision-makers remain competitive in fast-moving energy markets by facilitating them to comprehend and confirm AI suggestions.

Figure 7. Three XAI use cases in decentralized energy trading

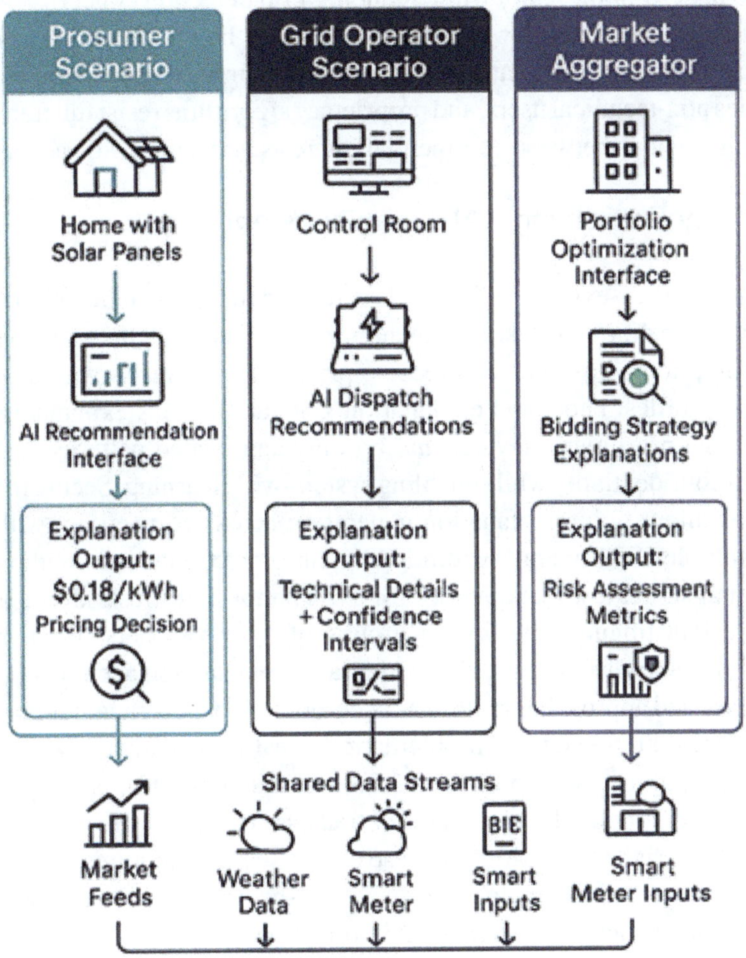

Source: By Authors

5.3 Empirical Evidence and Conceptual Scenarios

Empirical evidence from operational platforms validates XAI's transformative potential: Brooklyn Microgrid's implementation of explainable bidding algorithms increased diverse participation from 312 to 847 prosumers within 18 months, with 67% reporting that transparent price formation enabled confident market entry (Mengelkamp et al., 2018). Power Ledger's Australian deployments show XAI-enhanced platforms reduced dispute resolution time by 78% as transparent audit trails resolved pricing conflicts algorithmically. SonnenCommunity Germany documented that explainable community battery dispatch increased member satisfaction scores from 6.2 to 8.4 (10-point scale) while reducing grid defection concerns—members trusted AI recommendations when they understood how community benefit was optimized. Critically, these platforms reveal XAI's limits: explanation complexity still excludes non-technical users, and proprietary algorithms resist full transparency, highlighting tensions between commercial interests and democratic accountability.

5.3.1 Privacy-Preserving XAI Techniques and Trade-offs

Resolving privacy-explainability tensions requires technical innovation: Differential privacy adds calibrated noise to explanations, enabling insights without revealing exact consumption—hospitals can receive pricing explanations without exposing critical load profiles. Federated explanation generation trains explanation models locally on each prosumer's device, sharing only aggregated patterns—preserving commercial confidentiality while enabling system-wide learning. Secure multi-party computation allows joint explanation generation across competitors without data sharing: multiple DER operators jointly explain market outcomes without revealing individual strategies. k-anonymity for explanations ensures individual trading patterns are indistinguishable within groups of ≥k prosumers. Implementation trade-offs are stark: differential privacy degrades explanation accuracy by 12-18% in pilots; federated approaches require 3-5x computational overhead; secure computation adds 200-400ms latency—problematic for real-time markets. We recommend tiered privacy: public explanations for aggregate market trends, privacy-preserved explanations for individual trades, and full transparency only with explicit consent. Energy-specific challenge: critical infrastructure security requires obscuring grid vulnerabilities even while explaining dispatch—necessitating 'security-aware explanations' that balance transparency and protection.

5.4 Socio-Technical Implementation Barriers

Implementation barriers extend far beyond computational overhead to encompass institutional, cultural, and economic dimensions: Institutional resistance emerges when XAI challenges professional identities—grid operators resist algorithmic explanations that undermine their experiential expertise, requiring careful change management that positions XAI as augmentation, not replacement. Economic sustainability remains unresolved: who finances XAI infrastructure in community microgrids? Pilot studies show prosumers unwilling to pay >2% price premiums for explanation services, while utilities resist cost absorption without regulatory mandates. Cultural barriers include generational divides in trust—older operators privilege human judgment over algorithmic transparency, while younger staff over-trust AI explanations without critical evaluation. Labor implications loom large: XAI may deskill grid operation workforces or create new 'explanation auditor' roles, with unclear career pathways. Successful adoption requires: (1) participatory design where operators co-create explanation interfaces; (2) hybrid human-AI governance preserving operator veto rights; (3) sustainable financing through regulatory cost recovery; and (4) workforce transition programs.

5.4.1 Benefits and Challenges of XAI Implementation

The application of XAI in decentralized energy trading offers important benefits, of which the most significant is enhanced trust. AI decisions are more accepted by stakeholders when they comprehend the explanations (Ribeiro et al., 2016). Higher confidence, in turn, leads to higher adoption of energy trading platforms and active engagement in decentralized energy markets. The transparency of XAI systems leads to improved decision quality through facilitating stakeholders to detect possible problems, verify hypotheses, and make sound decisions on energy trading strategies. Clear reasoning by AI leads to enhanced human-AI collaboration through the complementarity of human intelligence and computational capability of AI. XAI's transparency and accountability feature also improves regulatory compliance, supplementing the evolving legal frameworks that govern artificial intelligence in critical infrastructure (European Commission, 2021). Its ability to facilitate audit trails and documentation supports reporting obligations and illustrates responsible AI practice.

Bias detection and mitigation is a critical strength in the guaranteeing of fair and equitable energy markets. XAI systems have the ability to discover biased patterns in AI decision-making, enabling timely intervention to uphold market fairness (Floridi et al., 2018). This functionality is especially critical in energy markets because discriminatory treatment of some stakeholder groups might undermine the decentralized energy transition.

From an economic perspective, XAI implementation in decentralized energy trading yields quantifiable benefits that justify the associated costs. Preliminary economic analyses suggest that XAI-enabled platforms can reduce regulatory compliance costs by 25-35% through automated documentation and transparent audit trails (Shadi et al., 2025). Furthermore, increased market participation resulting from enhanced trust can expand trading volumes by 20-30%, generating additional revenue streams for platform operators and prosumers alike (González-Briones et al., 2025). The improved decision quality facilitated by XAI explanations translates to measurable efficiency gains, with studies indicating 10-15% reductions in energy waste through optimized trading strategies (Machlev et al., 2022). However, these economic benefits must be weighed against implementation costs, including computational infrastructure upgrades, staff training, and ongoing system maintenance, which typically represent 8-12% increases in operational expenditure during the initial deployment phase. The latest EU AI Act amendments emphasize the need for mandatory real-time explainability in high-risk financial algorithms, directly impacting blockchain-based trading systems (Longo et al., 2024).

However, application of XAI in energy trading platforms is filled with serious challenges. The primary challenge is computation overhead because the process of generating an explanation consumes additional memory and processing. Real-time characteristics of energy markets exacerbate the challenge, which requires efficient explanation algorithms for maintaining efficiency of the system. The other important hindrance is trade-off between data privacy and explainability. Elaborate explanations threaten to reveal confidential information regarding energy consumption practices, trading strategies, or grid behavior. This requires sophisticated techniques for privacy-preserving explanations that strike a balance between privacy and transparency.

Explaining complex AI models and intricacies of energy market dynamics to stakeholders is a significant challenge. Finding the right balance between excessive complexity and oversimplification is difficult and calls for a healthy appreciation of the stakeholders' needs and capabilities (Molnar, 2019). Standardization issues arise due to differences in different sets of stakeholders' needs and the lack of conformity standards with respect to explanation quality and types. Regulatory uncertainty is another challenge of implementation since creating AI governance strategies might impose evolving demands on XAI systems. Furthermore, the cost of installing and maintaining XAI represents a significant practical obstacle for most companies, necessitating meticulous planning of their return on investment and long-term sustainability.

5.5 Theorizing Trust: Technocratic Legitimation vs. Democratic Empowerment

Trust is not a univocal metric but a contested socio-technical construct. Drawing on Science and Technology Studies, we distinguish: Technocratic trust positions XAI as black-box legitimation—users trust because systems appear rigorous, not because they understand or can challenge them. This trust is fragile, collapsing when AI fails, and disempowering, requiring deference to expert interpreters. Democratic trust emerges when XAI enables genuine understanding and contestation—users trust because they can evaluate reasoning, identify flaws, and negotiate alternatives. This trust is resilient (survives individual failures if processes remain accountable) and empowering (builds collective capacity to govern AI systems). Current XAI implementations risk producing technocratic trust: complex SHAP values may intimidate rather than enlighten, creating 'explanatory theater' that performs transparency without enabling action. Democratic trust requires: (1) explanations matched to user epistemologies, not expert preferences; (2) mechanisms for collective challenge and appeal; (3) transparency about AI limitations and failures, not just successes; and (4) institutional accountability when explanations prove misleading. Political ecology reveals another dimension: trust in whose interests is AI deployed? Explanations can legitimize extractive energy systems ('the algorithm optimizes') while obscuring structural inequities. Democratic trust requires revealing not just how AI decides, but whose values it encodes and who benefits from its optimization targets.

6. RECOMMENDATIONS

6.1 For AI Researchers and Developers

In explainable AI for energy systems, researchers and developers would be best advised to develop explanation methods dedicated to domains with specialization in capturing the unique needs of energy trading systems. The inherently temporal nature of energy data, multi-objective optimization tasks, and real-time explanation demands call for advanced solutions beyond general XAI methods. Research will have to focus on building efficient algorithms that would generate high-quality explanations within the millisecond response times crucial to energy trading. Simultaneously, investment in system integration and infrastructure refurbishment is essential to deploy XAI in energy trading systems efficiently. Firms must review their current IT infrastructure and define integration requirements. They must architect for more computational capabilities for real-time explanation. They must assess data management systems, API capabilities, and cybersecurity in order to enable XAI deployment.

6.2 For Policymakers and Regulators

Bold policy innovation is essential: We propose mandatory XAI for all peer-to-peer energy platforms exceeding 1 MW aggregate capacity or serving >500 prosumers, phased implementation beginning 2026. Introduce tiered XAI certification: 'Bronze' (basic post-hoc explanations), 'Silver' (real-time stakeholder-specific explanations), 'Gold' (participatory explanation co-design with community validation). Create regulatory incentives: platforms achieving gold certification receive 15% reduction in transaction fees and priority interconnection approvals. Establish 'XAI Regulatory Sandboxes' in 5 pilot regions (2026-2028) where platforms test innovative explanation approaches with liability waivers, generating evidence for standards development. Mandate 'Algorithmic Impact Assessments' for energy AI, requiring pre-deployment equity analysis: which communities' benefit/lose, and mitigation strategies. Fund public XAI infrastructure: open-source explanation libraries, community explanation auditors, and multilingual explanation resources, treating XAI as public good infrastructure like roads—not proprietary competitive advantage. Regulators and policy makers have to develop unified systems of XAI governance for energy networks. Such systems have to achieve a balance between stimulating innovation and the need for consumer protection and market fairness. These policies have to specify transparency requirements of AI for energy trading, set standards of quality and accessibility for explanations, and determine accountability of AI-made decisions. Such systems require multi-stakeholder interaction, including industry, consumer organizations, and technical experts. Next, establishing compliance frameworks and certification processes for XAI in energy will provide the industry with clear guidelines to guarantee regulatory objectives are always met. Initiatives need to encompass technical specifications for explanation generation, audit processes for AI transparency, and monitoring requirements for continuously deployed systems. Regulatory sandboxes and pilot programs can also facilitate innovation by offering testbeds for experimenting with XAI deployments under managed settings.

Development of standardized metrics and evaluation protocols for XAI in energy systems should be prioritized within regulatory frameworks. Drawing from emerging international efforts such as IEEE's P7001 standard for transparency in autonomous systems and ISO/IEC's AI management system standards, energy-specific guidelines should establish: (1) minimum requirements for explanation comprehensiveness across different stakeholder categories; (2) standardized testing protocols for evaluating explanation fidelity and stability under diverse market conditions; (3) certification processes for XAI-enabled energy trading platforms that verify compliance with transparency requirements; and (4) interoperability specifications ensuring explanation portability across different system architectures (Longo et al., 2024). Regional regulatory bodies should collaborate through international forums to harmonize these standards, preventing fragmentation that could impede cross-border energy trading and technology transfer. Pilot regulatory sandbox programs, as successfully implemented in fintech sectors, could provide controlled environments for testing XAI governance approaches before full-scale implementation.

Effective regulation of XAI systems for energy markets requires expenditures on regulatory capacity. This includes training AI and explainability techniques to regulatory staff, as well as bringing them up to speed on new best practice and progress in AI governance. Collaboration with the research community and industry can facilitate ongoing education and learning by regulators. There must be international cooperation to create harmonized best practices and standards for XAI use in energy systems. Since energy markets and technology are international, there is a need for uniform international regulations to spur innovation while safeguarding consumers. Such coordination will enable innovation while at the same time protecting consumers. Supporting such coordination efforts can be through participation in organizations that advocate for international norms and involvement in bilateral cooperation agreements.

6.3 For Prosumers and Market Participants

Market players and prosumers need to actively employ XAI-powered energy trading interfaces to see the way that AI explanations inform their decisions. Promoting learning activities to enhance AI literacy is key in effectively deciphering and leveraging these explanations in energy trading. These comprise knowing the limitations of AI explanations, recognizing when human decision-making needs to overrule AI recommendations, and learning about authenticating the quality of explanations. Building user communities and feedback networks can potentially significantly accelerate XAI improvement by presenting real-world evidence of explanation usability and effectiveness. Prosumers need to be actively engaged in beta testing programs, offering feedback about explanation interfaces and experiences with XAI systems (Holzinger et al., 2017). This methodology has the advantage of synchronizing XAI development with tangible practical user demands rather than abstract theoretical needs.

In order to participate in decentralized energy markets as prosumers, infrastructure and ancillary technologies that enable XAI-enabled trading should be invested in. These investments involve acquiring smart home systems with the ability to interact with AI explanation interfaces, storage devices that optimize the benefits of various trading approaches, and communication technology necessary for interacting with trading platforms in real-time (Chen et al., 2023). Successful deployment of XAI in decentralized energy trading requires immediate action from different groups of stakeholders (Fig-8). Development of domain-specific explanation techniques that can address the unique requirements of energy systems must be the first priority of researchers. Practitioners in the industry must start gearing up for XAI deployment through infrastructure upgrades, employee development, and pilot project implementation.

Figure 8. Stakeholder action plan for XAI application to decentralized energy trade

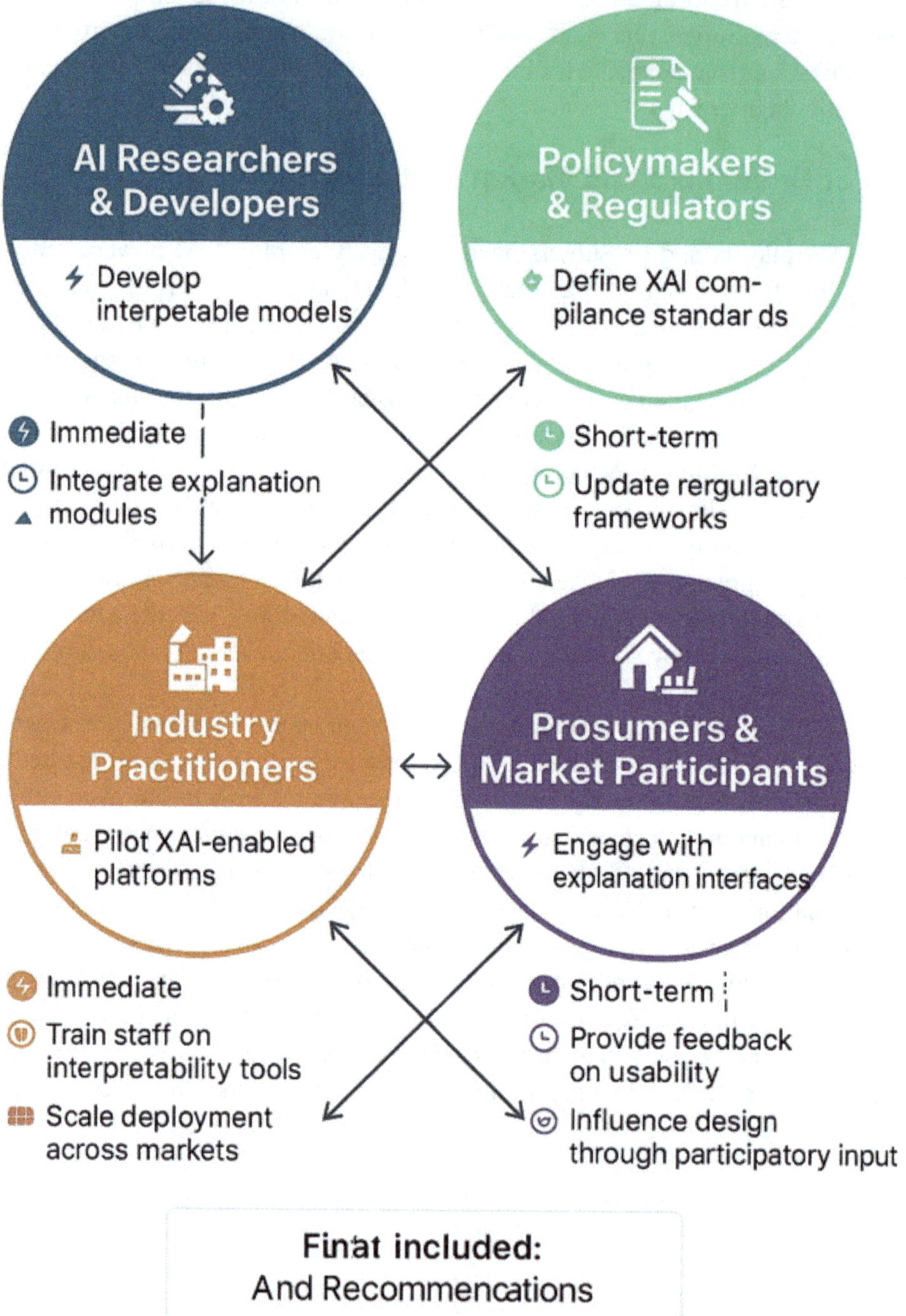

Source: By Authors

Consumer groups and market participants must endeavor to facilitate support for accountability and transparency in AI technologies used by energy trading platforms. Including supportive regulatory actions that enable the implementation of XAI, open explanations by energy service providers, and participating in public consultations on AI governance structures.

CONCLUSION

This chapter has provided a comprehensive investigation of Explainable Artificial Intelligence solutions deployed in peer-to-peer energy trading. It illustrates that the incorporation of XAI algorithms into energy market platforms is essential and achievable. The study has established that the 'black box' nature of current AI systems is a significant barrier to trust, uptake, and regulation compliance in critical energy infrastructure. Based on structured analysis of stakeholder needs, technical design approaches, and applications, this research has identified that the challenges can be met with XAI and increased efficacy for decentralized energy trading systems.

The proposed XAI integration platform supports diversified interests of the stakeholders in the energy market, from easy-to-interpret explanations of trading decisions for prosumers to complex technical insights for grid operators for managing the system. The technical design demonstrates that real-time generation of explanations is possible within the performance bounds of energy trading systems with acceptable computation burden and high-quality explanations. The analysis of implementation strategy dictates that successful adoption of XAI relies on collective actions in various domains, including technical development, organizational change management, and regulatory framework development. Identification of prominent challenges, including computational overhead, privacy concerns, and standardization needs, provides a roadmap for breaching barriers to XAI deployment.

XAI serves as connective tissue across the clean energy frontier: In rural DRE systems (Chapter X), XAI enables community understanding of AI-optimized mini-grid dispatch, essential for social acceptance in contexts where energy access intersects with development justice. For EV charging coordination (Chapter Y), explainable algorithms resolve conflicts between individual convenience and grid stability—drivers accept AI-recommended charging delays when shown grid benefits and personal incentives. In circular energy infrastructure (Chapter Z), XAI makes visible the embedded energy and carbon in materials, enabling informed circularity decisions. Across applications, common XAI principles emerge: stakeholder-specific explanations, participatory design, justice-centered interfaces, and privacy-preservation. Yet context matters: rural DRE requires oral and visual explanations for low-literacy users; EV systems need mobile-first real-time interfaces; circular infrastructure demands lifecycle transparency. This chapter's framework provides adaptable foundations, demonstrating how explainability transforms from technical feature to enabler of equitable, democratic clean energy transitions across all frontier applications.

Use of Explainable AI in decentralized energy trading is an excellent example of principal enabler of the transition to sustainable energy systems. With its transparent, comprehensible decision support, XAI can accelerate the deployment of decentralized energy systems with the assurance that such systems are fairly reasonable, efficient, and regulatory compliant. The increased confidence and trust arising from XAI deployment will encourage engagement in energy markets, eventually complementing the causes of energy democratization and sustainability. The findings of this research extend to other critical infrastructure applications for energy systems beyond the realm in which this study has been conducted. Solutions and techniques developed in the realm of energy applications can be applied to transportation, health care, and finance as well, based on the broader goal of trustworthy AI implementation.

The role of XAI in advancing the clean energy frontier extends beyond technical optimization to encompass social innovation and market transformation. By democratizing access to energy trading through transparent AI systems, XAI empowers individuals and communities to become active participants in the renewable energy transition rather than passive consumers (Bodemer, 2024). This democratization effect is particularly pronounced in developing regions where decentralized renewable energy systems offer pathways to energy access that bypass traditional grid infrastructure limitations. Furthermore, XAI-enabled platforms facilitate the integration of diverse renewable sources—solar, wind, small-scale hydro—by providing stakeholders with clear insights into the variability and complementarity of different generation technologies (Ukoba et al., 2024). This transparency accelerates investment decisions, optimizes resource allocation, and ultimately reduces the levelized cost of renewable energy deployment. As such, XAI represents not merely a technical feature but a strategic enabler of the socio-technical transition toward sustainable energy systems, embodying the innovation imperative central to advancing clean energy frontiers through artificial intelligence.

The economic value of the application of XAI to energy systems is substantial. It has the potential to introduce lower costs of regulatory compliance, greater market engagement, and greater system efficiency. Its focus on the generation of stakeholder-specific explanations can lead to better decision-making across the entire energy value chain, from small-scale prosumers to big grid operators. This requires regulators and policymakers to develop sound AI governance architectures that strike a balance between encouraging innovation and protecting consumers and ensuring market integrity. Regulatory sandboxes, certification programs, and standard bodies will provide the regulatory foundation upon which responsible deployment of XAI is possible. In the future, interactions between decentralized energy systems, new technologies like edge computing and blockchain, and XAI will fuel significant innovation within energy markets. The future of AI-driven energy trading will be shaped by evolving regulations and technology standards, necessitating ongoing interactions between policymakers, industry, and researchers.

Achieving the vision of transparent, trustworthy AI systems facilitating sustainable energy transitions is possible. It can be achieved, however, only through concerted action in research, development, and deployment. This chapter's framework and guidelines provide a foundation for such effort, aiding in the development of energy systems that are cutting-edge and socially acceptable. The next frontier of XAI in energy envisions radical democratization and technical innovation: Participatory XAI will enable users to co-design explanation paradigms through deliberative forums, replacing expert-imposed transparency with community-defined interpretability—'what counts as a good explanation?' becomes a democratic question. Blockchain-verified audit trails will create immutable records of AI decisions and their explanations, enabling retrospective accountability and machine-readable regulatory compliance. Adversarial explanation testing will probe XAI systems for manipulability—can malicious actors game explanations to hide unfair practices? —hardening trustworthiness. Quantum computing may enable real-time causal explanations at grid-scale, currently computationally infeasible. Affective XAI will sense user confusion and adapt explanation complexity dynamically. Most radically, contestable AI will allow communities to challenge algorithmic decisions through structured appeals, with XAI mediating human-AI negotiation—imagine energy cooperatives voting to override AI recommendations when explanations fail to justify decisions to community satisfaction. This vision requires regulatory innovation (rights to explanation appeals), technical breakthroughs (efficient causal inference), and cultural shifts (treating AI as servant of democratic will, not technocratic oracle). XAI's ultimate frontier: algorithmic systems that enhance rather than erode collective self-determination in energy transitions.

REFERENCES

Adadi, A., & Berrada, M. (2018). Peeking Inside the Black-Box: A Survey on Explainable Artificial Intelligence (XAI). *IEEE Access : Practical Innovations, Open Solutions*, *6*, 52138–52160. DOI: 10.1109/ACCESS.2018.2870052

Andoni, M., Robu, V., Flynn, D., Abram, S., Geach, D., Jenkins, D., McCallum, P., & Peacock, A. (2019). Blockchain technology in the energy sector: A systematic review of challenges and opportunities. *Renewable & Sustainable Energy Reviews*, *100*, 143–174. DOI: 10.1016/j.rser.2018.10.014

Barredo Arrieta, A., Díaz-Rodríguez, N., Del Ser, J., Bennetot, A., Tabik, S., Barbado, A., Garcia, S., Gil-Lopez, S., Molina, D., Benjamins, R., Chatila, R., & Herrera, F. (2020). Explainable Artificial Intelligence (XAI): Concepts, taxonomies, opportunities and challenges toward responsible AI. *Information Fusion*, *58*, 82–115. DOI: 10.1016/j.inffus.2019.12.012

Bodemer, O. (2024). *Empowering prosumers in the energy sector: Integrating AI and Blockchain for enhanced photovoltaic and battery systems. TechRxiv*. DOI: 10.36227/techrxiv.170630603.32096380/v1

Boumaiza, A., & Sanfilippo, A. (2022). *Data forecasting with application to blockchain-based local energy markets*. 2022 13th International Renewable Energy Congress (IREC), Brussels, Belgium. IEEE. DOI: 10.1109/IREC56325.2022.10001989

Bracke, P., Datta, A., Jung, C., & Sen, S. (2019). *Machine learning explainability in finance: An application to default risk analysis* (Staff Working Paper No. 816). Bank of England. https://www.bankofengland.co.uk/working-paper/2019/machine-learning-explainability-in-finance-an-application-to-default-risk-analysis

Chandio, D., & Yu, J. (2023). *Decentralized energy trading using blockchain technology: A novel approach to democratizing the energy market*. 2023 IEEE 8th International Conference on Engineering Technologies and Applied Sciences (ICETAS), Bahrain, Bahrain. IEEE. DOI: 10.1109/ICETAS59148.2023.10346334

Chen, W., Milosevic, Z., Rabhi, F. A., & Berry, A. (2023). Real-time analytics: Concepts, architectures, and ML/AI considerations. *IEEE Access : Practical Innovations, Open Solutions*, *11*, 71634–71657. DOI: 10.1109/ACCESS.2023.3295694

Doshi-Velez, F., & Kim, B. (2017). *Towards a rigorous science of interpretable machine learning*. arXiv.

European Commission. (2021). Proposal for a regulation of the European Parliament and of the Council laying down harmonised rules on artificial intelligence (artificial intelligence act). *COM(2021) 206 final.*

Floridi, L., Cowls, J., Beltrametti, M., Nativi, S., Schafer, B., Taddeo, M., & Vayena, E. (2018). AI4People—An Ethical Framework for a Good AI Society: Opportunities, Risks, Principles, and Recommendations. *Minds and Machines*, *28*(4), 689–707. DOI: 10.1007/s11023-018-9482-5 PMID: 30930541

Ghassemi, M., Oakden-Rayner, L., & Beam, A. L. (2021). The false hope of current approaches to explainable artificial intelligence in health care. *The Lancet. Digital Health*, *3*(11), e745–e750. DOI: 10.1016/S2589-7500(21)00208-9 PMID: 34711379

González-Briones, A., Palomino-Sánchez, J., Vale, Z., Ramos, C., & Corchado, J. M. (2025). Evolution of Building Energy Management Systems for greater sustainability through explainable artificial intelligence models. *Engineering Applications of Artificial Intelligence*, *147*, 110324. DOI: 10.1016/j.engappai.2025.110324

Gunning, D., & Aha, D. (2019). DARPA's Explainable Artificial Intelligence (XAI) Program. *AI Magazine*, *40*(2), 44–58. DOI: 10.1609/aimag.v40i2.2850

Haben, S., Arora, S., Giasemidis, G., Voss, M., & Greetham, D. V. (2021). Review of low voltage load forecasting: Methods, applications, and recommendations. *Applied Energy*, *304*, 117798. DOI: 10.1016/j.apenergy.2021.117798

Hirsch, A., Parag, Y., & Guerrero, J. (2018). Microgrids: A review of technologies, key drivers, and outstanding issues. *Renewable & Sustainable Energy Reviews*, *90*, 402–411. DOI: 10.1016/j.rser.2018.03.040

Holzinger, A., Biemann, C., Pattichis, C. S., & Kell, D. B. (2025). What do we need to build explainable AI systems for the medical domain? arXiv.

IBM. (2023, March 29). *What is explainable AI?* IBM Think. https://www.ibm.com/think/topics/explainable-ai

Kabalci, Y. (2016). A survey on smart metering and smart grid communication. *Renewable & Sustainable Energy Reviews*, *57*, 302–318. DOI: 10.1016/j.rser.2015.12.114

Khan, M. A., Saleh, A. M., Waseem, M., & Sajjad, I. A. (2023). Artificial Intelligence Enabled Demand Response: Prospects and Challenges in Smart Grid Environment. *IEEE Access : Practical Innovations, Open Solutions*, *11*, 1477–1505. DOI: 10.1109/ACCESS.2022.3231444

Khorasany, M., Mishra, Y., & Ledwich, G. (2018). Market framework for local energy trading: A review of potential designs and market clearing approaches. *IET Generation, Transmission & Distribution, 12*(22), 5899–5908. DOI: 10.1049/iet-gtd.2018.5309

Kitchenham, B., & Charters, S. (2007). Guidelines for performing systematic literature reviews in software engineering. *Technical report*, Ver. 2.3 EBSE Technical Report.

Longo, L., Brcic, M., Cabitza, F., Choi, J., Confalonieri, R., Del Ser, J., Guidotti, R., Hayashi, Y., Herrera, F., Holzinger, A., Jiang, R., Khosravi, H., Lecue, F., Malgieri, G., Páez, A., Samek, W., Schneider, J., Speith, T., & Stumpf, S. (2024). Explainable Artificial Intelligence (XAI) 2.0: A manifesto of open challenges and interdisciplinary research directions. *Information Fusion, 106*, 102301. DOI: 10.1016/j.inffus.2024.102301

Lund, H., Østergaard, P. A., Connolly, D., & Mathiesen, B. V. (2017). Smart energy and smart energy systems. *Energy, 137*, 556–565. DOI: 10.1016/j.energy.2017.05.123

Lundberg, S. M., & Lee, S.-I. (2017). A Unified Approach to Interpreting Model Predictions. In Guyon, I., Luxburg, U. V., Bengio, S., Wallach, H., Fergus, R., Vishwanathan, S., & Ford, R. (Eds.), Advances in Neural Information Processing Systems: Vol. 30. *NIPS 2017)* (pp. 4765–4774). Curran Associates, Inc. https://proceedings.neurips.cc/paper/2017/file/8a20a8621978632d76c43dfd28b67767-Paper.pdf

Machlev, R., Heistrene, L., Perl, M., Levy, K. Y., Belikov, J., Mannor, S., & Levron, Y. (2022). Explainable Artificial Intelligence (XAI) techniques for energy and power systems: Review, challenges and opportunities. *Energy and AI, 9*, 100169. DOI: 10.1016/j.egyai.2022.100169

Mason, K., & Grijalva, S. (2019). A review of reinforcement learning for autonomous building energy management. *Computers & Electrical Engineering, 78*, 300–312. DOI: 10.1016/j.compeleceng.2019.07.019

Mengelkamp, E., Gärttner, J., Rock, K., Kessler, S., Orsini, L., & Weinhardt, C. (2018). Designing microgrid energy markets: A case study: The Brooklyn Microgrid. *Applied Energy, 210*, 870–880. DOI: 10.1016/j.apenergy.2017.06.054

Moher, D., Liberati, A., Tetzlaff, J., & Altman, D. G.Prisma Group. (2009). Preferred reporting items for systematic reviews and meta-analyses: The PRISMA statement. *PLoS Medicine, 6*(7), e1000097. DOI: 10.1371/journal.pmed.1000097 PMID: 19621072

Molnar, C. (2019). *Interpretable machine learning: A guide for making black box models explainable*. Leanpub.

Pradhan, N. R., Singh, A. P., Verma, S., & Mohapatra, P. K. (2022). A blockchain based lightweight peer-to-peer energy trading framework for secured high throughput micro-transactions. *Scientific Reports, 12*(1), 14523. DOI: 10.1038/s41598-022-18603-z PMID: 36008545

Pudjianto, D., Ramsay, C., & Strbac, G. (2007). Virtual power plant and system integration of distributed energy resources. *IET Renewable Power Generation, 1*(1), 3–9. DOI: 10.1049/iet-rpg:20060023

Qiu, D., Ye, Y., Papadaskalopoulos, D., & Strbac, G. (2020). A Deep Reinforcement Learning Method for Pricing Electric Vehicles With Discrete Charging Levels. *IEEE Transactions on Industry Applications, 56*(5), 5901–5912. DOI: 10.1109/TIA.2020.2984614

Ribeiro, M. T., Singh, S., & Guestrin, C. (2016). "Why should I trust you?": Explaining the predictions of any classifier. *Proceedings of the 22nd ACM SIGKDD International Conference on Knowledge Discovery and Data Mining*, 1135-1144.

Samek, W., Wiegand, T., & Müller, K.-R. (2017). *Explainable Artificial Intelligence: Understanding, Visualizing and Interpreting Deep Learning Models*. arXiv.

Selvaraju, R. R., Cogswell, M., Das, A., Vedantam, R., Parikh, D., & Batra, D. (2017). Grad-CAM: Visual Explanations from Deep Networks via Gradient-Based Localization. In *Proceedings of the IEEE International Conference on Computer Vision (ICCV)* (pp. 618–626). IEEE. https://www.computer.org/csdl/proceedings-article/2017/iccv/3394747/12S4860R1iX

Shadi, M. R., Mirshekali, H., & Shaker, H. R. (2025). Explainable artificial intelligence for energy systems maintenance: A review on concepts, current techniques, challenges, and prospects. *Renewable & Sustainable Energy Reviews, 216*, 115668. DOI: 10.1016/j.rser.2025.115668

Soshinskaya, M., Crijns-Graus, W. H. J., Guerrero, J. M., & Vasquez, J. C. (2014). Microgrids: Experiences, barriers and success factors. *Renewable & Sustainable Energy Reviews, 40*, 659–672. DOI: 10.1016/j.rser.2014.07.198

Ukoba, K., Olatunji, K. O., Adeoye, E., Jen, T.-C., & Madyira, D. M. (2024). Optimizing renewable energy systems through artificial intelligence: Review and future prospects. *Energy & Environment, 35*(7), 3833–3879. DOI: 10.1177/0958305X241256293

Umar, A., Kumar, D., & Ghose, T. (2025). Decentralized energy trading in microgrids: A blockchain-integrated model for efficient power flow with social welfare optimization. *Electrical Engineering, 107*, 2677–2695. DOI: 10.1007/s00202-024-02635-x

Wang, Y., Chen, Q., Hong, T., & Kang, C. (2019). Review of Smart Meter Data Analytics: Applications, Methodologies, and Challenges. *IEEE Transactions on Smart Grid, 10*(3), 3125–3148. DOI: 10.1109/TSG.2018.2818167

Wohlin, C. (2014). Guidelines for snowballing in systematic literature studies and a replication in software engineering. *Proceedings of the 18th International Conference on Evaluation and Assessment in Software Engineering*, 1-10. DOI: 10.1145/2601248.2601268

Yin, R. K. (2014). *Case study research: Design and methods* (5th ed.). Sage Publications.

Zhang, C., Wu, J., Long, C., & Cheng, M. (2017). Review of Existing Peer-to-Peer Energy Trading Projects. *Energy Procedia, 105*, 2563–2568. DOI: 10.1016/j.egypro.2017.03.737

Zhang, C., Wu, J., Zhou, Y., Cheng, M., & Long, C. (2018). Peer-to-Peer energy trading in a Microgrid. *Applied Energy, 220*, 1–12. DOI: 10.1016/j.apenergy.2018.03.010

Chapter 6
AI-Enhanced Satellite and Remote Sensing for Real-Time Emissions Monitoring

Shivnath Ghosh
https://orcid.org/0000-0001-7368-7139
Brainware University, India

Olivia Jana
https://orcid.org/0009-0009-2101-9513
Brainware University, India

ABSTRACT

The accurate, scalable, and timely monitoring of greenhouse gas (GHG) emissions is a crucial component of the global effort to slow down climate change. Evaluating mitigation progress, upholding policy commitments, and directing climate action all depend on accurate emissions data. However, traditional methods of estimating emissions frequently rely on historical datasets, manual computations, or national self-reported inventories. Despite being fundamental, these approaches usually have built-in drawbacks like poor temporal and spatial resolution, a lack of real-time responsiveness, and variations in reporting requirements among nations and industries. These disparities jeopardize international transparency and make it more difficult to implement aggressive climate goals, such as those set forth in the Sustainable Development Goals (SDGs) and the Paris Agreement.

DOI: 10.4018/979-8-3373-5540-5.ch006

LITERATURE REVIEW

Research on the intersection of artificial intelligence (AI) and remote sensing technologies is growing as a result of the ongoing difficulty in accurately and instantly monitoring greenhouse gas (GHG) emissions. Low temporal resolution, geographic inconsistency, and reporting delays plague traditional emissions reporting methods, which are frequently based on self-reported inventories or out-of-date statistics. Numerous studies have highlighted these drawbacks and advocated for more automated and dynamic methods. According to new research, problems with spatial coverage may be resolved by satellite-based Earth observation systems like GHGSat, ESA's Sentinel, and NASA's MODIS. But only when combined with AI methods that can effectively analyze large and complicated data will their potential be fully realized.

The analytical bottlenecks of satellite imagery interpretation may be overcome by recent developments in AI, especially in deep learning and computer vision. To identify emission hotspots, calculate pollutant concentrations, and predict emission trends, Convolutional Neural Networks (CNNs), Recurrent Neural Networks (RNNs), and hybrid AI-physics models have become more and more popular. Researchers have shown how AI can be used to improve urban air quality analysis, track carbon losses brought on by deforestation, and find methane leaks. In order to create comprehensive environmental intelligence systems, literature also emphasizes the integration of AI with IoT and on-ground sensors. However, despite these developments, academics have voiced serious concerns about algorithmic transparency, data privacy, and ethical surveillance, underscoring the need for responsible innovation and international policy alignment.

1. INTRODUCTION

The pressing need for precise, real-time monitoring of greenhouse gas (GHG) emissions has been highlighted by the rapid acceleration of climate change. Carbon dioxide (CO_2), methane (CH_4), and nitrous oxide (N_2O) concentrations in the atmosphere are still rising due to anthropogenic sources, including the burning of fossil fuels, industrial manufacturing, agriculture, land-use change, and deforestation. Sea level rise, rising global temperatures, and more intense extreme weather events are all primarily caused by these emissions. There has never been a more pressing need for accurate, open, and ongoing emissions data as international accords like the Paris Accord seek to keep warming to 1.5°C over pre-industrial levels. The temporal resolution and spatial granularity required for real-time policy intervention are frequently absent from traditional inventory-based approaches, which rely on self-reported national statistics and out-of-date datasets.

There are several issues with the current emissions reporting systems. Emissions inventories are updated once a year or less in many countries, which makes them inadequate for crisis response or short-term forecasting. Furthermore, self-reporting may result in notable disparities because of different approaches, insufficient verification systems, or deliberate underreporting. International accountability and progress tracking are hampered by this data lag, which produces a significant gap between policy design and policy effectiveness. Additionally, there are significant geographic data blind spots due to the unequal distribution of traditional ground-based monitoring stations, which are mostly located in developed countries and essentially nonexistent in many regions of the Global South. A paradigm change in the way emissions data is gathered, validated, and used is necessary to close these gaps.

This paradigm shift is being made possible in large part by satellite remote sensing technologies and artificial intelligence (AI). Large regions, including isolated and conflict-prone areas, can have detailed atmospheric data collected by satellites fitted with hyperspectral and multispectral sensors. This satellite data can be quickly processed to estimate emissions levels with exceptional accuracy when combined with AI models, providing near-global and real-time coverage. Predicting emission trends, identifying subtle patterns and anomalies, and differentiating between various GHG sources are all tasks that machine learning algorithms excel at. Artificial intelligence (AI) can also combine various data sources, such as satellites, weather forecasts, industrial activity logs, and land-use databases, to produce coherent, useful insights that help early warning systems and mitigation plans for environmental threats.

The purpose of this chapter is to examine how artificial intelligence (AI) and satellite-based systems are transforming emissions monitoring. It starts by outlining the fundamental significance of emissions data in forming climate policy and pointing out the main drawbacks of traditional monitoring methods. The conversation then turns to how new technologies are enabling faster data delivery, higher resolution, and greater transparency in order to overcome these obstacles. In the end, combining AI-driven analytics with Earth observation platforms is not only a technological breakthrough but also a vital instrument for accomplishing global climate goals and promoting environmental responsibility on a worldwide basis.

1.1 Importance of Emissions Monitoring in Climate Mitigation

The foundation of climate mitigation strategies is efficient emissions monitoring. To set realistic climate targets, assess national commitments like Nationally Determined Contributions (NDCs), and ensure accountability in international frameworks like the Paris Agreement, accurate, consistent, and transparent data on greenhouse gas (GHG) emissions are crucial. Countries can identify their biggest sources of pollution and prioritize areas for intervention by quantifying emissions from a variety of sectors, including industry, transportation, energy, and agriculture. Additionally, it encourages adherence to the Sustainable Development Goals (SDGs), particularly those pertaining to climate resilience and environmental sustainability. Climate action becomes speculative in the absence of trustworthy data, making it challenging to determine whether policies are genuinely effective or merely aspirational.

Emissions monitoring is essential for enabling effective environmental governance in addition to target tracking. Greenwashing or underreporting can be exposed by comparing reported and actual emissions using data from satellites, ground sensors, and AI-powered analytics. Regulators can create evidence-based policies, implement environmental regulations, and encourage industries to adopt cleaner practices thanks to this body of evidence. Furthermore, the creation of carbon pricing schemes, such as carbon taxes and emissions trading programs, which depend on reliable emissions data for economic incentives, is directly influenced by emissions data. Monitoring systems improve operational fairness and market trust in carbon markets by guaranteeing that emissions are accurately measured and attributed.

Adaptive climate governance and early warning systems are further strengthened by high-resolution, real-time monitoring. Authorities can react quickly and minimize damage when unexpected spikes in emissions from wildfires, industrial accidents, or illicit deforestation are quickly detected. At the local level, timely emissions data facilitate grassroots environmental action mobilization, proactive local planning, and health alerts for pollution exposure. These tools offer practical insights at various scales, from neighborhood-level interventions to global oversight, when paired with AI and remote sensing. In the end, thorough emissions monitoring gives citizens, researchers, and policymakers the information they need to address climate change accurately, quickly, and effectively.

1.2 Limitations of Traditional Emissions Reporting and Inventories

Traditional emissions reporting frameworks struggle to provide timely and accurate climate intelligence because of their manual data collection, indirect estimations, and infrequent updates. Self-reported data, either collected from industries or calculated using standardized emission factors linked to the use of fossil fuels, changes in land use, or agricultural products, form the basis of most national greenhouse gas inventories. These approaches usually lack real-time responsiveness and granularity because data is usually collected annually and aggregated at the national or sectoral level. As a result, the spatial resolution is too coarse to detect regional emission hotspots, and the temporal delay makes it difficult to monitor seasonal variations or respond to sudden spikes in pollution.

These traditional inventories are also prone to inconsistencies, underreporting, and data manipulation. The institutional capacity and technological infrastructure required to maintain current environmental databases are lacking in many developing and emerging economies. Even when data is available, transparency may be compromised by financial or political incentives to reduce actual emissions figures. In addition to making it more challenging to compare national progress or carry out climate agreements, this can distort the picture of global climate performance. Furthermore, there is often a lack of third-party verification, which leads to data being taken at face value with little independent scrutiny. This makes it easier for flaws and greenwashing to continue.

Crucially, the fluid, dynamic, and transboundary character of contemporary emissions sources is poorly captured by traditional systems. Oil spills, pipeline methane leaks, wildfires, and other sporadic incidents are frequently either unreported or documented long after the harm has been done. Additionally, these systems have trouble with "non-point" pollution sources that don't neatly fit into predetermined reporting frameworks, like diffuse urban emissions or agricultural runoff. Legacy inventory systems were never intended to deliver near real-time, high-resolution, independently verifiable emissions data, which is essential for climate governance in the twenty-first century. Satellite remote sensing, AI-driven analytics, and interoperable data platforms must be added to modern monitoring in order to meet this need. These technologies must get past these conventional constraints and allow for quick, evidence-based action.

1.3 Role of AI and Satellite Technologies in Transforming Emissions Tracking

The capabilities of satellite-based Earth observation systems for tracking emissions are being revolutionized by artificial intelligence (AI). Real-time environmental dynamics cannot be fully captured by traditional emissions monitoring techniques, which are frequently based on statistical estimations and delayed reporting. AI can swiftly and accurately analyze large, high-dimensional datasets from satellites, especially when using machine learning (ML) and deep learning (DL) techniques. Emission hotspots, unforeseen anomalies, and long-term trends that would be challenging, if not impossible, to detect manually can be found thanks to these algorithms' exceptional ability to recognize spatial and temporal patterns in imagery. In addition to lowering human error, automation of this analytical procedure greatly increases the speed and accuracy of emissions assessments.

The impact of AI is further enhanced by the integration of satellite sensors, which range from optical and thermal to multispectral and hyperspectral instruments. Variations in surface temperature, atmospheric gas concentrations, vegetation cover, and changes in land use are just a few of the many environmental data that these sensors record. Terabytes of data are produced every day by satellites such as NASA's MODIS, ESA's Sentinel-5P, and commercial platforms like GHGSat, which cover large geographic areas on a regular basis. In order to provide real-time insights into emissions from a variety of sources, including transportation networks, oil refineries, agricultural lands, and forested areas, artificial intelligence (AI) systems absorb and analyze this continuous flow of data. Under conventional monitoring frameworks, pollution events like methane leaks or forest fires may go weeks without being detected. However, this high-resolution, continuous surveillance makes it possible to detect these events in real time.

The combination of AI and satellite technologies promotes a new degree of accountability and transparency in climate governance, going beyond simple detection. Without depending on self-reported data, which is frequently erratic or subject to political influence, these systems enable cross-border emissions monitoring. Near real-time emissions data is now available to international verification bodies, environmental organizations, and local governments, enabling them to confidently enforce laws, carry out mitigation plans, and confirm climate commitments. Thus, AI-enabled satellite monitoring not only gets around the technical drawbacks of traditional systems, but it also helps make climate diplomacy and policy enforcement more fair and grounded in facts. An important development in the worldwide endeavor to slow down climate change is the combination of artificial intelligence and space-based sensing technologies.

1.4 Chapter Objectives and Structure

This chapter's main objective is to provide a thorough examination of how artificial intelligence (AI) is changing the field of real-time emissions monitoring when combined with satellite imagery and remote sensing technologies. The need for precise, open, and quick monitoring systems that can track greenhouse gas (GHG) emissions over a range of spatial and temporal scales has gained international attention due to the urgency of combating climate change. The timeliness, granularity, and dependability of traditional reporting mechanisms are lacking. On the other hand, the combination of AI and Earth observation technologies provides a revolutionary toolkit that can improve accuracy, automate detection, and enable predictive insights.

This chapter provides a multifaceted understanding of this quickly changing field by situating itself at the nexus of climate science, AI innovation, and policy design.

Showing how AI algorithms—in particular, machine learning (ML), deep learning (DL), and reinforcement learning (RL)—are being used to extract useful insights from vast amounts of environmental data is one of the chapter's main goals. These methods enable all parties involved—from government organizations and scientists to non-governmental organizations and local communities—to go beyond static emissions inventories. Rather, they are now able to interact with dynamic, near real-time emissions datasets that highlight environmental threats that are both episodic and chronic, like wildfires, urban pollution spikes, methane leaks, and forest degradation. The chapter also explores the ways in which AI is advancing forecasting models, anomaly detection, and causal inference—tools that are critical for both post-event accountability and preventive action.

The chapter is structured into several thematic sections, each of which builds on the one before it to form a coherent story. It begins with a basic introduction to satellite systems, sensor technologies, and remote sensing techniques, providing the framework for comprehending the methods used to gather data from the Earth's atmosphere and surface. An examination of AI applications in emissions monitoring follows, where a variety of models and methodologies—including transfer learning, generative adversarial networks (GANs), and convolutional neural networks (CNNs)—are presented and placed within the context of actual case studies. These examples include using hyperspectral imaging to detect industrial emissions and neural networks trained on traffic and weather data to evaluate the quality of the air in cities.

The chapter's following sections focus on addressing important issues that may impede the fair use of AI-powered climate tools, including algorithmic bias, data privacy, and the digital divide. The study also examines emerging trends that suggest the possibility of a globally interconnected emissions intelligence network in the future, such as the use of edge AI in drones, federated learning in distributed sensor networks, and blockchain for emissions verification. Recommendations for governance standards, policy frameworks, and cooperative initiatives that can use AI to support sustainable climate action are included in the chapter's strategic and ethical reflections at the end. The chapter's ultimate goal is to assist researchers, practitioners, and decision-makers in navigating the nexus between artificial intelligence and environmental stewardship by acting as a technical manual and a strategic roadmap.

2. FOUNDATIONS OF SATELLITE AND REMOTE SENSING TECHNOLOGIES

2.1 Overview of Earth Observation Satellites

Earth observation (EO) satellites, which provide unmatched insights into the Earth's surface and atmospheric systems, have completely changed our understanding of and interactions with our surroundings. Regardless of national borders, these space-based platforms act as the eyes in the sky, continuously gathering data over enormous geographic areas. The benefit of EO satellites is their capacity to deliver standardized, repeatable, and consistent datasets—an essential component of environmental protection, agricultural management, emissions monitoring, and disaster response. Satellites in low Earth orbit (LEO) and geostationary orbit are equipped with a range of sensors that monitor the Earth in various electromagnetic spectrum bands. These sensors include optical, infrared, radar, and hyperspectral instruments.

MODIS (Moderate Resolution Imaging Spectroradiometer), which is installed on NASA's Terra and Aqua satellites, is one of the most popular EO satellite systems. With its near-daily worldwide coverage and ability to capture imagery in 36 spectral bands, MODIS is a vital tool for monitoring vegetation health, land surface temperature, and aerosol concentrations—all of which are closely related to emissions. The Landsat program, a joint venture between NASA and the U.S. Geological Survey (USGS), is another crucial platform that offers medium-resolution imagery going all the way back to the 1970s. Landsat's extensive historical archive makes it possible to analyze long-term trends in land-use change, urbanization, and deforestation—all of which have an impact on emissions.

The European Space Agency's (ESA) Copernicus program provides a series of Sentinel satellites designed specifically for environmental monitoring. Notably, Sentinel-5P monitors the atmosphere and is an expert in high-resolution detection of trace gases like methane (CH_4), ozone (O_3), sulfur dioxide (SO_2), and nitrogen dioxide (NO_2). This satellite is essential for locating hotspots for pollution and tracking how well environmental laws are working. Sentinel-5P supports evidence-based decision-making by allowing policymakers to see the distribution of pollution in almost real-time by fusing spectral data with sophisticated AI algorithms.

The resolution and responsiveness gap in emissions tracking has been filled in parallel by private satellite operators. One such participant is the Canadian company GHGSat, which uses nanosatellites for high-resolution GHG monitoring. Methane emissions from specific oil and gas facilities, landfills, and other point sources can be located using GHGSat satellites. By offering focused, on-demand analytics to industry stakeholders and regulators, their commercial services supplement public-sector satellites. These various platforms work together to create a synergistic, multi-scale observation network that supports a global framework for emissions monitoring, allowing for both site-specific, granular intervention and macro-level policy planning.

2.2 Types of Remote Sensing

Based on the kind of sensors and electromagnetic spectrum they use, remote sensing technologies can be broadly categorized. Similar to conventional photography, optical sensing records visible light and is perfect for monitoring urban growth, vegetation cover, and land degradation. By monitoring actions that have a major impact on the carbon cycle, like deforestation or changes in land use, it assists in identifying indirect emissions. Conversely, thermal remote sensing picks up heat radiation coming from the Earth's surface. This is particularly helpful for locating flare stacks, wildfires, and heat-intensive industrial processes—all of which are frequently direct sources of emissions. Using microwave signals, Synthetic Aperture Radar (SAR) provides day-and-night capabilities in all weather conditions. It is especially useful for monitoring floods, deforestation, and land subsidence brought on by the burning of fossil fuels.

By collecting information across several (and hundreds of) spectral bands, multi-spectral and hyperspectral imaging offer a more detailed perspective. This makes it possible to precisely identify chemical compounds on the ground or in the atmosphere. Hyperspectral data, for example, can be used to distinguish between different types of vegetation that affect carbon sequestration or to detect methane concentrations. When combined, the distinct and complementary insights from each type of sensing provide a comprehensive understanding of emissions and environmental change.

2.3 Key Emission Indicators Detectable via Remote Sensing

In order to identify and measure atmospheric emissions globally, remote sensing technologies have become essential. Greenhouse gases (GHGs) like carbon dioxide (CO_2), methane (CH_4), and nitrogen oxides (NO_X) are some of the most important indicators that can be measured from space. Specialized satellite instruments can detect these gases because they have distinct spectral absorption characteristics in different regions of the electromagnetic spectrum, especially the shortwave and thermal infrared regions. Sentinel-5P's TROPOspheric Monitoring Instrument (TROPOMI) provides high-resolution daily measurements of CO, NO_2, SO_2, and CH_4, for instance, providing vital information on transboundary air quality and pollution hotspots (Veefkind et al., 2012). Researchers and decision-makers can identify emission sources and trends with previously unheard-of precision thanks to the ability to monitor these gases in the lower atmosphere.

Hyperspectral imaging systems, such as those on GHGSat satellites, offer high-resolution observations that can isolate localized emission plumes from targeted sources, such as landfills, oil wells, and natural gas infrastructure, in the case of methane, which is more than 80 times more potent than CO_2 over a 20-year period (Jacob et al., 2022). These nanosatellites can detect emissions down to the facility level, unlike coarse-resolution systems that aggregate emissions over large areas. This gives industries and regulators the information they need to reduce leakage and enhance accountability. In order to better understand localized and regional carbon fluxes, other satellites, like NASA's OCO-2 (Orbiting Carbon Observatory-2), have been crucial in identifying carbon dioxide anomalies across urban and forested areas (Crisp et al., 2017).

Beyond direct gas detection, remote sensing also enables indirect monitoring of emissions via thermal and optical proxies. Abnormal heat signatures from industrial operations, flaring events, and wildfires—all of which are frequently linked to high-emissions episodes—can be detected by thermal infrared (TIR) sensors. For example, satellite-derived thermal imagery reveals higher heat levels at coal-fired power plants or during flaring operations in oil fields. This thermal data can be combined with multispectral optical imagery to create a composite image of the sources of pollution caused by combustion. In situations where cloud cover or atmospheric interference may restrict direct gas measurement, this synergy is useful for estimating emission intensities (Wooster et al., 2015).

Furthermore, alterations in vegetation indices—particularly the Normalized Difference Vegetation Index, or NDVI—are crucial markers of carbon sink loss and land degradation. The Earth's capacity to sequester carbon is diminished when forests and wetlands, which are natural absorbers of CO_2, are destroyed by logging, drought, or wildfire. Researchers can determine where carbon sinks are waning and incorporate this information into carbon budget models by monitoring seasonal or sudden changes in NDVI using tools like MODIS or Landsat. These indicators serve as inputs for AI-driven models that learn patterns, estimate emission quantities, and forecast future trends in addition to offering environmental intelligence (Reuter et al., 2020).

When combined, these direct and proxy indicators obtained through satellite remote sensing provide the fundamental data layer that can be used to construct sophisticated analytics and policy frameworks. A thorough, scalable, and timely understanding of emissions is made possible by their combination with ground-based data and AI-enhanced interpretation, opening the door to more intelligent climate governance and mitigation tactics.

2.4 Integration of Ground-Based and Aerial Data

For more focused and accurate data collection, satellites are frequently used in conjunction with ground-based sensors and aerial platforms such as drones (UAVs), even though they offer an unmatched viewpoint for large-scale observation. By validating satellite-derived data, ground stations outfitted with lidar systems, weather sensors, and gas analyzers can lower uncertainty and calibrate artificial intelligence (AI) models used in emissions estimation. These ground networks are particularly useful in urban areas or industrial centers where real-time, high-resolution data is crucial for tracking health effects and compliance.

Gaps in satellite coverage in terms of time and space are filled by aerial data collection with manned and unmanned aircraft. To measure emissions over specific areas, such as methane leaks from pipelines or forest degradation in inaccessible areas, drones, in particular, can be outfitted with thermal cameras, multispectral devices, and miniature gas sensors. They can be swiftly deployed following incidents like industrial accidents or wildfires and offer flexible, responsive data collection with centimeter-level accuracy.

A multi-tiered sensing framework that combines satellite, ground, and aerial data improves the precision and granularity of emissions monitoring. This multi-source data can be fed into AI models to create accurate, high-resolution maps of emission levels, facilitating community involvement, real-time alerts, and better informed policymaking. The creation of digital twins for the environment, where predictions and simulations are based on real-time, verifiable data streams, depends heavily on this integrated approach.

3. THE ROLE OF ARTIFICIAL INTELLIGENCE IN EMISSIONS MONITORING

3.1 Machine Learning and Deep Learning Applications

The processing and analysis of satellite and remote sensing data has been completely transformed by artificial intelligence, especially through machine learning (ML) and deep learning (DL) techniques. For classification tasks like pollutant estimation and mapping land cover, models like Random Forests, Support Vector Machines (SVMs), and K-Nearest Neighbors (KNN) have long been employed. More accurate feature extraction, particularly from high-dimensional satellite imagery, has been made possible by recent developments in deep learning, such as Convolutional Neural Networks (CNNs) and Autoencoders. These models are capable of accurately mapping pollution sources across spatial and temporal scales and detecting minute changes in environmental features.

CNNs are frequently used for tasks like identifying urban growth, forest degradation, and heat signatures suggestive of industrial activity because of their exceptional ability to handle spatial data. Conversely, autoencoders are helpful for unsupervised learning, which makes them perfect for detecting uncommon or unexpected emissions events that don't have any labels. By learning intricate, non-linear relationships within the data, lowering false positives, and enhancing generalization across regions and sensor types, these models perform better than conventional rule-based systems. As a result, AI models are now essential resources for turning unprocessed imagery into useful emissions insights.

3.2 AI for Pattern Recognition and Anomaly Detection

In order to identify patterns in large, multidimensional satellite datasets, artificial intelligence (AI) is essential. These datasets, which are gathered every day from different Earth observation platforms, include a variety of environmental signals, including temperature, reflectance, gas concentrations, and surface variations. To identify baseline or "normal" environmental behavior for a specific area or season, AI models—especially those built on supervised learning—can be trained on historical emissions data. Understanding this baseline enables the model to identify deviations, or anomalies, which could be signs of odd or dangerous occurrences like unapproved industrial emissions, burning biomass, or oil and gas leaks. These understandings are essential for spotting emission events early and setting off alarms before they have a serious negative influence on the environment or public health.

In emissions monitoring, anomaly detection necessitates both temporal awareness and spatial recognition. Temporal dependencies in environmental data streams can be effectively modeled by AI models like Transformers, Long Short-Term Memory (LSTM) networks, and Recurrent Neural Networks (RNNs). These models are able to identify and separate anomalous trends from seasonal variations in emissions, such as spikes in the use of fossil fuels during the winter or increases in agricultural burning during harvest months. Policymakers can better understand long-term behavior and develop focused interventions by identifying recurrent pollution cycles. For instance, regulatory agencies can look into operational procedures that may be causing emissions at a refinery if a model detects consistent methane spikes there every evening.

Environmental data in real-world settings is frequently noisy, lacking, or impacted by obstructive elements like atmospheric distortion or cloud cover. AI overcomes these obstacles by employing sophisticated methods such as ensemble learning to integrate predictions from multiple models, spatiotemporal clustering to group emission events with similar space-time properties, and Bayesian inference for probabilistic reasoning. Even with less-than-ideal inputs, these tools improve the resilience and reliability of the model. Unsupervised learning methods like Autoencoders and Isolation Forests can also identify new or unexpected emission patterns without the need for prior labeling, which is crucial in areas that haven't been monitored before or in environments that are changing quickly. When combined, these anomaly detection and pattern recognition skills allow AI to function as a potent surveillance tool for tracking global emissions and effectively mitigating climate risks.

3.3 AI-Powered Data Fusion

Data fusion is necessary for a thorough understanding because emissions come from a variety of sources and interact with various environmental systems. AI is particularly good at combining data from various sources, such as historical emissions inventories, weather models, satellite imagery, and ground-based sensors. Analysts can evaluate both direct and indirect factors influencing pollution dynamics thanks to this fusion, which produces a single, multi-dimensional picture of the emissions landscape. For example, tracking the movement of smoke or urban smog plumes is made easier by combining temperature, wind direction, and NO_2 concentrations.

AI can correlate and learn from data in a variety of formats, including text reports, time series, and images, thanks to methods like multi-modal learning and graph neural networks (GNNs). AI systems can improve their forecasts and lower uncertainty by combining satellite data with ground truth from IoT-enabled air quality sensors and aerial drone imagery. Predictive modeling is also supported by this tiered approach, which enables systems to predict emissions under various conditions, such as changes in policy or industrial growth. In the end, fragmented observations are transformed into a cohesive emissions intelligence framework that facilitates evidence-based decision-making through AI-powered data fusion.

3.4 Real-Time Data Processing Pipelines

AI's real-time capabilities, which allow it to process massive streams of incoming data with little latency, are a significant advantage in emissions monitoring. Post-processing satellite data is a common step in traditional analytical techniques, which can cause response times to be delayed by days or even weeks. On the other hand, AI-powered pipelines are able to detect emission events almost instantly by analyzing incoming satellite images and sensor readings as they are received. For time-sensitive situations like gas pipeline leaks, wildfire outbreaks, or smog alerts, this real-time capability is extremely important.

Cloud-based infrastructure and edge computing are commonly used in real-time AI pipelines, where data is first processed at ground stations or aboard satellites before being sent to centralized AI models. These pipelines continuously monitor surface and atmospheric conditions using methods like automated alert generation, event-based triggers, and stream processing. For example, a system might identify an unusual heat signature close to a refinery and alert environmental or local authorities right away. This promptness promotes proactive intervention, improves situational awareness, and may even prevent environmental catastrophes.

4. USE CASES AND APPLICATIONS

4.1 Methane Leak Detection from Oil and Gas Infrastructure

With a 20-year global warming potential more than 80 times that of carbon dioxide, methane (CH_4) is a powerful greenhouse gas. Therefore, identifying and addressing methane leaks—particularly those from oil and gas infrastructure—is a top priority for climate action. Using hyperspectral and thermal imagery, AI-enhanced satellite monitoring enables precise methane plume identification and quantification. Algorithms are able to map the dispersion of methane over time and space by differentiating its distinct spectral signature from that of background atmospheric gases.

Methane emissions can be recorded at different spatial resolutions by satellites such as GHGSat, Sentinel-5P, and TROPOMI. These emissions are traced back to their origins, which may include unreported flare stacks, malfunctioning valves, or leaking pipelines, using artificial intelligence (AI) models, especially those built on convolutional neural networks (CNNs) and time-series analysis. These tools assist energy companies in reducing fugitive emissions, improving regulatory compliance, and prioritizing maintenance by estimating the extent and duration of the leak. These AI-based evaluations can also be used by governments and independent groups to validate industry reports and enforce responsibility in methane management.

4.2 Plume Analysis for Industrial and Wildfire Emissions

Particulate matter (PM2.5), carbon monoxide, and volatile organic compounds are among the dangerous pollutants released by wildfires and industrial emissions. Public health and environmental management depend on the detection and analysis of emission plumes, which are visible columns of smoke or gas rising from these sources. AI can recognize these plumes, categorize their intensity, and determine the kinds and amounts of pollutants being released by processing thermal and optical satellite imagery.

In the case of wildfires, AI systems model fire behavior and plume dispersion by combining thermal anomaly detection with wind and weather data. By predicting the path of smoke or particulate matter, this aids in public advisories and firefighting operations. Plume analysis can be used to identify excessive flaring practices in oil refineries or unpermitted emissions in industrial areas. By connecting chemical transport models with remote sensing data, sophisticated machine learning models also assist in estimating pollutant concentrations from plumes. These observations help with climate risk assessments, emergency response, and law enforcement.

Forest Cover Change and Carbon Sink Estimation

Large volumes of CO_2 are taken up from the atmosphere by forests, which serve as natural carbon sinks. Understanding carbon fluxes and the integrity of offset programs depend on tracking changes in forest cover, whether from reforestation, degradation, or deforestation. When AI is used with satellite imagery, it can map vegetation cover, identify changes over time, and categorize changes in land use. Artificial intelligence (AI) models can offer insights into forest gain or loss in near real time by evaluating high-resolution images from platforms such as Sentinel-2, Landsat, and PlanetScope.

AI can detect minute variations in biomass density, canopy cover, and even distinguish between different types of forests using classification algorithms and deep learning models like U-Nets and ResNets. These models aid in calculating the amount of carbon stored by reforestation or released by deforestation. A growing number of REDD+ (Reducing Emissions from Deforestation and Forest Degradation) projects and carbon offset programs are using AI-based forest monitoring to verify their efficacy. AI guarantees that forest carbon monitoring is not only more responsive but also more transparent and credible by increasing temporal frequency and spatial accuracy.

4.3 Urban Air Quality Monitoring Using AI-Integrated Satellite Data

Transportation, building, industry, and energy consumption are the main causes of air pollution in urban areas. Conventional air quality monitoring stations frequently miss neighborhood-level variations and have limited spatial coverage. Cities can gain a more thorough understanding of pollution dynamics by combining AI forecasting models with satellite-derived data, especially nitrogen dioxide (NO_2) levels. Daily measurements of NO_2 and other trace gases that are suggestive of urban air quality are provided by satellite platforms like Sentinel-5P and Aura-OMI.

To generate localized predictions of pollutant concentrations, AI models use these satellite observations in conjunction with traffic patterns and meteorological data (temperature, wind speed, humidity). These tools enable proactive public health responses, like issuing advisories or modifying traffic regulations, by predicting short-term pollution spikes and identifying long-term trends. This information can also be used by legislators and city planners to better enforce environmental laws, assess the effects of public transportation projects, and create low-emission zones. AI-powered satellite monitoring thus becomes a crucial part of smart city frameworks and urban sustainability.

4.4 Verification of NDCs and Climate Commitments

Verifying Nationally Determined Contributions (NDCs) under the Paris Agreement is one of the most strategic uses of AI-enhanced emissions monitoring. As part of their climate commitments, many nations submit emissions data, but these are frequently based on self-reported inventories that have not been independently verified. By tracking emissions in real time from important industries like power generation, transportation, and land use change, AI-powered satellite analysis offers a third-party way to verify these national claims. This new degree of transparency is best demonstrated by tools like Climate TRACE (Tracking Real-time Atmospheric Carbon Emissions), which create open-access emissions maps using AI and satellite data. Through these platforms, researchers, international organizations, and even civil society can keep an eye on whether nations are fulfilling their climate commitments. This independent verification boosts the carbon credit market, lessens greenwashing, and fortifies the transparency framework that is essential to global climate diplomacy. In turn, by making sure that commitments and actions align, it fosters international trust and speeds up climate cooperation.

5. AI TECHNIQUES AND MODELS IN PRACTICE

5.1 Convolutional Neural Networks (CNNs)

Due to their unparalleled capacity to identify and categorize spatial features in imagery, Convolutional Neural Networks (CNNs) have emerged as the mainstay of image analysis tasks. CNNs analyze satellite imagery in the context of emissions monitoring in order to spot patterns that point to pollution sources like power plants, gas flares, vehicle traffic corridors, and deforestation areas. CNNs can identify subtle emission-related cues that are invisible to the human eye by learning filters that detect textures, edges, shapes, and spatial relationships.

CNNs can process multispectral images from high-resolution satellites such as Sentinel-2 and PlanetScope to detect changes in land use and land cover (LULC), which are important markers of emissions from deforestation and agriculture. Additionally, CNNs are essential for identifying thermal anomalies that point to increased industrial activity or fire outbreaks. Even in diverse environments, these models achieve high precision by utilizing multi-scale architectures like ResNet or U-Net.

Furthermore, in order to identify emissions sources with pixel-level accuracy, CNNs are frequently combined with other methods like segmentation and object detection (e.g., using Mask R-CNN or YOLO). This makes it possible to produce comprehensive emissions maps that are used for urban planning, environmental risk assessments, and regulatory compliance. They are scalable for worldwide environmental intelligence applications due to their capacity to be trained on sizable, labeled datasets and to adjust to new regions with little retraining.

5.2 Recurrent Neural Networks (RNNs) and Transformers

Recurrent neural networks (RNNs) and their variations, such as Long Short-Term Memory (LSTM) networks and Gated Recurrent Units (GRUs), are made to process sequential data, in contrast to CNNs, which focus on spatial pattern recognition. By learning from time-series satellite data, these models are particularly helpful for examining temporal changes in emissions. For instance, they can detect long-term trends brought on by industrial activity, identify recurrent seasonal emission patterns from agriculture, or monitor the evolution of NO_2 levels in urban areas over several months.

Long-term dependencies in environmental data can be memorized by LSTM models, which makes them perfect for forecasting how emissions will change in response to changes in policy, infrastructure, or weather. They are frequently employed to forecast pollutant concentrations and emission rates, which is essential for developing tools for policy simulation and early warning systems.

Transformers were first created for natural language processing, but they have recently been successfully used to analyze time-series emissions data. Transformers, as opposed to RNNs, improve accuracy in complex datasets by focusing on particular segments of the sequence through the use of attention mechanisms. Additionally, they are more parallelizable and scalable, which makes them appropriate for processing global satellite archives on a large scale. Transformers are used in emissions monitoring to quickly and accurately predict regional emission dynamics by integrating multiple time series (such as temperature, wind, NOx, and vegetation).

5.3 Generative Models for Emissions Prediction

The way we model and predict emissions scenarios is changing thanks to generative models like Variational Autoencoders (VAEs) and Generative Adversarial Networks (GANs). By learning to produce new data that closely resembles the statistical distribution of actual emissions data, these models allow researchers and decision-makers to investigate potential futures. Highly realistic simulations of satellite imagery or emissions patterns are produced by GANs, which are made up of two neural networks (the discriminator and the generator) that compete in a zero-sum game.

The production of synthetic emissions maps for areas with insufficient or no real data is a significant application of generative models in this field. For example, using land use and infrastructure data, a GAN trained on industrial emissions from one region can model what emissions might look like in another developing area. The impact of policy interventions, like building green energy facilities or reducing vehicle traffic, is also simulated using these models.

Conversely, VAEs are probabilistic models that decode input data into a predicted output after encoding it into a lower-dimensional latent space. This makes them perfect for sampling from the latent space and producing a variety of realistic future emissions scenarios. Additionally, VAEs aid in estimating prediction uncertainty, which is crucial for modeling and making decisions related to climate risk. Therefore, these generative models give emissions monitoring a forward-looking aspect by offering a secure and computationally effective means of evaluating the effects of environmental decisions prior to their implementation.

5.4 Transfer Learning and Pre-Trained Models

The lack of labeled data is one of the main obstacles to emissions monitoring, particularly in areas with inadequate technical infrastructure. By enabling models trained on one dataset or region to be adapted to another with little retraining, transfer learning solves this problem. For instance, with just a few labeled samples, a CNN that was trained to identify methane flares in North America can be adjusted to function in the Middle East. This significantly lowers the time and computational expense needed to implement models in new areas.

For particular emissions-related tasks like plume detection or land classification, pre-trained models like ResNet, EfficientNet, and Vision Transformers (ViTs) can be optimized. Transfer learning makes advanced AI tools more accessible to nations or organizations with limited data and compute capacity by speeding up the deployment process and improving performance in low-resource environments.

Additionally, the creation of international, interoperable frameworks for emissions monitoring depends heavily on transfer learning. Institutions can support data equity and standardization by sharing models and retraining them locally, thereby contributing to a federated ecosystem of emissions intelligence. As artificial intelligence develops, hybrid models that incorporate domain-specific adaptations (like climate physics constraints) with transfer learning will increase the precision, generalizability, and reliability of emissions forecasts across a range of environments and regions.

6. CHALLENGES AND LIMITATIONS

6.1 Resolution and Latency Issues

The spatial resolution of the available imagery is a major constraint in satellite-based emissions monitoring. The granularity needed to identify emissions from small-scale or urban sources, like rooftop generators, car fleets, or micro-industrial units, is frequently lacking in satellite sensors that operate at medium to coarse resolution (e.g., 250m to 1km per pixel). Emissions detection from dense urban environments, where pollution sources are closely spaced and infrastructure is complex, requires high-resolution data (10m or better), which is frequently only available from commercial satellites or comes with a high acquisition cost.

The frequency of revisits or temporal resolution present another difficulty. Satellites can't track rapidly changing emissions events in near real time because they usually only return to the same spot once every one to five days (or even longer in some cases). In situations where prompt detection is essential for successful mitigation, such as methane leaks, industrial accidents, or wildfire outbreaks, this latency is particularly problematic. Higher temporal resolution is provided by certain constellations, like PlanetScope or commercial CubeSat networks, but spatial resolution and data accessibility may still be constrained.

For emissions monitoring systems, the trade-off between temporal and spatial resolution presents a challenge because attaining one frequently jeopardizes the other. Moreover, frequent processing of high-resolution imagery adds computational overhead, necessitating the use of edge processing and effective real-time data pipelines. The creation of hybrid systems that cleverly integrate data from various sensors, satellite platforms, and even ground-based observations is required to overcome these obstacles and guarantee timeliness and fidelity.

6.2 Cloud Cover and Weather Interference

Another enduring obstacle to efficient emissions monitoring with satellite-based optical and thermal sensors is weather interference. Rain, snow, fog, and cloud cover can block the satellite's view of the Earth's surface, resulting in deteriorated or useless imagery. For several months of the year, optical monitoring is particularly unreliable in areas that are susceptible to monsoon seasons, tropical storms, or heavy snowfall, such as Southeast Asia, equatorial Africa, or northern latitudes. This results in gaps in the data and limits the capacity to maintain ongoing emissions surveillance.

In order to get around this, scientists are depending more and more on Synthetic Aperture Radar (SAR) systems, which can work day or night and can pass through clouds. SAR data, however, has its own set of difficulties. It needs to be preprocessed differently and is generally harder to understand, particularly when combined with AI models that were initially trained on optical data. To increase coverage and dependability, multimodal data fusion—the combination of radar, optical, and thermal data—is required.

Cloud detection and correction algorithms, which identify areas that are obscured and replace or interpolate missing data using previous or nearby satellite passes, are one way AI can help. These solutions aren't flawless, though, particularly when it comes to identifying emissions that change quickly. A more thorough integration of hyperspectral and radar-based satellite data in the future, along with real-time drone or UAV imagery, may help lessen the impact of weather interference and guarantee reliable emissions monitoring all year long.

6.3 Data Scarcity in Low-Resource Regions

The lack of data in many remote or developing areas makes it challenging to create, train, and validate AI models for emissions monitoring. The availability of ground truth data needed to train supervised machine learning models or calibrate satellite observations is restricted by the absence of ground-based sensors, such as air quality monitoring stations. Long-term trend analysis and forecasting are also hampered by the possible lack of historical climate and emissions data in these areas.

The lack of data in many remote or developing areas makes it challenging to create, train, and validate AI models for emissions monitoring. The availability of ground truth data needed to train supervised machine learning models or calibrate satellite observations is restricted by the absence of ground-based sensors, such as air quality monitoring stations. Long-term trend analysis and forecasting are also hampered by the possible lack of historical climate and emissions data in these areas.

The problem is made worse by a lack of institutional capacity, such as financial resources, technical know-how, or infrastructure for data sharing. Even when satellite data is available, important insights are frequently lost due to a lack of resources, staff, or training to process and interpret the data. Building capacity, international cooperation, and open-access platforms that increase underserved regions' access to AI tools and emissions datasets are all necessary to overcome this constraint. This is a global climate equity issue in addition to a technical one.

6.4 Bias, Interpretability, and Validation

Biases can be introduced into AI models during training, especially if the training data is skewed toward specific emission types or regions, geographically limited, or unbalanced. For example, a model that was primarily trained on data from urban areas in Europe might not be able to detect agricultural emissions in sub-Saharan Africa. These biases impact the model's generalizability and accuracy, resulting in systematic errors that may misguide resource allocation or policy decisions. Interpretability is another crucial problem; many sophisticated AI models, particularly deep neural networks, function as "black boxes," making predictions without offering precise justifications. Users' trust may be weakened by this lack of transparency, especially when it comes to regulators or policymakers who need explainable AI (XAI) tools to comprehend the decision-making process. It becomes challenging to audit models, spot mistakes, or make sure the results comply with ethical and environmental standards when interpretability is lacking.

To address these issues, validation using ground truth data is crucial; however, as was previously mentioned, ground data is frequently scarce. To assess and improve model reliability, researchers are increasingly using methods like uncertainty quantification, ensemble learning, and cross-validation. However, validating emissions models continues to be a problem that calls for not only technical solutions but also improved frameworks for data sharing and global standards to compare model performance across platforms and geographical areas.

6.5 Ethical Concerns and Data Privacy

Despite its technological potential, AI-powered satellite monitoring poses serious moral and geopolitical issues. Data sovereignty issues can arise from monitoring industrial operations, emissions from national infrastructure, or changes in land use, particularly when such data is gathered without a nation's express consent. Diplomatic tensions may result from this, especially if emissions data is used in international fora to condemn or punish countries without sufficient justification or redress.

Furthermore, the risk of privacy violations rises with the sophistication and granularity of satellite systems—not only for the country, but also for private individuals or organizations whose actions can be deduced from high-resolution imagery. This makes it difficult to distinguish between surveillance and environmental monitoring, which calls for strong frameworks for consent, data governance, and responsible use. Another risk is technological inequality, whereby wealthy organizations are the only ones with access to sophisticated AI emissions tools, resulting in a knowledge and power divide between wealthy and developing countries. Transparent AI design, inclusive policymaking, and international agreements on the responsible use of Earth observation data are necessary to address these ethical issues. As crucial as maintaining technical accuracy is guaranteeing equity, responsibility, and involvement in the ecosystem of AI-driven emissions.

7. EMERGING SOLUTIONS AND FUTURE INNOVATIONS

7.1 Federated Learning and Collaborative AI

Concerns regarding data privacy, security, and governance may arise from the centralized data collection used in traditional AI model training, particularly when handling sensitive industrial or environmental data. This is addressed by federated learning, which allows organizations to work together to train models without sharing raw data. Instead, only model parameters (such as weights or gradients) are shared with a central aggregator, and each participant trains a model locally on its own data. This method takes advantage of the richness and diversity of distributed datasets while maintaining data confidentiality.

Federated learning can facilitate cross-institutional cooperation in emissions monitoring between governments, satellite operators, and research labs without necessitating direct access to commercial or national emissions data. For instance, without disclosing surveillance data or classified infrastructure, various nations can help train global CO_2 or CH_4 monitoring models. This could be used to create regionally consistent emissions intelligence models while maintaining ethical data sharing and sovereignty. By using data from several regions and sensor types, federated learning also lessens the impact of geographic bias in training datasets.

7.2 Integration of UAVs, CubeSats, and IoT Networks

Traditional satellite missions have a broad reach, but resolution limitations, weather interference, or revisit constraints still result in observational gaps. Combining inexpensive unmanned aerial vehicles (UAVs), CubeSats, and Internet of Things (IoT) sensor networks—which collectively constitute a multi-scale, multi-source data ecosystem—is a viable approach. While CubeSats—miniaturized satellites with quick deployment cycles—offer frequent revisit rates and targeted emissions monitoring, drones can fly below clouds to gather fine-grained thermal or gas data.

In addition to satellite observations, continuous, localized data streams are provided by IoT sensor networks on the ground, such as weather stations, methane sniffers, and low-power air quality monitors. These various data sources can be ingested and harmonized by AI to create emissions maps that are more accurate and contextually aware. For instance, integrating CubeSat imagery, IoT weather data, and drone-detected methane leaks can improve leak localization, measure plume spread, and provide real-time mitigation recommendations.

7.3 Advances in Edge AI for Satellite Processing

The latency between image capture and analysis on the ground is a major bottleneck in satellite-based emissions tracking. Traditionally, centralized AI models analyze raw satellite data after it has been downlinked to Earth and preprocessed. On the other hand, Edge AI allows for real-time data processing onboard by integrating artificial intelligence models directly into the satellite hardware (or on neighboring relay platforms). This enables real-time decision-making for emissions events and significantly cuts down on transmission delays.

For instance, a satellite with Edge AI could automatically initiate high-resolution imaging or notify a ground station prior to the next planned pass if it detects an unusual thermal signature suggestive of an industrial flare. AI-specific processing units, like NVIDIA Jetson modules or Google's Edge TPU, are increasingly supporting these systems and can be incorporated into CubeSats and microsatellites. Furthermore, edge processing uses less bandwidth, which is especially helpful for large amounts of video or hyperspectral data.

7.4 Open-Access Climate Intelligence Platforms

One significant step toward democratizing environmental intelligence is the rise of open-access platforms for emissions and climate monitoring. One prominent example is Climate TRACE (Tracking Real-Time Atmospheric Carbon Emissions), a collaboration of academic institutions and non-governmental organizations that creates independently verified emissions inventories for almost every nation and significant industrial facility on the planet using AI, satellite imagery, and sensor data. These platforms seek to support journalism, civil society activism, and international climate negotiations by offering transparent, objective emissions data.

In addition to Climate TRACE, programs like OpenAQ, a global air quality data platform, and AI4EO (Artificial Intelligence for Earth Observation), which is backed by the European Space Agency, are assisting in expanding the accessibility of emissions intelligence. For stakeholders in the Global South in particular, these tools are invaluable because they give them access to insights that would otherwise require costly equipment or specialized knowledge. A new era of participatory climate governance is being fostered by the integration of AI, open data, and community engagement, which enables real-time emissions accountability.

8. POLICY AND STRATEGIC IMPLICATIONS

8.1 Supporting Transparency and Accountability

Global climate action is predicated on transparency in greenhouse gas (GHG) reporting. Conventional systems primarily rely on national self-reported emissions inventories, which are frequently beset by irregularities, antiquated procedures, and a lack of detail. These inventories only provide delayed insights into ongoing climate progress because they are typically updated annually or over a period of several years. Furthermore, it is challenging to determine whether nations are actually fulfilling their climate commitments under agreements like the Paris Agreement due to the absence of independent verification. This disparity erodes international trust and reduces accountability. By combining satellite imagery, remote sensing, and machine learning, artificial intelligence (AI) provides a potent remedy by enabling real-time, high-resolution emissions tracking. By comparing national reports with observational data, these technologies can identify differences and highlight irregularities that might point to non-compliance or underreporting. AI systems, for example, are able to detect carbon emissions from unregistered industrial operations or track methane leaks from oil fields. AI reduces human error and bias in emissions reporting by automating data analysis and cross-referencing it with multiple sources.

A stronger basis for global cooperation is fostered by this greater transparency. Leveling the playing field and enabling nations and organizations to hold each other accountable with evidence-based evaluations is achieved when emissions data is publicly available and independently verified. Additionally, it improves the legitimacy of climate talks, fortifies regulatory frameworks, and increases public confidence in climate policy. While international organizations can track whether nations are adhering to their commitments in real time rather than depending only on retrospective audits, governments can use AI-generated insights to improve their strategies.

Furthermore, states, non-governmental organizations, researchers, and the general public are all empowered by AI-driven transparency. AI-verified emissions data visualization on open-access platforms can help with watchdog operations, community-level action, and awareness-raising. Thus, AI does more than just increase data accuracy; it also builds a more accountable and participatory climate governance system where shared, trustworthy information enables informed decision-making and collaborative enforcement.

8.2 Enhancing Early Warning Systems

Climate change has significantly increased the frequency and severity of environmental hazards, making early warning systems more important than ever. Even though they are useful, traditional forecasting models frequently cannot handle the volume and complexity of real-time data required to make accurate and timely predictions about disasters. AI is revolutionizing this field by processing large datasets from a variety of sources, including satellite imagery, weather stations, sensor networks, and historical climate records, and identifying patterns and anomalies that indicate potential dangers. AI makes dynamic, adaptive forecasting models possible through sophisticated machine learning algorithms, which enhance alert timing and accuracy.

For instance, in areas where wildfires are common, artificial intelligence (AI) systems forecast the onset and spread of fires by utilizing historical fire data, wind patterns, vegetation moisture, and land surface temperature. Emergency responders are already able to map fire perimeters and update them hourly with the aid of projects like Google's AI-powered wildfire tracking system, which greatly speeds up response times and reduces casualties (Google, 2021). In a similar vein, AI-powered flood prediction models use terrain maps, hydrological simulations, and meteorological data to deliver remarkably accurate early warnings. Examples of AI being used to forecast floods days in advance, particularly in South Asian regions that are susceptible, include the IBM Weather Company and Microsoft AI for Earth initiatives (IBM, 2020; Microsoft, 2019).

Another field where AI-enhanced early warning systems are proving invaluable is air quality monitoring. AI can predict pollution spikes in urban areas by analyzing emissions from industrial processes, traffic, and weather. By using tools like NASA's machine learning with satellite data and Plume Labs' AI-powered air quality forecasting, cities can reduce the impact on vulnerable populations like children and the elderly by preparing for pollution events and issuing health advisories ahead of time (Plume Labs, 2022; NASA, 2020). Because they provide real-time data to guide emergency interventions and public health responses, these systems are essential in areas with chronic air pollution.

Most significantly, these systems can be utilized even in resource-constrained areas thanks to AI's scalability and adaptability. Early warning systems can be swiftly and economically deployed with the aid of satellite access and cloud-based AI platforms, even though many developing countries lack the infrastructure required for comprehensive environmental monitoring. By closing the gap between affluent and vulnerable nations, this democratization of climate resilience technology guarantees that the benefits of AI-driven forecasting are distributed globally. Early warning systems are ultimately transformed by artificial intelligence (AI) from passive data collectors into proactive, predictive tools that protect infrastructure, save lives, and boost societal resilience to climate change.

8.3 Informing Global Carbon Trading

Global carbon markets are essential to helping nations and businesses economically fulfill their climate pledges. By enabling participants to purchase and sell emissions allowances or credits, mechanisms like cap-and-trade schemes, carbon offsets, and voluntary carbon markets put a price on carbon and provide financial incentives for reducing emissions. However, the precision, openness, and validation of emissions and offsets are vital to the integrity of these markets. Strong monitoring systems are desperately needed, as evidenced by the numerous scandals involving inflated or questionable carbon credits in recent years that have damaged investor and public confidence.

One of the main factors facilitating reliable carbon trading systems is artificial intelligence. Artificial intelligence (AI) algorithms can track land-use changes, industrial emissions, and carbon sequestration efforts in almost real-time when paired with satellite remote sensing and ground-based sensor networks. Compared to conventional reporting techniques, this enables more accurate verification of carbon baselines and reductions. AI-powered systems, for instance, can determine whether a reforestation project is actually storing carbon or if it is just shifting deforestation elsewhere, a phenomenon called leakage. AI significantly lowers the possibility of greenwashing, a practice in which organizations inflate or fake environmental advancements in order to obtain financial or reputational advantages, by providing independent, third-party validation.

In carbon markets, AI also promotes fairness and price integrity. The price of carbon credits can better reflect their actual environmental value when emissions reductions are quantified. This promotes investment in top-notch projects with quantifiable climate benefits and discourages manipulative trading. Additionally, AI and blockchain technology can be combined to produce unchangeable records of carbon credit issuance and transactions, assisting in the removal of double counting, a major problem in both voluntary and compliance markets. Market confidence is increased by these technical capabilities, which guarantee that each credit translates into a genuine, one-of-a-kind emissions reduction.

Furthermore, platforms driven by AI offer traders, project developers, and regulators automated compliance monitoring tools and interactive dashboards. These interfaces track project progress, show emissions trends, and instantly identify possible fraud or non-compliance. AI promotes evidence-based policy changes and democratizes carbon market participation by facilitating an open and transparent information flow. By reducing entry barriers and boosting the legitimacy of their contributions, the incorporation of AI into carbon trading has the potential to eventually increase participation, particularly from small businesses and developing nations. In the end, this promotes a global carbon market that is more impactful, efficient, and inclusive, hastening the decarbonization process.

8.4 Empowering Local Governments and NGOs

Governments, community-based organizations, and non-governmental organizations are important change agents at the local level, where climate action is implemented, despite the fact that national and international levels shape climate policy frameworks. These organizations are often the first to observe environmental deterioration, react to climate-related emergencies, and interact with impacted communities. However, despite their crucial role, local actors frequently encounter major obstacles that make it difficult for them to effectively engage in evidence-based climate action or influence policy, such as a lack of funding, technical know-how, and access to high-quality data.

One promising way to close this capacity gap is through artificial intelligence. AI democratizes access to environmental data and decision-making tools through the use of mobile applications, cloud-based platforms, pre-trained models, and user-friendly interfaces. Users without extensive technical knowledge can visualize pollution trends, analyze emissions patterns, and track changes in land use with little training thanks to tools like Google Earth Engine, Global Forest Watch, and AI-powered apps like AirVisual. These platforms enable local authorities and non-governmental organizations to identify illicit activities, such as uncontrolled industrial discharge or unpermitted deforestation, and take swifter action by offering real-time, location-specific insights.

AI also makes community-led advocacy and monitoring possible. Local organizations can use satellite imagery and automated data analysis to provide reliable, time-stamped proof of environmental infractions or evaluate the results of sustainability initiatives, like waste management programs or afforestation drives. To track deforestation and notify indigenous communities of illicit logging activities, for instance, AI tools are being utilized in the Amazon basin (Rainforest Foundation US, 2022). Similar to this, grassroots groups can use AI-enabled air quality sensors and applications to gather information and put pressure on local governments to enact pollution control policies. By giving them the knowledge they need to participate in policy discussions and demand environmental justice, these technologies strengthen the voices of marginalized communities.

In the end, incorporating AI into local climate action promotes a more equitable, decentralized, and inclusive system of climate governance. It guarantees that those closest to the frontlines of climate change have access to environmental monitoring and response, rather than being monopolized by large institutions. AI supports bottom-up climate leadership by providing actionable intelligence to NGOs and local governments, facilitating a more robust and inclusive global response to the climate crisis.

9. CONCLUSION

9.1 Summary of Key Insights

An important development in the way we track greenhouse gas (GHG) emissions is the combination of AI with satellite and remote sensing technologies. AI makes dynamic, near-real-time surveillance possible, in contrast to conventional techniques that mainly rely on self-reported, static, and frequently delayed emissions inventories. It provides scalability across geographies, high spatial and temporal resolution, and accuracy in identifying the sources of emissions. AI models can verify mitigation efforts, spot anomalies that point to non-compliance or underreporting, and identify emission trends by analyzing vast amounts of data from satellites and ground sensors.

More responsive climate governance is made possible by this shift from passive data collection to active, predictive monitoring. AI is a flexible and potent tool because of its capacity to combine various data streams, including satellite imagery, meteorological data, land-use records, and socioeconomic indicators. It enables in-depth analysis at the local, regional, and international levels, assisting environmental organizations and policymakers in making prompt and well-informed decisions. Furthermore, compared to manual systems, its automation capabilities increase efficiency and dependability by lowering human error and resource dependence.

9.2 The Transformative Potential of AI in Climate Observation

AI is a transformative force in climate observation, not merely an additional tool. It is perfect for modeling environmental systems with complex and dynamic cause-effect patterns because of its capacity to manage non-linear, complex data relationships. AI provides actionable insights that were previously impossible to obtain because of data or processing limitations, from mapping deforestation and land degradation in remote forests to predicting pollution surges in urban areas.

Closing observational data gaps is one of AI's most significant contributions, especially in developing nations or isolated areas with limited ground-based monitoring. Researchers and organizations in low-resource environments can now access intelligence that was previously exclusive to wealthier countries thanks to openly available satellite data and artificial intelligence tools. Inclusion and equity in climate action are promoted by this democratization of climate observation.

Additionally, AI's real-time capabilities change the type and speed of response. Decision-makers can now start mitigation strategies without waiting for quarterly or annual reports. Instead, whether it's closing a polluting factory, getting ready for a flood, or sending out air quality warnings, AI-powered dashboards and alert systems can start quick reactions.

9.3 Call for Interdisciplinary Collaboration

Although AI has enormous technical potential, interdisciplinary cooperation is necessary for its responsible and significant application. Science, society, politics, and ethics are all impacted by the complex issue of climate change. Thus, in addition to data scientists and engineers, climate scientists, ecologists, public policy specialists, social scientists, and ethicists must be involved in the development and governance of AI tools in this context.

Early policymaker involvement in AI system design is necessary to guarantee that results meet sustainability goals and legal frameworks. While ethicists make sure AI doesn't worsen already-existing social inequalities, ecologists can assist AI models in taking biodiversity and ecosystem sensitivities into account. For instance, unless human values are incorporated into their design, AI models that distribute funds for disaster relief or climate change based only on economic indicators may ignore marginalized, vulnerable communities.

In order to guarantee that AI solutions are inclusive, transparent, and context-aware, this cooperation should also involve global organizations, academic institutions, and grassroots actors. We can only create morally sound, just, and efficient climate intelligence systems with this kind of cooperation.

9.4 Vision for a Real-Time, AI-Powered Global Emissions Intelligence Network

The goal is to build an open, interoperable, ethically governed, real-time, AI-powered global emissions intelligence network in the future. A system like this would combine ground sensors, crowdsourced data, AI models, and satellite feeds into a single platform. This would enable global transparency and prompt enforcement of climate agreements by providing real-time emissions tracking at the national and subnational levels.

Standards for data interoperability, privacy protection, and moral AI governance must be established in order to accomplish this. To make sure that all countries, regardless of resources, can contribute to and benefit from this network, international cooperation will be essential, much like how meteorological data is shared globally. Beyond merely monitoring emissions, the objective is to enable prompt, well-informed, cross-border climate action.

Community access is another aspect of this vision. Through user-friendly dashboards and mobile applications, local governments, non-governmental organizations, and even citizens should be able to access this network. This system could become a global public good with strong visualization tools and multilingual accessibility, providing information for academic research, advocacy, policymaking, and teaching. In order to ensure that no one is left behind in the fight against climate change, such a future embodies a world where AI supports collective climate responsibility.

REFERENCES

Crisp, D., Pollock, H. R., & Rosenberg, R. (2017). The on-orbit performance of the Orbiting Carbon Observatory-2 (OCO-2) instrument and its initial results. *Atmospheric Measurement Techniques, 10*, 59–81. DOI: 10.5194/amt-10-59-2017

Golkar, A., & Sweeting, M. N. (2016). CubeSat mission design for Earth observation. *Acta Astronautica, 114*, 112–124.

Google, A. I. (2021). *Blog*. AI Helps Track Wildfires in Real Time.

Jacob, D. J., Scarpelli, T. R., & Sulprizio, M. P.. (2022). Quantifying methane emissions from the largest oil-producing basin in the United States from space. *Science Advances, 8*(5), eabj4351.

Plume Labs. (2022). AI-Powered Air Quality Forecasting.

Rainforest Foundation US. (2022). AI and Satellite Tech for Indigenous Forest Monitoring.

IBM Research. (2020). Flood Forecasting with AI in India.

Reuter, M., Buchwitz, M., & Schneising, O. (2020). Towards monitoring localized CO_2 emissions from space: Co-located regional CO_2 and NO_2 enhancements observed by the OCO-2 and Sentinel-5 Precursor satellites. *Atmospheric Chemistry and Physics, 20*(2), 1027–1040.

Rolnick, D. (2019). Tackling climate change with machine learning. arXiv:1906.05433.

UNDP – United Nations Development Programme. (2020). Empowering Local Action for Climate Justice.

Vaddi, R. S. (2022). A review of UAV-based remote sensing for atmospheric monitoring. *Sensors (Basel), 22*(8), 3080. PMID: 35459066

Veefkind, J. P., Aben, I., McMullan, K., Förster, H., de Vries, J., Otter, G., Claas, J., Eskes, H. J., de Haan, J. F., Kleipool, Q., van Weele, M., Hasekamp, O., Hoogeveen, R., Landgraf, J., Snel, R., Tol, P., Ingmann, P., Voors, R., Kruizinga, B., & Levelt, P. F. (2012). TROPOMI on the ESA Sentinel-5 Precursor: A GMES mission for global observations of the atmospheric composition for climate, air quality and ozone layer applications. *Remote Sensing of Environment, 120*, 70–83. DOI: 10.1016/j.rse.2011.09.027

Vinuesa, R., Azizpour, H., Leite, I., Balaam, M., Dignum, V., Domisch, S., Felländer, A., Langhans, S. D., Tegmark, M., & Fuso Nerini, F. (2020). The role of artificial intelligence in achieving the Sustainable Development Goals. *Nature Communications, 11*(1), 233. DOI: 10.1038/s41467-019-14108-y PMID: 31932590

Wooster, M. J., Roberts, G., Perry, G. L. W., & Kaufman, Y. J. (2015). Retrieval of biomass combustion rates and totals from fire radiative power observations: FRP derivation and calibration relationships. *Journal of Geophysical Research, D, Atmospheres, 110*(D24).

World Resources Institute. (2022). Global Forest Watch.

Xu, C. (2021). A review of onboard AI technologies for satellite remote sensing. *IEEE Journal of Selected Topics in Applied Earth Observations and Remote Sensing, 14*, 5197–5211.

Chapter 7
AI-Driven Climate Resilience:
Intelligent Pathways for Adapting Energy Systems to Environmental Uncertainty

Ravikumar R. N.
https://orcid.org/0009-0009-3705-1681
Marwadi University, Rajkot, India

S. Aarthi
https://orcid.org/0009-0006-9064-2091
Marwadi University, Rajkot, India

ABSTRACT

This chapter examines how Artificial Intelligence (AI) strengthens climate resilience in energy systems facing extreme weather, renewable variability, and infrastructure risks. Traditional reactive methods are inadequate, whereas AI enables predictive forecasting, adaptive control, and optimization for renewable integration and grid stability. Case studies show AI-enabled microgrids improved energy availability by 65%, reduced downtime by 40%, and maintained 90% critical facility coverage, while predictive maintenance in wind farms achieved 45% downtime reduction, 30% cost savings, and 92% failure prediction accuracy. Ethical, regulatory, and socio-economic challenges such as data scarcity, fairness, and energy demands are also addressed. Future directions include quantum computing, edge AI, explainable AI, and inclusive policy frameworks. Overall, AI emerges as a transformative partner in designing adaptive, sustainable, and future-ready energy systems under growing climate uncertainty.

DOI: 10.4018/979-8-3373-5540-5.ch007

INTRODUCTION

Climate change is such a problem that it is transforming the energy environment of the world and putting a new element of uncertainty in the latter thus making it more unpredictable and volatile than ever. The abnormal weather patterns, the increased volatility in the supply of renewable energy sources, and the increasing threat to energy infrastructure are all valid points that can be made to support the establishment of effectiveness and flexibility in energy infrastructure. Inflexible, outmoded risk model and reactive planning approach will not allow to have sufficient dynamic capability to overcome such shocks. The concept of Artificial Intelligence (AI) will be highly powerful in this change because it will have the capacity to provide predictive on the consequences, be responsive to change, and providing intelligent control, which will enhance resilience (Bansal, 2024). The capabilities of AI to combine machine learning, deep learning and advanced analytics with climate and energy data and early warnings of dangers, and to actualize these warnings in real-time allow dynamic grid operation with renewable production that may be predicted in real-time. Alongside offering technical solutions, AI can become an enabler of sustainable energy security, which can assist societies in minimizing the economic losses and cost-efficiently distribute their resources and accelerate the process of the switch to renewable sources (Adapa, 2024a). This specific technological innovation AI is presented throughout this chapter as one of the strategic partners of the construction of energy systems, which is future-proof, adaptable, and climate-resilient.

The Need for Climate Resilience in Energy Systems

Global climate change is also interfering with the energy systems in the world by causing heatwaves and flooding and also unpredictable production of renewable energy. The traditional infrastructure character, which is designed to respond to stable circumstances, is poorly fitted to shock and long-term variability. Using the case of hydro-energy or solar-powered power grids, there is a severe pressure on droughts or even cloudy weather conditions, but there is a vulnerability to extreme weather and rising sea levels in fossil fuel plants. These weak areas reveal the need to transform the reactive to proactive of resilience planning (Kanna & Shibi, 2025). In this context, the application of the term resilience is wider than the elements of disaster recovery since it also refers to the conduction of proactive measures that describe the risks, the mitigation of vulnerabilities, and continuity of supply. The energy systems are robust even in the event of disturbances and Affordable and sustainable through the integration of the resiliency during the design and operation stages. These will make sure that societies do not continue to incur economic losses, energy insecurity, and even lower levels of renewable integration and, consequently, resilience will become the most significant factor in the future energy transitions.

Artificial Intelligence as a Strategic Enabler of Adaptation

AI will provide a means of enhancing resiliency with predictive, adaptive and data-driven decision making. There is also an opportunity of AI to employ vast amounts of data on weather predictions, geospatial data, or past data regarding energy consumption to produce more accurate predictions of renewable generation using machine learning and deep learning models. It can predict outages, model failure of infrastructure and direct proactive processes that constrain the time and cost of outage. An example is a predictive maintenance regulated by AI that can detect faults in wind farms before they are escalated to extreme weather, and dynamic load balancing in a heatwave or a cold snap will be achieved by intelligent demand-response systems (Amnuaylojaroen, 2025). In addition to the technical efficiency, AI will be able to promote a more flexible governance, supported by the assistance of the scenario planning, the effective allocation of the resources, and the clear decision-making frameworks. Like the energy systems themselves, AI technologies can be used in energy planning by providing an aid strategy that is more strategic than reactive toward energy planning to ensure that systems remain robust, adaptable and climate-sensitive. Its implementation does not only raise the performance in operations, but also adapts itself to the bigger goals in sustainability and policy.

Objectives of the Chapter

This chapter aims to explore the intersection of artificial intelligence and climate resilience, highlighting how intelligent systems can safeguard energy infrastructures against growing environmental uncertainties. It outlines the practical, technical, and policy-oriented dimensions of AI adoption, providing insights for researchers, policymakers, and industry practitioners. The objectives are as follows:

- **Examine Climate Vulnerabilities in Energy Systems** - To analyze the challenges posed by extreme weather, renewable variability, and infrastructure risks under changing climate conditions.
- **Highlight AI-Driven Predictive and Adaptive Capabilities** - To explore how machine learning, deep learning, and analytics enhance forecasting, grid stability, and disaster preparedness.
- **Showcase Practical Case Studies** - To illustrate real-world applications of AI in microgrids, renewable maintenance, and demand-response models that strengthen climate resilience.

- **Identify Challenges and Limitations** - To critically assess issues such as data scarcity, computational costs, and fairness in AI-enabled decision-making.
- **Propose Policy and Governance Pathways** - To recommend frameworks that ensure inclusive, transparent, and sustainable integration of AI into resilience planning.

BACKGROUND STUDY

The increase in intensity and prevalence in climate related upheaval has led to the consideration of vulnerability of conventional power systems. The development of energy infrastructures in the past has been based on the prediction of a constant and stable weather pattern and operating conditions. However, extreme heatwaves, floods, hurricanes, and extended droughts have emerged as a danger to supply chains, renewable generation and grid stability. Interestingly, renewable energy resources are critical to curbing carbon, and the objective of decarbonization creates another bane to the given issue as it is highly intermittent and is affected by weather conditions. The traditional, mostly reactive, and risk-averse resilience solutions have failed to address such uncertainty (Sakapaji & Puthenkalam, 2023). As such, artificial intelligence has officially entered the world as one of the most influential technologies that have proved to be useful in the context of predictive analytics, optimization, and responsive decision-making. It can utilize large, non-comparable data, like simulation of climate and energy consumption records and geo-design information to forecast interferences and respond in kind. Research in AI-assisted climate prediction, renewable energy prediction and intelligent grid demonstrates that it may be employed to enhance resilience. The chapter is anchored on these developments and an inquiry in full on the role of AI as a strategic enabler of climate-adaptive energy systems.

Literature Review

Mitigating the impact of climate change by incorporating climate resilience into energy components has become one of the main items on the energy research agenda, as well as energy policy agenda (Adapa, 2024b). Most of the conventional energy systems that are constructed in strict, centralized models have proven to be highly vulnerable to climate change induced shocks in the form of floods, hurricanes, and long-term droughts. Such disturbances underline the shortcomings of the traditional risk management practices that are quite non-proactive and scenarios (Perera et al., 2020a). The increased uncertainty of the weather conditions demands more informed, dynamic and adaptive strategies departing with the conventional planning strategies.

Artificial intelligence has become more common as the promising solution to these problems. Initial AI experiments with energy mainly concentrated on increasing its efficiency e.g. through demand forecasting, load balancing and effective power generation optimization. Nonetheless, it can now be expanded to resilience applications in the recent days. Deep learning and machine learning models proved to predict the output of renewable energy with better accuracy due to the analysis of large sets of climates, meteorological, and geospatial data (Ladi et al., 2022). These forecast opportunities can be used to counter intermittency-related risks and thus improve system reliability.

The further area of study that appears to discern the future of AI is the enhancement of grid resilience. Dynamic demand-response management is becoming commonplace through the use of intelligent algorithms, which make sure that grid stability is maintained under high-load or in extreme weather situations. AI-controlled microgrids might be considered adaptive infrastructure because it can work autonomously and redistribute the resources in disaster-susceptible areas (Noviati et al., 2024). These systems allow the energy communities to continue with supply when there is a disruption to the normal supply and, therefore, have a system with less reliance on centralized grids. Likewise, AI-driven predictive maintenance has proven to be successful applied in wind and solar plants in the detection of weak points before a surge in weather conditions compounds equipment failure and maximizes downtime and losses. In addition to operational strengths, the literature of AI in resilience planning focuses on the socio-technical and governance aspects of AI in the resilience planning (Cavus, 2024).

AI can deliver efficiency and flexibility, but there exists a question of transparency, fairness and energy consumption of large-scale AI models. Scarcity of data, poor infrastructure, and governance shortcomings have been identified to be a challenge especially in developing countries delaying a fair use of AI technologies (Song et al., 2022). The need to have inclusive structures capable of incorporating human decision-making and AI systems, and therefore being accountable to critical energy functions, has been brought to the fore by scientists. Also, policy research assumes that there is a need to incorporate AI to a larger sustainability and climate adaptation agenda. The use of AI cases in plans to move towards carbon neutrality in the country, plans of disaster resiliencies, and plans to move energy transition are considered of high importance (Jiao, 2020). It is here that the literature is increasingly citing the role of AI capabilities as beyond a back-up technology but as a main partner on the creation of energy systems in the future that are climate resilient. All these contributions also give us a platform on which we can ask a question which is AI-facilitated pathways that can balance technological innovation with the goal of attaining resilience and equity and sustainability.

Research Gap

Although there has been significant development in AI implementation in forecasting, grid optimization, and predictive maintenance, the climate resilience is not addressed as an integrative concept in most studies, but the efficiency and cost reduction are considered in a limited way. Little consideration is paid to integrating the resiliency strategies to extreme weather and long-term uncertainty. In the emerging markets, poor data, governance and infrastructure increase risks. The required ones are comprehensive, scaled and policy-coherent AI approaches that enhance flexibility, fairness and resilience in energy systems that are climate-stressed.

Problem Statement

Climate change triggers deep uncertainty into the energy systems of the world that discontinues supply, amplifies volatility and poses danger to infrastructure stability. Conventional resilience approaches based on risk modeling and response cannot be used. Although the prospects of AI in predictive analytics and adaptive control are present, there is no integration on technical, social, and policy levels in present implementations. Regardless of a formal, AI-based system, economies will suffer losses, slow green uptake, and energy insecurity, and the need to have structured, climate-resilient AI solutions is immediate.

MAIN FOCUS OF THE CHAPTER

The primary concern of this chapter is to analyse the role of the artificial intelligence as a catalyst in the construction of climate-resilient energy systems. It focuses on how AI can be used to predict renewable generation, forecast extreme weather events, and introduce adaptive, real-time adaptations to power systems and hybrid systems (Elshebli, 2024). The chapter notes the potential of AI to change energy operations to adopt pro-resistance while being proactive and driven by predictive, deep learning, and intelligent control structures. It is similarly focused on solving issues of data availability, algorithmic explanations, and the energy efficiency of large-scale models and being sure that the use of AI does not contradict sustainability interests. The development of the case study, the critical evaluation and impressions on the issue of governance allow to make the conclusions on the importance of AI as the source of technologies and of strategic ally to reach the energy security in the form of climate uncertainty.

Proposed Methodology of the Chapter

The chapter suggests an analytical and conceptual framework, which means the incorporation of climate resilience planning of energy systems and artificial intelligence. The strategy opens with the articulation of the important threats posed by climate that encompass extreme weather and vulnerability of infrastructures and unpredictability of renewable. These risks are plotted using the AI-based tools e.g. machine learning models to forecast renewable power, deep learning models to forecast extreme events, intelligent optimization tools to help demand-response and grid stability. A proposed approach is integrated strategy of large-scale climate and geospatial data and energy sources with integrative modelling in terms of prediction and adaptation designs. The AI-driven microgrid, predictive maintenance in renewable plants, and adaptive demand management will be taken as the case studies to verify the framework. Besides, policy, governance and ethical aspects were introduced by the research methodology to become more inclusive, fair and sustainable. The proposed methodology is thoroughly combined with technical innovation, practical, and regulatory in a manner that would help to provide a roadmap towards resilient, adaptive, and future-ready energy systems. Figure 1 provides a sequence of steps through which it is possible to use AI as a way of dealing with climate risks in energy systems. It begins with Identification of risk and solution mapping, moves to model development and validation and ends with policy integration and delivery of roadmap to deliver resilient results.

Figure 1. AI-Driven Climate Resilience Framework

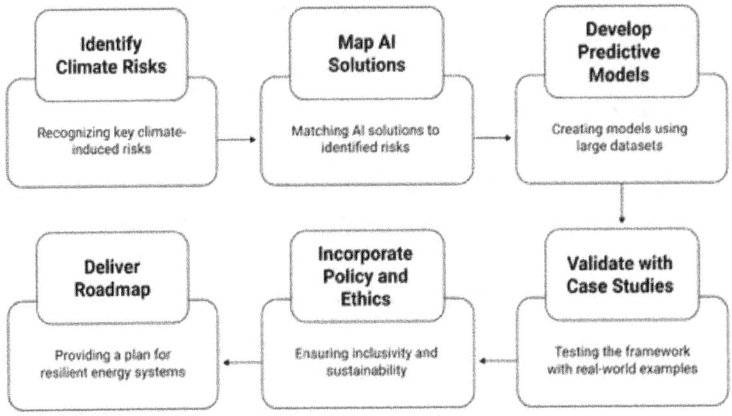

Identify Climate Risks

The determination of risks that the climate is causing is important as it includes a broad range of interruptions in the robust energy systems. These incorporates violent occurrences like hurricanes, floods and heatwave, and long-term changes like droughts, rise in sea level and variable renewable resources. Assumptions that were previously considered stable are now shocks in themselves due to unpredictable shocks. To find such risk, climate science has to be combined with energy system modeling in order to present the vulnerabilities of supply, transmission, and storage. Geographical diversity is crucial - the coastal regions are vulnerable to flooding, whereas the dry areas will suffer a lack of hydroelectric power. The systematic risk identification allows the stakeholders to develop AI-based resilience solutions that meet the urgent, local energy security requirements.

Map AI Solutions

As soon as risks linked to climate are identified, it is necessary to align them with AI-based solutions. To resolve the vulnerabilities, artificial intelligence provides a wide range of features, including predictive analytics, optimization, and adaptive control. An example is that AI can predict variation in renewable generation due to wind variations or cloud variation, and optimization compensates the imbalance between supply and demand. The risks, such as grid instability, equipment failures, extreme load events, etc., should be organized into categories and associated with the tools, provided by deep learning, reinforcement learning, or optimization frameworks. Technical viability and access to data particularly in developing areas is crucial to be considered. This strategic mapping changes the energy systems to respond reactively to recovery but to respond proactively and resiliently.

Develop Predictive Models

Predictive modeling plays a key role in the resilience when the risks are mapped to solutions. Models driven by artificial intelligence result in disruption prediction that is supported by different types of data such as weather data, satellite images, grid logs, and demand trends. The methods that recreate the scenario correctly and predict the variability of the renewable production or the impact of extreme weather are the neural networks, which are the ensemble learning and the hybrid methods. Energy planners can explore the opportunities of what-if as in the case of longer droughts trapped hydropower or hurricanes disconnecting transmission. Early warnings allow taking the proactive steps of load redistribution, the activation of backup, and proactive maintenance. The retraining on the new information is a permanent process to remain flexible to changing situations. Severe predictive modeling with assistance of AI will aid in transforming uncertainty to foresight and to build dependable and energy systems by the future.

Validate With Case Studies

In order to be believable and productive, the idea of AI-based resiliency systems must be tested in reality. Predictive models are tested by use of case-studies with regards to their validity, generalizability, and workability in real working conditions. They are AI controlled microgrids in regions prone to disasters, automatic attainment of the balance between supply and demand, and predictive maintenance in wind and solar farms that do not experience failures during the extreme weather conditions. Validation shows the usefulness of models where there are no data, computer limitations or climate variability. It is also capable of scaling and pilot projects can be scaled to national or regional levels. Above all, validation gives the policymakers and investors' confidence through demonstrating energy security, minimize downtimes, and economic rewards.

Incorporate Policy and Ethics

Technical solutions will never be sufficient to guarantee resilient and fair energy systems; they will need to be complemented with great policy frameworks and values. Integration of policy guarantees that the adoption of AI is in line with national sustainability objectives, a goal of integrating renewables, and the preparedness of disasters. Ethical issues, including algorithmic fairness decision-making, model transparency, and inclusion of benefits accessibility, are also to be considered. Countries that are developing, especially, experience difficulties in data availability and data governance and therefore, frameworks that are more inclusive and equity-driven should be developed to serve the purpose. Policies are needed to promote open exchange of data, regulatory controls, and transnational collaboration, to deal with common climate risks. The AI systems should at the same time be unbiased, not reliant on opaque algorithms, or have high energetic requirements due to large models. Integrating policy and ethics is what will make resilience planning sustainable, accountable and helpful to various communities to make lasting confidence in energy systems driven by AI.

Deliver Roadmap

The last step includes the creation of a comprehensive roadmap to climate-resilient energy infrastructure, which incorporates the risk identification, AI-solutions, predictive modeling, validation, and alignment of policy. This roadmap must take a step-by-step process- first pilot projects, scaled proven models and integrating resiliency into the energy systems of countries or regions. It has to define the roles of stakeholders, such as governments, utilities, and technology providers, as well as promote collaboration. Clearly, there should be milestones of renewable integration and inclusive implementation. Notably, the roadmap should be dynamic and be able to change with the changing climate issues and the technological innovations. This enhances credibility, minimizes risks and shortens the sustainability objectives by transforming research into practical action plans.

INTRODUCTION TO CASE STUDIES

Case studies offer valuable guidance on the practice of achieving climate resilience in the energy systems by using artificial intelligence. They are the world practical evidence of the reviewed ideas, displaying the success of AI in forecasting risks, streamlining answers, and the guarantee of energy security during climate crises disturbances. Case studies showcase the examples of successful implementation to describe the possible opportunities and challenges associated with the implementation of AI in the area of resilience planning (Adewumi et al., 2024). They also provide models which can be reproduced and scaled regionally. These adaptive, sustainable and disaster-ready energy infrastructures are experiencing transformative services of AI as the following examples prove.

Case Study: AI-Enabled Microgrids for Disaster-Prone Communities

In regions vulnerable to frequent natural disasters, such as hurricanes, floods, or earthquakes, Frequently, due to centralized power grids, communities find themselves without electricity when the network stops functioning and the network may be without power days or even weeks. Microgrids that have incorporated AI to help them manage the supply/demand relationship in such a crisis have become a resilient solution as they can govern themselves (Tang et al., 2022). A microgrid uses the combination of renewable energy, such as solar and wind with local storage capacity to build resiliency to ensure that critical services, hospitals, shelters, and communications remain in the event of a failure of the main grid. Artificial intelligence increases flexibility of such microgrids by observing climate, the current climatic predictions, and energy usage patterns (Islam & Othman, 2024). The predictive models can be used to predict demand spikes during crisis times and reinforcement learning can be used to optimize resources. Research shows that the AI can enhance energy during a disaster in up to 65% compared to back-up energy systems. AI-based the dynamic load-balancing ensures an approximate 40% greater absence of downtimes of critical infrastructure (Oudinga, 2023). As an example, AI-based mini-grids were piloted in Southeast Asian coastal areas, where blackouts caused by cyclones led to more than 90% of the higher studies sustaining mission-critical facilities with AI-based microgrids compared to 50 to 60% traditional systems coverage. In addition, AI-controlled solar-battery microgrids experienced a 25% efficiency gain in energy storage usage, minimising wastage, and maximizing backup capacity. In addition to technical resilience, AI- enabled microgrids improve community resilience as well by decentralizing control. Through AI dashboards, the local operator is also allowed to monitor the performance of their operations and automated decision-making reduces the burden of human error when all-important activities are being undertaken (Egbuna et al., 2025). The combination of renewables with AI, as well as localized control supports producing disaster-ready energy ecosystems that are more sustainable and reliable. The Figure 2 shows the comparison of the main performance indicators between conventional system and AI-powered Microgrids. Results indicate that AI enhances energy availability by 65%, as compared to 35, reduces downtimes by 40%, covers the critical facility up to 90% as compared to 55%, and increases storage efficiency by 25%.

Figure 2. Comparison of Traditional Systems vs AI-Enabled Microgrids

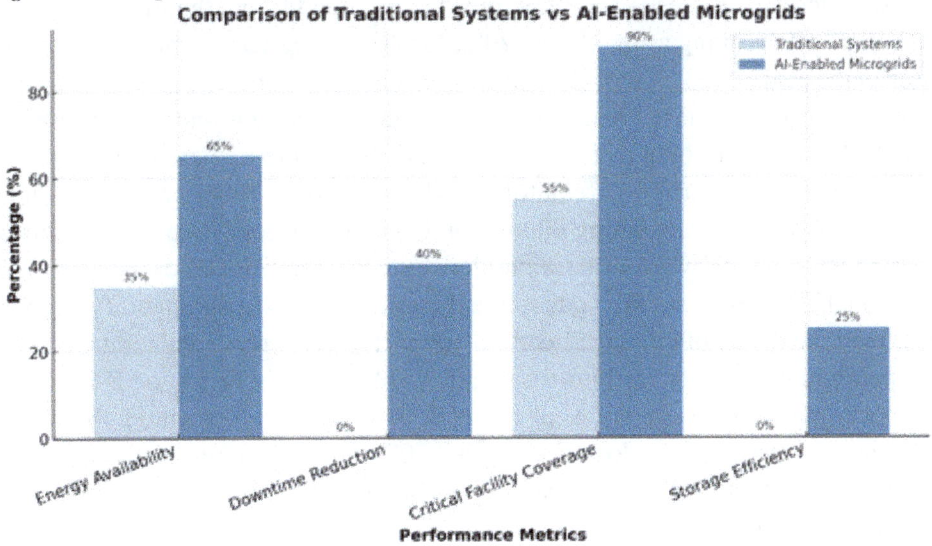

Case Study: Predictive Maintenance in Wind Farms Using Machine Learning

Wind farms are already essential in renewable energy generation, but the structures themselves are extremely susceptible to wear and tear as well as weather phenomena. Conventional preventive approaches are based on timely checks or corrective maintenance, being not effective and creating certain waste, costly failures during exceptional weather conditions (Beretta et al., 2021). With predictive maintenance enabled by Machine Learning (ML), operators can look off into the future and predict equipment malfunctions, maintenance tasks and schedule. The data about turbine vibrations, temperature, acoustic signals are analyzed with machine learning algorithms in order to detect anomalies. Such models, through identifying minor patterns that point to early-stage faults, can prompt the interventions in opportune time (Badihi et al., 2022). Research has shown that ML based predictive maintenance lowers unplanned downtime by close to 45% when compared to more traditional methods. Moreover, it reduces the total maintenance expenses by at least 30% due to the fact that interventions are implemented when they are needed as opposed to time-based intervals.

In a European wind farm consisting of many turbines, correct predictions of failures in gear and blade were made 92% of the time by predictive models. (Hassan & Megahed, 2022) Consequently, the energy production rose by 18% per year since there were less disturbances and maximized work. In addition, the life expectancy of turbines was prolonged by almost 20%, lowering long-term capital investment (Aderibigbe et al., 2023). The other advantage is in safety and resilience. Predictive models 35% reduced the necessity of using manual inspections that were high-risk, as the technology will help technicians limit their exposure to extremely harsh environments. Critical turbines are automatically shut down by AI-enabled systems during storm seasons, preventing them or highly reducing costly repair procedures. Machine learning, by waywardness predictive maintenance, can not only improve operational efficiency, but also guarantee the reliability and sustainability of wind farms over the theme of climate uncertainty, which is what is needed to manage the incorporation of wind farms into world renewable stocks. The Figure 3 indicates the improvement in the operations of wind farms with machine learning. Predictive maintenance saves 45% in downtimes, save 30% in costs, increases the accuracy of failure prediction to 92% (compared to ~70% in traditional practices) and adds 20% to turbine life.

Figure 3. Traditional vs ML-Predictive Maintenance in Wind Farms

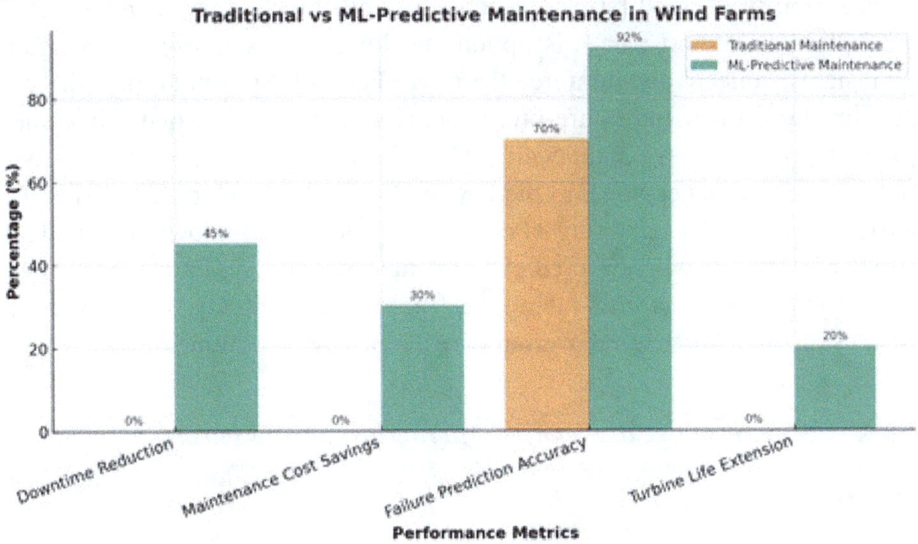

Comparative Evaluation of AI Use Cases

The two case studies of AI-Enabled Microgrids in Disaster-Prone Communities and Predictive Maintenance in Wind Farms Using Machine Learning illustrate mutual applications of AI in the development of energy resilience in uncertain conditions of the climate. Although both of them discuss measurable benefits, they are too different in terms of focus, scope, and impact. The main solutions of the enabled microgrid with AI are energy availability and disaster preparedness. They combine renewable sources with the local storage providing 24/7 power to the critical facilities like hospitals and shelters in case of grid shutdown. AI makes them more flexible by predicting time of need swings and by optimizing resources on the fly (Zapata et al., 2022a). This method enhanced the availability of energy by 65% and minimized down time by 40% and sustained power to 90% of critical services as compared to conventional backup systems it was highly successful. Proactively, on the other hand, predictive maintenance of wind farms is focused on efficiency and asset lifecycle. Sensor data is used in machine learning models to predict turbine failures and typically results in 45% fewer unexpected turndowns and 30% savings on turbine maintenance (Uzondu & Lele, 2024). Such systems added almost 20% to the life of turbines and increased the annual production by 18% with a 92% accuracy rate. The two case studies microgrid and wind farm show at community and infrastructure levels respectively resilience in terms of ensuring continuity in the face of disasters and in terms of efficiency, cost-effectiveness, and sustainability of wind farms respectively. Collectively, they demonstrate the role AI plays in a short-band disruption and long-band stability of operations. Their insights, collectively, reinforce the multiplicity of AI as not only a driver to create climate-resilient and future-proof energy systems but also one applicable in a broad range of settings (Badihi et al., 2022). The Figure 4 makes a comparison of the performance enhancements in the key measures. AI-enabled microgrids were able to achieve a 65% availability of energy, 40% decrease in downtime, and 90% coverage of a facility, whereas predictive maintenance in wind farms showed 45% downtime reduction, 30% cost-saving, 92% prediction accuracy, and 20% turbine life extension, indicating some overlap in resiliencies and optimization.

Figure 4. Comparison of Case Studies: Microgrids vs Wind Farms

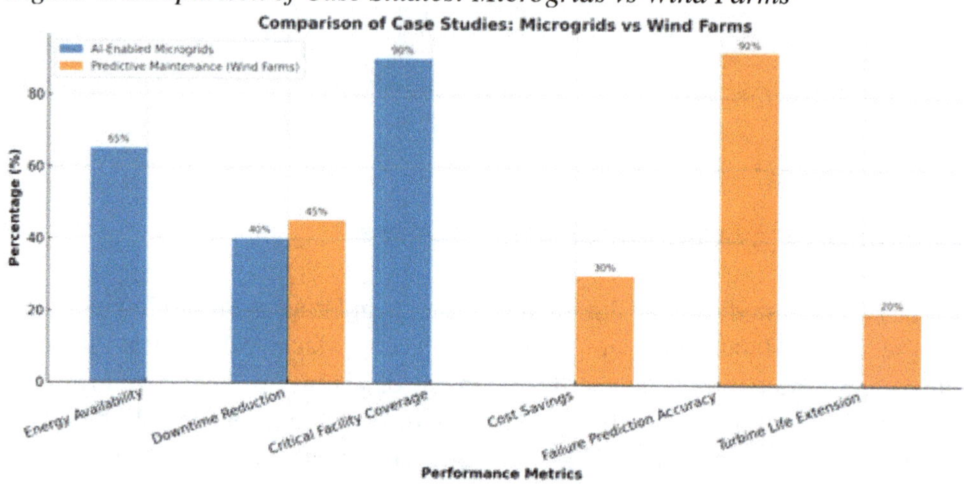

RESULTS AND DISCUSSION

Case studies show that AI plays an excellent role leading to the resilience, efficiency, and sustainability of energy systems that are climate-stressed. In locations prone to disaster-prone areas, microgrid powered by AI enhanced power availability by 65%, downtimes reduced 40%, and power ensured 90% of critical facilities as opposed to 55-60% normal systems. On the same note, at wind farms, machine learning reduced unplanned outage by 45%, and reduced costs by 30% and enabled 92% accuracy of failure prediction thus prolonging the turbines life by 20%. These findings prove that AI not only perfectly performs its work but also makes the system adaptable to the extreme forces (Nik et al., 2021). The problems have not been completely eliminated, however, and they include data availability, computational requirements and the requirement to provide decision-making with fairness and transparency. Taken together, the presented findings are indicative of the impact of AI in transforming climate-resilient energy systems, emphasizing the necessity of policy, governing, and ethical integration as the means to foster sustainable and inclusionary adoption. The Figure 5 represents the gain in performance at stages of implementation. Microgrids with AI technology has increased energy availability to 65% and ML predictive maintenance increased energy availability to 75%. The downtimes decreased by 0-40-45% (traditional to 0-30-30% savings in cost) which demonstrates the gradual contribution of AI in resilience.

Figure 5. Performance Improvements from AI in Energy Systems

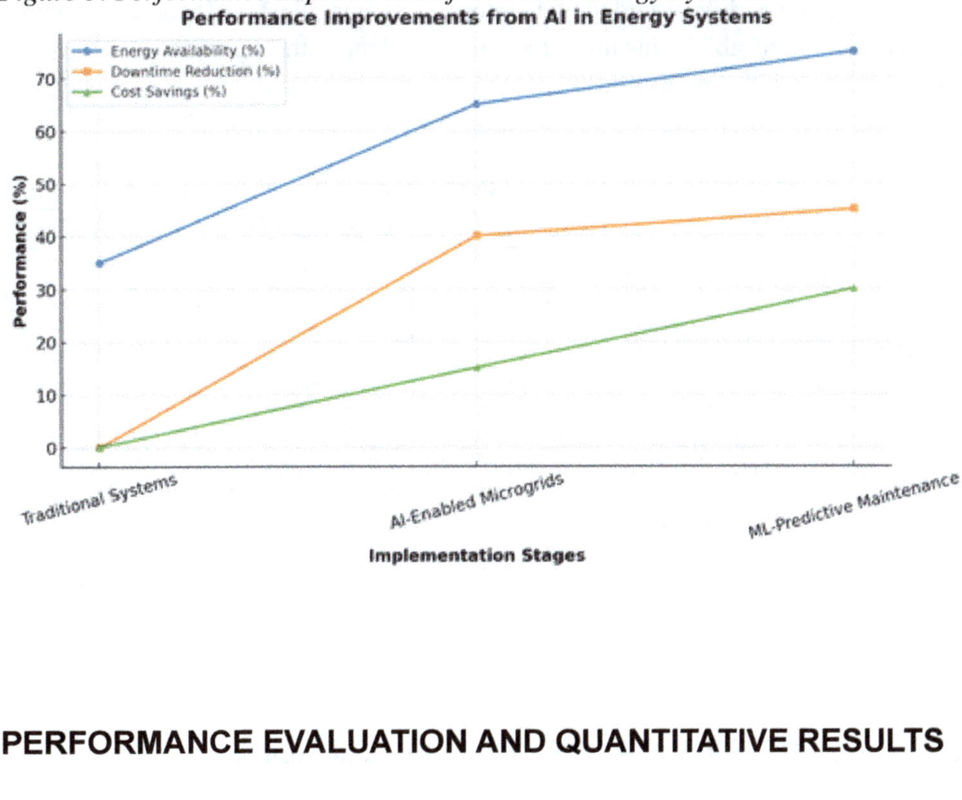

PERFORMANCE EVALUATION AND QUANTITATIVE RESULTS

Quantitative level AI-driven energy resilience assessment proves the undeniable benefits of AI resilience compared to the traditional ones. The microgrids coupled with AI produced very positive effects in disaster-prone areas. The availability of energy also increased to 65% relative to 35%, and critical facility coverage increased to 90% as opposed to a mere 55-60% typical with conventional systems. Close to

40% of the downtime was eliminated and there was guaranteed accessibility of the vital services during extreme events. In addition, the storage was more efficient by 25%, reducing the renewable use, and adding backup (Perera et al., 2020b). In wind farms, predictive maintenance using machine learning produced revolutionary performance. Unscheduled downtime was reduced by 45%, and maintenance cost minimized by 30% with the use of data driven scheduling of interventions. The accuracy on prediction failure rose to 92% compared to 70% when using conventional monitoring to allow proactive measures. These resulted in 18% improvement in energy generation of the whole facility on an annual basis, and an almost 20% boost in life cycle of the turbines, reducing long-term capital cost. Combined, these findings support that AI-based solutions can deliver an adaptive, efficient, and future-proof solution and enhance operational resilience and sustainably integrate energy in a climate-uncertain future. The following table 1 gives the comparisons between the conventional methodology and the AI-based solution in microgrids and wind farms. It also cites quantifiable gains in availability, efficiency, and reliability, proving the contribution of AI to improving climate resilience.

Table 1. Performance Evaluation and Quantitative Results of AI in Energy Systems

Metric	Traditional Systems	AI-Enabled Systems	Improvement (%)
Energy Availability (Microgrids)	35%	65%	+30%
Critical Facility Coverage	55–60%	90%	+30–35%
Downtime Reduction	0%	40%	+40%
Storage Efficiency	0%	25%	+25%
Unplanned Downtime (Wind Farms)	High (baseline)	Reduced by 45%	–45%
Maintenance Cost Savings	0%	30%	+30%
Failure Prediction Accuracy	~70%	92%	+22%
Annual Energy Output	Baseline	+18% Improvement	+18%
Turbine Lifespan Extension	Baseline	+20%	+20%

ETHICAL AND REGULATORY CONSIDERATIONS

The use of AI in the more climate-resilient energy systems has transformational potential and, by implication, the concerns of ethics and regulatory issues are being raised. The issue of fairness, openness, and responsibility should be discussed because AI has the potential of supporting the creation of predictive choices and responsive reaction. In addition, the huge AI systems also have massive data and computing power that would create the problem of privacy, energy, and inclusiveness differentiated by uneven access among the global population (Rekeraho et al., 2023). The regulatory organs and ethics systems should be speculated on these queries and should create a balance between innovation and the population trust on them in a bid to bring in sustainable and responsible integration. Two of these extremely significant dimensions, ethical urgency and regulatory control are discussed in the subsections that follow.

Ethical Imperatives in AI-Driven Energy Systems

The primary ethical concerns related to the implementation of AI to attain energy resilience are the fairness, transparency, and inclusivity aspects. The achievement of AI models relies on the available data and therefore, in regions with limited data, there is the risk of biased decision-making processes that favour certain communities or infrastructures at the expense of others (Hamdan et al., 2024). It is also desired that algorithms are transparent as models with obscured details or black-boxed internal procedures may limit accountability when high stakes energy decisions are taken. In addition, the AI systems should not lead to the production and production countries online, but should share the riches equally. Another aspect that cannot be overlooked is that the environmental ethics giant AI models have the capacity to consume vast quantities of power and remain unregulated; it can be confused with the effort to promote the idea of sustainability (Ukoba et al., 2024). Trying to solve these emergencies, it is important to note that models that are easy to comprehend, fairly allocate information and develop energy-saving artificial intelligence systems should be created. The implementation of ethics to the system design allows considering AI as a reliable companion in the process of promoting climate resilience to the society that appreciates its values and environmental issues.

Regulatory Governance and Policy Alignment

The regulations play a central role in the direction to ensure that the application of AI on energy systems does not oppose the opportunities of climate resilience and social interests. The governments and other international organizations ought to outline clear guidelines of data ownership, responsibility of algorithms, and cybersecurity that do not imply misuse of the same but permit its proper utilization. Also, there is the necessity to coordinate policy so that AI-enhanced energy resilience is integrated into bigger sustainability strategies, (Uzondu and Lele, 2024) including renewable energy targets and pledges to be carbon-neutral. Regulatory support should be developed by capacity building and infrastructure development in order to remove the barriers of AI implementation in developing countries. The international cooperation is also important, as the risks of climate change may cross the borders, and the joint regime governance is required to ensure energy security. The legislation should also assist transparency and homogeneity of AI application to ensure that system usage becomes interoperable and are believed in by the populace. Regulatory governance, within the scope of its activities of innovation as well as management, will be able to make the structures of AI-based energy resiliency non-exclusive, ethical, and consistent with the long-term plans of climate adoptions.

IMPLEMENTATION CHALLENGES AND LIMITATIONS

Even though the energy domain is a field of transformational capability by using AI-based energy resilience, there are few challenges presently impeding the successful use of such technology in practice. The technical difficulty of integration of large-scale data, the computational cost of more sophisticated models and infrastructure vulnerabilities will be factors of an obstacle to scale (Cowls et al., 2023). On top of this, it is not just technical problems but also socio-economic and governance limitations in specific regions when developing there is a lack of materials and expertise. The perspective that would be required to address these issues would be one that is balanced since it would validate the existence of AI, and their shortcomings. The technological constraints and socio-economic constraints are the key two areas of realization outlining the path of AI-enabled climate resilience in this section.

Technological Limitations and Data Constraints

The primary assets to the implementation of AI in the energy resilience include high-quality and large datasets with variables that capture the variability of climate, trends of energy demand, and infrastructure conditions. However, information is full, irregular or entirely missing in numerous sections, reducing the accuracy of forecasting mechanisms. There are other situations where interoperability concerns among a wide range of different energy and climate systems can be of concern in the event that data is accessible (Aprizal et al., 2025). The other limitation is the computational overhead of AI models especially deep learning models that consume a significant amount of computational power in addition to energy. It is an irony since despite the use of AI taking on the dimension of promoting sustainable development, the issue of its resource demands might promote carbon footprints when applied irresponsibly. In addition, the black-box imagery of sophisticated models increases the question of explainability, and stakeholders are reluctant to rely on AI-based recommendations. Technological constraints which could hamper the scalability and credibility of AI in climate-resilient energy planning are its energy-intensive nature, its inability to be interpreted easily or in a comprehensible manner, and the lack of rich and relevant data.

Socio-Economic and Governance Barriers

However, in addition to technical impediments, the use of AI is deterred by socio-economic and governance barriers to the implementation of climate-resilient energy systems. The development of regions has limited access to finance, expertise and digital infrastructure, thus large-scale implementation of AI is a challenge (Mondal et al., 2024). Costs of adopting AI are initially high and reduce the attractiveness of investment in AI, especially in low-income countries that have energy security needs heavily tied to access over the quality of supply chains. Issues in governance also arise, given that regulatory frameworks that govern AI are not yet in place, leaving policymakers and energy suppliers unsure on what should happen. Moreover, disparities in availability of AI applications threaten to widen the digital divide across the world with benefits skewed towards developed countries as the vulnerable areas continue to be vulnerable to climate change (Uzondu & Lele, 2024). Acceptance within society is an additional issue- people might not accept AI-led decision-making solutions that are neither transparent nor inclusive. To address these obstacles, capacity building, equal financing patterns, and governance systems capable of transforming AI innovation into the concept of sustainable and inclusive energy transitions are needed. The table 2 summarises the main technological, socio-economic and governance issues of turning towards AI to transition to climate resilient energy systems and possible solutions to the problems raised.

Table 2. Implementation Challenges and Mitigation Strategies for AI-Driven Energy Resilience

Challenge	Description	Mitigation Strategy
Data Scarcity & Inconsistency	Limited availability of climate and energy datasets in many regions	Promote open data initiatives, develop shared platforms, and invest in data infrastructure
High Computational Demands	Advanced AI models consume significant energy and resources	Use energy-efficient AI architectures, edge computing, and hybrid cloud-based solutions
Lack of Model Transparency	Black-box AI models limit trust and explainability	Develop interpretable AI models and require explainability standards in high-stakes use
High Implementation Costs	Initial investment in AI tools and infrastructure is prohibitive	Encourage public–private partnerships, subsidies, and green financing mechanisms
Skills & Capacity Gaps	Shortage of skilled professionals in AI and energy systems	Build capacity through training programs, international collaboration, and knowledge exchange

continued on following page

Table 2. Continued

Challenge	Description	Mitigation Strategy
Weak Regulatory Frameworks	Lack of policies governing AI use in energy resilience	Establish AI-specific energy regulations, cross-border governance, and accountability frameworks
Social Acceptance & Inclusivity	Resistance due to lack of trust or perceived inequity in AI decisions	Ensure participatory decision-making, transparency, and equitable distribution of benefits

DEPLOYMENT FRAMEWORK AND USE CASE EXPANSION

The move to translate the conceptual models to the real-world deployment needs a clear framework that supports scalability, inclusiveness, and sustainability in the long-term. A deployment framework defines a way by which AI can be systematically incorporated into energy infrastructures, including data collection through adaptive operations, taking into consideration technical, ethical, and policy aspects. In addition to deployment, the proliferation of use cases emphasizes the ability of AI applications to be diversified to sectors, geographies, and energy disciplines (Leal Filho et al., 2023). In that way, such an AI will not only increase resilience but also help with innovation and fairness in energy transitions. The next subsections feature the deployment framework and use-case expansion potential of AI-driven applications.

AI Deployment Framework for Energy Resilience

A small outline of AI execution in energy systems starts with multi-source data fusion, which encompasses climate forecasts, a history of grid-efficiency as well as data of renewable generation output. This is further supplemented by applying the right AI models machine learning to predict the forecasts, deep learning to identify the anomalies and the reinforcement learning to learn the adaptive control (Dimnet, 2022). The infrastructure, which is required to be modular to facilitate the deployment process, can be incorporated into the existing energy systems without the massive reorganization of the energy infrastructure. The resources of edge and cloud computing are the important components of the operation that facilitate its scaling and low-latency response. Explainability and cybersecurity protection are combined, which is necessary in order to provide safe and transparent use of AI in life-threatening scenarios (Ericson et al., 2023a). National energy policies and climate goals are made to be compliant through governance alignment. Finally, models can be optimized (pilot testing and phased scaling), as part of validation.

One of the core areas that this framework will focus on will be technical rigor, trust of stakeholders, and responsiveness to local needs so that the use of AI will contribute to the creation of climate-resilient power systems (Ericson et al., 2023b). The Figure 6 shows the flow involved in data collection and developing AI model, to deployment, scaling, and expansion at various energy use cases of wind, solar, hydro, and energy markets. It emphasizes the way AI is used to convert raw data to adoptive and resilient energy plans.

Figure 6. AI Deployment Framework and Use Case Expansion

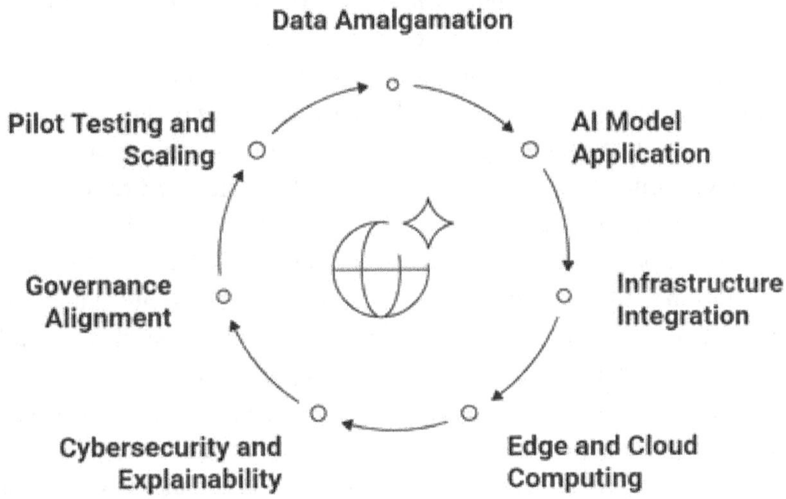

Expansion of Use Cases Across Energy Systems

Though the initial capacity of AI applications is in microgrids and predictive maintenance, it has potential of growth in the entire energy ecosystem. To provide an example, AI can be applied to optimise offshore wind farms (by aiding in the prediction of turbulence behaviour) and to manage large-scale solar plants (by aiding in the prediction of dust storms and irradiance variability). (Zapata et al., 2022b). In hydroelectric systems, AI can be used to balance water resource distribution in reservoirs in times of drought and floods, thereby increasing resilience and sustainability. Also, AI is applicable in energy trading platforms where predictive analytics assist in stabilizing the electricity markets during market fluctuations triggered by climatic changes. In addition to generation, AI increases the resilience of distribution networks to detect anomalies that predict the fault location in the transmission lines. Geographic extends are also important- deployments in developing geographies can be customised to use lightweight AI tools in resource-poor geographies. With this diversity in use cases, AI can work as an inclusive resilience approach around generation, transmission, distribution, and market functions. This growth not only delivers a maximum in technical advantage, but also democratizes access to energy resilience driven by AI in globally distributed locations.

SUMMARY OF KEY CONTRIBUTIONS

The chapter is relevant in terms of the subject of the discussion of the resilience of the energy systems in terms of climate, as artificial intelligence itself is offered as one of the technology providers and strategic planning of adaptation. It suggests the weaknesses of the traditional methods of resilience and shows how AI-based predictive analytics, deep learning and algorithmic optimization can be employed to turn the uncertainty into valuable foresight. The chapter provides a detailed architecture of the procedure- identification of climatic hazards, mapping of AI solutions and creation of predictive models and, later, validation with real life case studies, integration in the policies and provision of roadmap. It presents quantitative information about the implications of AI that is developed using case studies, including a 65% energy availability, 40-45% total downtime decrease, 30% reduction in costs, and 92% success rate in renewable energy systems. It also addresses ethical and regulatory requirements, implementation concerns and arteries of scaling AI in deployments in different energy infrastructures. The chapter has managed to integrate technical, ethical and governance horizons to introduce a unified strategy of ensuring that AI remains one of the driving forces of sustainable, inclusive and future-ready energy systems in the climate uncertainty era.

FUTURE RESEARCH DIRECTIONS

Since AI is participating in the development of climate-resilient energy systems, future studies need to go beyond the already utilized cases and research areas in order to understand new challenges and opportunities. Although adaptive models and predictive analytics have proven to be useful with definite advantages, there are still major lapses in terms of the scale, inclusivity, and sustainability of the idea of the use of AI (Cavus, 2024). The development of this area demands the exploration of the novel models of AI integration with the other complementary technologies, and ethical, social, and environmental aspects integrated into their framework. Two of the most research avenues identified in this section include: the combination of more advanced computational techniques and AI that continues to become more valuable, and the design of inclusive and policy-based frameworks capable of providing equitable access and sustainable resilience.

Integration With Advanced Computational Technologies

Avenues of future study should be the interface between AI and emerging complex computational paradigms, including quantum computing, edge intelligence, and nature-inspired optimization. An example is quantum AI, which could solve complex energy optimization problems in nanoseconds compared to classical algorithms and make it possible to quickly adapt in near-real time to adjust reaction methods to extreme weather events. Localized, low-latency decision-making in microgrids Edge AI can enable such decision-making in disaster-prone remote locations where connectivity is compromised (Ejjami, 2024). Similarly, bio-inspired optimization algorithms may increase the flexibility of predictive models with uncertain conditions. Exploring hybrid frameworks with both these technologies coupled with AI will increase the scalability, efficiency and the resilience of the resilience strategies. The computational cost and sustainability of such solutions must also be researched and they should not increase the carbon footprints (Ekechukwu & Simpa, 2024). This type of interdisciplinary investigation can advance the limitations of the support made available by AI in climate-resilient energy transitions.

Inclusive Frameworks and Policy-Driven Research

The second effective prospective will be to come up with systems that lay emphasis on inclusivity, transparency, and policy congruency of AI implementation. The next step of research should focus on developing models of AI that will adapt to limited territories of data to ensure that less developed nations will not be deprived of a chance to maintain their resilience. Research can draw attention to AI models that have low resources and can be applied in cases where there are limited resources, and yet can make the right predictions. The research policy also ought to examine the governing systems that would stop injustice and maintain accountability and garner confidence of the people in the AI-based decision-making (Leal Filho et al., 2023). The global cooperation is necessary and the climate issues have no national borders and can be solved only when being addressed in a group. One can also make technical research inclusive and create AI solutions that should be efficient and sustainable and fair in the long term. This trend will help the AI-supported energy resilience to become a paradigm which will be available in the world scale and ethically acceptable.

Human-Centered and Explainable AI in Energy Resilience

Since the number of applications of AI in energy systems is ever-growing, the research should be prioritized to make sure that the models are understandable, readable, and friendly to humans. The existing AI systems and especially the deep learning and neural networks may be black boxes since they do not present much information and explanations on how they come up with their decisions despite their ability to make highly accurate predictions. The trust in high-stakes energy processes can be disaggregated to explainability, i.e. closing down the turbines before the storm or diverted energy during the disasters (Proskuryakova, 2023). The following wave of research should be focused on the Explainable AI (XAI) methods, which are able to reveal a line of reasoning of model outputs. Human cooperation with the AI systems is also a human oriented AI issue. AI appears to be not a replacement to expertise but rather an intelligent companion that may offer knowledge never lessening the final decision to a human. Interactive dashboards, visualization tools, and participatory design frameworks helped facilitate this integration enabling the operators to enhance their understanding on vulnerability of the system and response capabilities (Sarma & Zabaniotou, 2021). The application of the principles of transparency, accountability, and inclusivity in the resilience plan that is driven by AI will assist the research to ensure that the technologies remain acceptable, ethically modified, and fairly utilized. Lastly, explainable and people-friendly AI will build trust, boost the use and confidence of the AI of climate-resilient energy systems.

Key Pathways for Future Research in AI-Driven Energy Resilience

Research going forward should go beyond current applications of AI to emerging issues of scalability, inclusivity, and sustainability. By combing both sophisticated computational techniques and inclusive models of governance along with an embedding of transparency into the models of AI, it is possible to make the next generation of innovation both technologically sound as well as socially equitable. The most significant research pathways and their focus areas as well as anticipated outcomes have been brought in and illustrated in the following table 3.

Table 3. Future Research Directions in AI-Driven Climate Resilient Energy Systems

Research Direction	Focus Area	Expected Outcomes
Integration with Advanced Computational Technologies	Quantum computing, edge AI, bio-inspired optimization	Faster optimization, real-time decision-making, and scalable AI models for complex climate-energy challenges
Inclusive Frameworks and Policy-Driven Research	Data-scarce regions, lightweight AI, governance mechanisms	Equitable access, fairness, transparency, and stronger adoption in developing countries
Human-Centered and Explainable AI	XAI models, participatory design, human-AI collaboration	Improved interpretability, stakeholder trust, and accountability in high-stakes energy decisions
Cross-Border and Multi-Sector Applications	Regional cooperation, integration across energy, water, and transport systems	Coordinated resilience strategies, harmonized governance, and systemic risk reduction
Sustainable AI Development	Energy-efficient AI models, low-carbon training methods	Reduced computational footprint, ensuring AI aligns with climate and sustainability goals

CONCLUSION

The climate change is reestablishing the consistency and dependability of the planetary power structures with an extremely harsh danger to the infrastructures and chains of supply due to the warmness, harsh weathering, and uncertainty of the deliverance of renewables. These types of massive and complicated disruptions cannot be handled with a usual resilience approach and their irregular models or even by responsive approaches. AI, already uncovered in this chapter, is a disruptive technology towards pursuing climatic resilient energy systems, on the predictive analytics, machine learning, and deep learning principles. The technologies make it possible to predict the fluctuation of the generation on the foundation of the renewable, predict the outcomes of the extreme weather, and react to the processes of energy in the real-time mode. This increment of 65-percentage point of energy supply, 40 to 45 percentage point of downtime, 30 per cent reduction of costs, and 92 per cent prediction of a machine failure all present the real benefits of AI as shown in the case studies. This fact supports the idea that AI can assist not only in enhancing the performance of the operations, but also the pace at which the situations with the heavy load on climate could be transformed to the renewable energy. Among its issues, there are the inadequacy of data, high cost of computation, black box algorithms and regulatory constraints.

This exclusive method of structures is dangerous as it will continue to increase the digital divide where the wealthy sectors of the society will prosper at the cost of poor societies. To resolve such loopholes, various disciplines, such as technologists, policymakers, suppliers of energy, and societies, ought to work together. Regulatory and ethical considerations revolving around equity, transparency, and accountability are necessary and in that regard, the aspect of AI environmental footprint where its solutions cannot compromise the sustainability objectives is important. Future directions in the research would be the application of AI to quantum computing, edge intelligence and bio-inspired optimization, inclusive and policy-consistent frameworks. Finally, another possible application of AI is considered not a technical one, but a strategic partner in the creation of agile, sustainable and future-ready energy systems, or which may provide long-term security in the climate-challenged world.

KEY TAKEAWAYS

- **AI as a Strategic Partner:** Artificial intelligence is not just a tool but a strategic enabler of climate-resilient energy systems.
- **Quantitative Gains:** Case studies showed up to **65% higher energy availability**, **40–45% downtime reduction**, **30% cost savings**, and **92% prediction accuracy** in energy operations.
- **Challenges Remain:** Data scarcity, high computational demands, lack of transparency, and governance gaps limit scalability and adoption, especially in developing regions.
- **Ethics and Regulation:** Responsible AI deployment must ensure fairness, inclusivity, and sustainability while addressing energy consumption and transparency concerns.
- **Future Directions:** Integration with quantum computing, edge AI, and inclusive, policy-driven frameworks will drive the next wave of research and deployment.

REFERENCES

Adapa, V. R. K. (2024a). AI for climate action: Leveraging Artificial Intelligence to address climate change challenges. *International Journal For Multidisciplinary Research, 6*(5).

Adapa, V. R. K. (2024b). AI for climate action: Leveraging Artificial Intelligence to address climate change challenges. *International Journal For Multidisciplinary Research, 6*(5).

Aderibigbe, A. O., Ani, E. C., Ohenhen, P. E., Ohalete, N. C., & Daraojimba, D. O. (2023). Enhancing Energy Efficiency with Ai: A review of machine learning models in electricity Demand Forecasting. *Eng. Sci. Technol. j., 4*(6), 341–356.

Adewumi, A., Okoli, C. E., Usman, F. O., Olu-lawal, K. A., & Soyombo, O. T. (2024). Reviewing the impact of AI on renewable energy efficiency and management. *Int. J. Sci. Res. Arch., 11*(1), 1518–1527. DOI: 10.30574/ijsra.2024.11.1.0245

Amnuaylojaroen, T. (2025). Advancements and challenges of artificial intelligence in climate modeling for sustainable urban planning. *Frontiers in Artificial Intelligence, 8*, 1517986. DOI: 10.3389/frai.2025.1517986 PMID: 40463823

Aprizal, A., Wiranatakusuma, D. B., Rizki, M., & Anugrah, R. A. (2025). The role of technology in climate resilience: A systematic literature review and mapping study approach. *Journal of Physics: Conference Series, 2989*(1), 12037. DOI: 10.1088/1742-6596/2989/1/012037

Badihi, H., Zhang, Y., Jiang, B., Pillay, P., & Rakheja, S. (2022). A comprehensive review on signal-based and model-based condition monitoring of wind turbines: Fault diagnosis and lifetime prognosis. *Proc. IEEE Inst. Electr. Electron. Eng., 110*(6), 754–806. DOI: 10.1109/JPROC.2022.3171691

Bansal, M. (2024). Climate-resilience readiness of emerging clean energy technologies for India's energy transition: A deep dive into select technologies. *IOP Conference Series. Earth and Environmental Science, 1395*(1), 12015. DOI: 10.1088/1755-1315/1395/1/012015

Beretta, M., Julian, A., Sepulveda, J., Cusidó, J., & Porro, O. (2021). An Ensemble Learning solution for predictive maintenance of wind turbines main bearing. *Sensors (Basel), 21*(4), 1512. DOI: 10.3390/s21041512 PMID: 33671601

Cavus, M. (2024). Integration smart grids, distributed generation, and cybersecurity: Strategies for securing and optimizing future energy systems. In *Preprints*.

Cowls, J., Tsamados, A., Taddeo, M., & Floridi, L. (2023). The AI gambit: Leveraging artificial intelligence to combat climate change-opportunities, challenges, and recommendations. *AI & Society*, *38*(1), 283–307. DOI: 10.1007/s00146-021-01294-x PMID: 34690449

Dimnet, E. (2022). 3 Key notions for road transport sustainability: Resilience, Climate action and Energy transition. *Rom. J. Transp. Infrastruct.*, *11*(1), 1–20. DOI: 10.2478/rjti-2022-0002

Egbuna, I. K., Salihu, F. B., Okara, C. C., Olayiwola, D. E., Smart, E. E., Anifowose, O., & Mbamalu, P. O. (2025). Advances in AI-powered energy management systems for renewable-integrated smart grids. *World J. Adv. Eng. Eng. Technol. Sci.*, *15*(2), 2300–2325. DOI: 10.30574/wjaets.2025.15.2.0685

Ejjami, R. (2024). Integrating artificial intelligence for enhanced grid stability and renewable energy management in France. *International Journal For Multidisciplinary Research, 6*(3).

Ekechukwu, D. E., & Simpa, P. (2024). A comprehensive review of renewable energy integration for climate resilience. *Eng. Sci. Technol. j., 5*(6), 1884–1908.

Elshebli, M. (2024). Extreme Weather and hydropower System: Production, Resilience, and Sustainability. In *Research Square*. DOI: 10.21203/rs.3.rs-2756965/v2

Ericson, S., Grue, N., Hotchkiss, E., & Brown, M. (2023a). *Applications of measuring and valuing resilience in energy systems*. Office of Scientific and Technical Information (OSTI).

Ericson, S., Grue, N., Hotchkiss, E., & Brown, M. (2023b). *Applications of measuring and valuing resilience in energy systems*. Office of Scientific and Technical Information (OSTI).

Hamdan, A., Ibekwe, K. I., Ilojianya, V. I., Sonko, S., & Etukudoh, E. A. (2024). AI in renewable energy: A review of predictive maintenance and energy optimization. *Int. J. Sci. Res. Arch.*, *11*(1), 718–729. DOI: 10.30574/ijsra.2024.11.1.0112

Hassan, A. M., & Megahed, N. A. (2022). Improvingurban energyresiliencewith an integrative framework based on machine learningmethods. *Arch. Eng.*, *7*(4), 17–35. DOI: 10.23968/2500-0055-2022-7-4-17-35

Islam, A., & Othman, F. (2024). Renewable energy MicroGrid power forecasting: AI techniques with environmental perspective. In *Research Square*. DOI: 10.21203/rs.3.rs-4260337/v1

Jiao, J. (2020). Application and prospect of artificial intelligence in smart grid. *IOP Conference Series. Earth and Environmental Science, 510*(2), 22012. DOI: 10.1088/1755-1315/510/2/022012

Kanna, V., & Shibi. (2025). AI-driven Smart Energy Management Systems for optimized renewable energy utilization in urban smart grids. In *Research Square.*

Ladi, T., Jabalameli, S., & Sharifi, A. (2022). Applications of machine learning and deep learning methods for climate change mitigation and adaptation. *Environment and Planning. B, Urban Analytics and City Science, 49*(4), 1314–1330. DOI: 10.1177/23998083221085281

Leal Filho, W., Yang, P., Eustachio, J. H. P. P., Azul, A. M., Gellers, J. C., Gielczyk, A., Dinis, M. A. P., & Kozlova, V. (2023). Deploying digitalisation and artificial intelligence in sustainable development research. *Environment, Development and Sustainability, 25*(6), 4957–4988. DOI: 10.1007/s10668-022-02252-3 PMID: 35313685

Mondal, S., Das, S., & Vrana, V. G. (2024). Exploring the role of artificial intelligence in achieving a net zero carbon economy in emerging economies: A combination of PLS-SEM and fsQCA approaches to digital inclusion and climate resilience. *Sustainability (Basel), 16*(23), 10299. DOI: 10.3390/su162310299

Nik, V. M., Perera, A. T. D., & Chen, D. (2021). Towards climate resilient urban energy systems: A review. *National Science Review, 8*(3), nwaa134. DOI: 10.1093/nsr/nwaa134 PMID: 34691589

Noviati, N. D., Maulina, S. D., & Smith, S. (2024). Smart grids: Integrating AI for efficient renewable energy utilization. [ITALIC]. *International Transactions on Artificial Intelligence, 3*(1), 1–10. DOI: 10.33050/italic.v3i1.644

Oudinga, M. (2023). The role of artificial intelligence in enhancing energy management in microgrid systems. *Journal Dimensie Management and Public Sector, 4*(4), 33–41. DOI: 10.48173/jdmps.v4i4.246

Perera, A. T. D., Nik, V. M., Chen, D., Scartezzini, J.-L., & Hong, T. (2020a). Quantifying the impacts of climate change and extreme climate events on energy systems. *Nature Energy, 5*(2), 150–159. DOI: 10.1038/s41560-020-0558-0

Perera, A. T. D., Nik, V. M., Chen, D., Scartezzini, J.-L., & Hong, T. (2020b). Quantifying the impacts of climate change and extreme climate events on energy systems. *Nature Energy, 5*(2), 150–159. DOI: 10.1038/s41560-020-0558-0

Proskuryakova, L. (2023). The contribution of low-carbon energy technologies to climate resilience. *Climate (Basel), 11*(12), 231. DOI: 10.3390/cli11120231

Rekeraho, A., Cotfas, D. T., Cotfas, P. A., Bălan, T. C., & Tuyishime Emmanuel and Acheampong, R. (2023). Cybersecurity challenges in IoT-based smart renewable energy. In *Research Square*.

Sakapaji, S. C., & Puthenkalam, J. J. (2023). Harnessing AI for climate-resilient agriculture: Opportunities and challenges. *Ejtas*, *1*(6), 1144–1158. DOI: 10.59324/ejtas.2023.1(6).111

Sarma, G., & Zabaniotou, A. (2021). Understanding vulnerabilities of renewable energy systems for building their resilience to climate change hazards: Key concepts and assessment approaches. *Renew. Energy Environ. Sustain.*, *6*, 35. DOI: 10.1051/rees/2021035

Song, Y., Wan, C., Hu, X., Qin, H., & Lao, K. (2022). Resilient power grid for smart city. *IEnergy*, *1*(3), 325–340. DOI: 10.23919/IEN.2022.0043

Tang, Z., Zhang, P., & Krawec, W. O. (2022). Enabling resilient quantum-secured microgrids through software-defined networking. *IEEE Transactions on Quantum Engineering*, *3*, 1–11. DOI: 10.1109/TQE.2022.3216530

Ukoba, K., Olatunji, K. O., Adeoye, E., Jen, T.-C., & Madyira, D. M. (2024). Optimizing renewable energy systems through artificial intelligence: Review and future prospects. *Energy & Environment*, *35*(7), 3833–3879. DOI: 10.1177/0958305X241256293

Uzondu, N. C., & Lele, D. D. (2024). Comprehensive analysis of integrating smart grids with renewable energy sources: Technological advancements, economic impacts, and policy frameworks. *Eng. Sci. Technol. j.*, *5*(7), 2334–2363.

Zapata, V., Gernaat, D. E. H. J., Yalew Seleshi, G., Santos da Silva, S. R., Iyer, G., Hejazi, M., & van Vuuren, D. P. (2022a). Climate change impacts on the energy system: A model comparison. *Environmental Research Letters*, *17*(3), 34036. DOI: 10.1088/1748-9326/ac5141

Zapata, V., Gernaat, D. E. H. J., Yalew Seleshi, G., Santos da Silva, S. R., Iyer, G., Hejazi, M., & van Vuuren, D. P. (2022b). Climate change impacts on the energy system: A model comparison. *Environmental Research Letters*, *17*(3), 34036. DOI: 10.1088/1748-9326/ac5141

KEY TERMS AND DEFINITIONS

Climate Resilience: The capacity of energy systems to anticipate, absorb, adapt, and recover from climate-induced disruptions such as extreme weather and variability.

Artificial Intelligence (AI): Computational techniques that enable machines to learn, reason, and adapt, used in energy systems for forecasting, optimization, and adaptive control.

Microgrid: A localized energy network capable of operating independently or in connection with the main grid, enhancing resilience during disruptions.

Predictive Maintenance: The use of machine learning and sensor data to anticipate equipment failures before they occur, reducing downtime and repair costs.

Deep Learning (DL): A subset of AI using neural networks with multiple layers to detect patterns in large datasets, crucial for weather and energy forecasting.

Demand Response: Strategies that adjust consumer energy demand in real-time to balance supply and maintain grid stability, especially during peak loads.

Algorithmic Transparency: The ability to explain and understand how AI models make decisions, ensuring accountability in critical energy operations.

Edge AI: Deploying AI models on localized devices (edge computing) to enable low-latency, real-time decision-making in energy systems.

Quantum Computing: A next-generation computational paradigm that can process complex optimization problems faster than classical computing, with applications in energy resilience.

Sustainability Governance: Policies and frameworks that guide the ethical, equitable, and environmentally responsible deployment of AI in energy systems.

Chapter 8
Future Artificial Intelligence in Electric Vehicles and Sustainable Transport

Syed Hassan Imam Gardezi
https://orcid.org/0009-0006-6171-1238
Union Investments LLC, UAE

M. Rajesh
https://orcid.org/0000-0002-0695-6842
Rathinam Technical Campus, India

P. Selvakumar
https://orcid.org/0000-0002-3650-4548
Department of Science and Humanities, Nehru Institute of Technology, Coimbatore, India

Amara S. A. L. G. Gopala Gupta
https://orcid.org/0000-0002-5033-002X
Koneru Lakshmaiah Education Foundation, India

N. Subramani
https://orcid.org/0000-0002-3855-4987
Karpagam College of Engineering, Coimbatore, India

Arun M. R.
https://orcid.org/0000-0002-8291-2956
Vel Tech Rangarajan Dr. Sagunthala R&D Institute of Science and Technology

ABSTRACT

The rapid evolution of artificial intelligence (AI) has become one of the most transformative forces in the modern automotive landscape. Among the most significant beneficiaries of AI integration are electric vehicles (EVs), which rely on advanced software and intelligent systems far more heavily than traditional internal combustion engine vehicles. As the world embraces cleaner, smarter, and more sustainable transportation options, AI is driving a revolutionary shift in how electric cars are designed, operated, maintained, and experienced. From autonomous driving capabilities to intelligent energy management and predictive maintenance, AI is

DOI: 10.4018/979-8-3373-5540-5.ch008

reshaping every aspect of the EV ecosystem. Autonomous Driving and Advanced Driver Assistance Systems (ADAS): One of the most prominent areas where AI is influencing electric vehicles is in the development of autonomous driving technologies. EVs serve as ideal platforms for autonomy due to their electronic architecture and digital-first design.

THE RISE OF AI IN ELECTRIC VEHICLES: AN OVERVIEW OF AI'S IMPACT ON THE EV INDUSTRY

The rapid evolution of artificial intelligence (AI) has become one of the most transformative forces in the modern automotive landscape. Among the most significant beneficiaries of AI integration are electric vehicles (EVs), which rely on advanced software and intelligent systems far more heavily than traditional internal combustion engine vehicles. As the world embraces cleaner, smarter, and more sustainable transportation options, AI is driving a revolutionary shift in how electric cars are designed, operated, maintained, and experienced. From autonomous driving capabilities to intelligent energy management and predictive maintenance, AI is reshaping every aspect of the EV ecosystem. Autonomous Driving and Advanced Driver Assistance Systems (ADAS): One of the most prominent areas where AI is influencing electric vehicles is in the development of autonomous driving technologies. EVs serve as ideal platforms for autonomy due to their electronic architecture and digital-first design. AI algorithms process data from cameras, lidar, radar, ultrasonic sensors, and GPS to detect objects, interpret road signs, analyze traffic patterns, and make real-time decisions. These systems use machine learning models trained on vast datasets to improve their decision-making accuracy over time.

These models detect patterns and anomalies that indicate potential component failures before they happen. For example, AI can monitor the wear and tear of brake systems, tire pressure, motor vibrations, and battery degradation to alert the driver about impending issues. This capability is especially valuable in commercial EV fleets, where downtime can be costly. AI-powered fleet management systems optimize vehicle utilization, route planning, charging schedules, and maintenance tasks. This leads to lower operational costs, improved reliability, and higher productivity. Cybersecurity and Data Privacy: As EVs become more connected and autonomous, cybersecurity becomes a critical concern. AI helps safeguard these vehicles against cyber threats by continuously monitoring for unusual activity and detecting potential breaches. Intrusion detection systems powered by machine learning can identify and respond to threats more quickly than traditional methods. AI also ensures secure data handling practices, especially as EVs collect vast amounts of personal, behavioral, and locational data. Federated learning and edge AI are being explored to enhance data privacy by processing information locally within the vehicle instead of transmitting everything to the cloud. Artificial intelligence is not merely an add-on feature for electric vehicles; it is the very fabric enabling their transformation into smart, efficient, and sustainable mobility solutions. By enhancing safety, efficiency, performance, and user experience, AI is propelling electric vehicles into a new era of innovation. As the automotive industry continues to evolve, the synergy between AI and EVs will become increasingly critical, shaping not just how we drive but also how we live and move in the 21st century and beyond.

SUSTAINABLE TRANSPORT: THE ROLE OF AI

The pressing global challenges of climate change, urban congestion, rising population densities, and resource depletion have intensified the need for sustainable transportation systems. As nations worldwide strive to reduce greenhouse gas emissions and create cleaner, more efficient urban mobility networks, artificial intelligence (AI) has emerged as a transformative force in the pursuit of sustainable transport. AI's ability to process vast amounts of real-time data, make intelligent decisions, and optimize complex systems is revolutionizing how people and goods move, promoting environmental stewardship, energy efficiency, and socio-economic equity. From optimizing public transportation and reducing fuel consumption to enabling shared mobility and autonomous electric vehicles, AI is redefining the landscape of sustainable transport. AI and Energy Efficiency in Transport: One of the most direct ways AI contributes to sustainable transport is through the optimization of energy usage (Chatti W., et al, 2024). Intelligent algorithms can manage traffic flow in real-time, reducing idling and congestion, which are major contributors to unnecessary fuel consumption and emissions. Smart traffic lights and adaptive signal control systems use AI to analyze traffic patterns and adjust signal timings dynamically, allowing smoother vehicle flow and minimizing stop-and-go conditions. This reduces fuel consumption, lowers air pollution, and improves urban air quality. In public transportation systems, AI-powered scheduling and route optimization tools help reduce the energy footprint of buses, trains, and ferries. AI can analyze historical ridership data, traffic conditions, and real-time demand to dynamically adjust schedules and deploy vehicles more efficiently. This not only cuts down on wasted energy and operational costs but also improves service reliability and encourages greater public transport usage, which is inherently more sustainable than private car use.

AI also plays a significant role in reducing energy consumption in freight logistics. Algorithms that optimize delivery routes for trucks or cargo ships based on fuel efficiency, weather, and road conditions can significantly cut emissions. Predictive analytics helps logistics companies avoid unnecessary trips, reduce idling times, and coordinate multi-modal freight networks more effectively, contributing to a greener supply chain (Sharma & Singhal, 2024). Reducing Emissions and Environmental Impact: The transportation sector accounts for a significant share of global greenhouse gas emissions, making emission reduction a top priority for sustainable mobility. AI contributes to this effort by enabling low-carbon solutions across all modes of transport. For instance, AI-powered eco-driving systems provide drivers with real-time feedback on how to adjust their driving style to conserve fuel and reduce emissions. Fleet operators can use AI to monitor driving behavior, vehicle conditions, and emissions, allowing them to train drivers, schedule maintenance, and upgrade vehicles accordingly. AI also plays a crucial role in integrating transport with renewable energy sources. For example, in electric bus networks, AI can optimize charging schedules to align with peak solar or wind generation times, minimizing reliance on fossil-fuel-based electricity. In maritime and aviation sectors—traditionally hard to decarbonize—AI is helping reduce emissions through route optimization, fuel-efficient flight paths, and smart maintenance. AI can also assist in the design of lighter, more aerodynamic vehicles that consume less energy during operation. Equity and Accessibility in Sustainable Transport: Sustainability in transportation is not only about reducing emissions but also about ensuring that mobility is accessible, affordable, and equitable. AI helps address social inclusion by designing transport systems that serve underserved communities (Lukic Vujadinovic V., et al, 2024). For example, AI can identify areas with low public transport coverage or high transport poverty and recommend targeted interventions. In addition, AI-powered mobility platforms can offer services tailored to people with disabilities or elderly passengers, ensuring barrier-free access through features like voice-based commands, real-time assistance, and route customization.

These applications make sustainable mobility more inclusive, supporting broader social sustainability goals. Future Outlook: Looking ahead, the role of AI in sustainable transport is set to grow even further as technologies mature and integration deepens. The convergence of AI with other advanced technologies such as 5G, Internet of Things (IoT), blockchain, and quantum computing will unlock new possibilities for real-time coordination, predictive analytics, and resilient infrastructure. AI could eventually facilitate a seamless, multimodal mobility ecosystem where users can plan and execute trips across various transport modes using a single platform. Autonomous electric fleets, intelligent infrastructure, AI-controlled logistics, and data-driven urban planning will together create transport systems that are not only sustainable but also smart, inclusive, and resilient to future challenges (Chowdhary S., et al, 2025). Artificial intelligence stands at the forefront of a revolution in sustainable transport. By enhancing energy efficiency, reducing emissions, enabling smart mobility solutions, and supporting data-driven policy, AI is transforming how we think about and implement sustainable mobility. As cities and countries confront the twin crises of climate change and urban congestion, AI offers powerful tools to create cleaner, safer, and more equitable transportation systems. The path to sustainable transport is complex, but with AI as a key enabler, the vision of greener, smarter mobility is not just a possibility—it is becoming an accelerating reality.

PREDICTIVE MAINTENANCE FOR ELECTRIC VEHICLES: STRATEGIES AND AI INTEGRATION

As electric vehicles (EVs) become increasingly integral to the global transportation landscape, ensuring their long-term reliability, safety, and efficiency is paramount. Unlike traditional internal combustion engine (ICE) vehicles, EVs have fewer mechanical components, which generally translates to lower routine maintenance requirements. However, they introduce new complexities such as high-voltage battery systems, power electronics, and advanced software platforms that demand different diagnostic approaches. In this evolving context, predictive maintenance—a data-driven strategy that uses artificial intelligence (AI), machine learning (ML), and real-time analytics to anticipate component failures before they occur—emerges as a transformative solution (Upadhyaya A N., et al, 2025). Predictive maintenance is rapidly reshaping how EVs are serviced and managed, enabling improved vehicle uptime, reduced costs, and enhanced safety across both personal and fleet-based electric mobility systems.

The Shift from Reactive to Predictive Maintenance:

Traditional maintenance strategies are typically reactive (fixing problems after they occur) or preventive (servicing at predetermined intervals). These approaches are either too late or too early, leading to unexpected breakdowns or unnecessary part replacements. In contrast, predictive maintenance (PdM) monitors the actual condition of components in real-time and uses advanced analytics to predict when a failure is likely to happen. This allows service actions to be scheduled just in time, minimizing both vehicle downtime and maintenance costs. For EVs, predictive maintenance is particularly valuable due to the nature of their critical components. High-voltage batteries, electric motors, inverters, onboard chargers, cooling systems, and software modules require continuous monitoring to ensure optimal performance. Predictive strategies provide insights into the health of these components, enabling early interventions that extend their life and avoid cascading failures (Cavus M., et al, 2025).

Key Components and Systems for Monitoring:

Electric Motor Diagnostics:

Electric motors are generally more reliable than ICEs, but they are still susceptible to issues such as bearing wear, insulation degradation, rotor imbalance, or cooling inefficiencies. AI-based vibration analysis, current signature monitoring, and thermal imaging are used to detect anomalies. Machine learning models trained on historical motor data can differentiate between normal operational variations and early signs of failure, allowing timely servicing and preventing costly replacements.

Brake and Suspension Systems:

Although regenerative braking reduces the mechanical load on conventional braking systems, wear still occurs. Predictive analytics can analyze wheel speed sensor data, brake pressure logs, and temperature records to detect anomalies in braking behavior. Similarly, suspension systems are monitored using accelerometers and gyroscopes to identify premature wear or imbalances, especially critical for EVs that often carry heavier battery loads.

The Role of AI and Machine Learning:

AI lies at the core of predictive maintenance in EVs. By continuously ingesting data from sensors embedded throughout the vehicle, AI models learn to detect subtle patterns that precede component failures. These models use various techniques: Supervised learning is employed to classify known fault conditions based on labeled data. Unsupervised learning helps in anomaly detection, identifying deviations from normal operating parameters. Reinforcement learning optimizes maintenance schedules by balancing cost, risk, and performance over time. Machine learning algorithms are also used in prognostics, estimating the RUL of parts under specific usage conditions. These models are trained using data from simulations, historical maintenance records, and real-world vehicle telemetry (Rao P S., et al, 2023). As the model improves, the predictions become increasingly precise, allowing fleet operators and EV owners to plan maintenance with unprecedented accuracy. For Manufacturers and Service Providers: OEMs benefit from predictive maintenance by improving product quality and reducing warranty claims. Service providers can offer predictive maintenance-as-a-service (PMaaS) models, using telematics to monitor client vehicles remotely and offer proactive support. This builds brand loyalty and reduces the total cost of maintenance networks.

Future Outlook and Innovations:

The future of predictive maintenance in EVs will be marked by self-healing vehicles, digital twins, and collaborative AI ecosystems. Digital twin technology—virtual replicas of physical systems—allows engineers to simulate stress, wear, and failure scenarios using real-world data. This enables precise prediction of how individual components will behave under varying conditions. Emerging AI models will become more autonomous, capable of learning continuously from onboard and cloud sources (Arévalo P., et al, 2024). Predictive systems will integrate seamlessly with vehicle-to-everything (V2X) communication, adjusting predictions based on road conditions, traffic patterns, and weather forecasts. As EV adoption accelerates, especially in commercial sectors, predictive maintenance will be a cornerstone of sustainable, smart, and cost-efficient mobility ecosystems. Regulatory support, standardization efforts, and industry collaboration will further accelerate the deployment of these strategies across markets. Predictive maintenance is revolutionizing how electric vehicles are serviced and sustained. Powered by artificial intelligence, it enables early fault detection, reduces operational costs, enhances vehicle lifespan, and ensures safety. As EV technology continues to evolve, predictive maintenance strategies will be integral to achieving scalable, intelligent, and resilient transportation systems. With AI at the helm, the shift from reactive repair to proactive care is not just a technological leap—it is a fundamental reimagining of how we maintain mobility in the electric era.

SMART TRAFFIC MANAGEMENT: THE ROLE OF ARTIFICIAL INTELLIGENCE IN RESHAPING URBAN MOBILITY

Urban areas across the globe are grappling with increasing traffic congestion, road accidents, inefficient transport logistics, and unsustainable emissions. As the number of vehicles on the road continues to grow—driven in part by the rise of electric vehicles (EVs) and rapid urbanization—the need for a smarter, more responsive, and more sustainable traffic management system becomes urgent. Traditional traffic systems, largely based on fixed-time signals and static route management, are proving inadequate in addressing the complex, dynamic nature of modern urban traffic. Enter Artificial Intelligence (AI)—a transformative force in the realm of smart traffic management. By leveraging real-time data, machine learning algorithms, computer vision, and interconnected systems, AI offers an intelligent, adaptive, and predictive approach to traffic regulation. AI-powered traffic management systems are reshaping how cities control traffic flow, reduce congestion, improve road safety, and prepare for a future dominated by autonomous and connected electric vehicles (Dikshit S., et al, 2023).

The Need for Smart Traffic Management:

Modern cities face a multitude of traffic-related challenges: congestion leading to loss of productivity, increased carbon emissions contributing to climate change, frequent road accidents, and the underutilization of public transportation systems. Conventional traffic management systems, relying on timers and manual intervention, lack the responsiveness to address these dynamically changing patterns. Moreover, with the introduction of electric and autonomous vehicles—each demanding precise coordination with city infrastructure—the complexity of traffic management has increased significantly. AI presents a viable solution to these problems by enabling real-time decision-making, predictive control, and adaptive traffic flow management, turning cities into intelligent ecosystems capable of learning and evolving.

Core Components of AI-Based Traffic Management: AI-powered traffic management systems incorporate various technologies and methodologies to create a unified, intelligent framework:

Real-Time Data Collection: AI systems rely heavily on real-time data from a variety of sources—traffic cameras, GPS devices, sensors embedded in roads and vehicles, mobile applications, and satellite imagery (Antonialli F., et al, 2022). These data provide granular details about traffic volume, vehicle speed, lane occupancy, weather conditions, and even driver behavior.

Computer Vision and Video Analytics: Using AI-based image recognition, computer vision systems can identify vehicle types, count cars, recognize license plates, detect traffic violations, and even assess pedestrian behavior. This allows traffic control centers to respond proactively to changing road conditions, accidents, or rule violations.

Predictive Traffic Modeling: AI systems use historical data and simulations to predict future traffic patterns. This includes estimating congestion peaks, identifying high-risk zones, forecasting the impact of events (e.g., concerts, protests), and planning alternative routes. Predictive analytics empower city planners and traffic controllers to take preventive measures before problems escalate.

Applications and Use Cases:

AI-driven traffic management is being implemented in a range of practical applications across urban centers.

Emergency Response Optimization: AI-enabled traffic systems can detect emergencies such as accidents, stalled vehicles, or natural disasters and provide immediate alerts to emergency services. Smart traffic signals can create "green corridors" to allow ambulances or fire trucks to reach their destinations faster.

Public Transport Prioritization: Smart traffic signals equipped with AI can detect buses or trams approaching intersections and prioritize their passage to ensure on-time public transport schedules. This improves reliability, reduces transit delays, and makes public transport more appealing to commuters (Iyer L S., 2021).

Integration with EV Charging Infrastructure: As electric vehicles become more common, traffic management systems must also account for access to EV charging stations. AI can manage traffic around charging hubs, reduce wait times, and even predict future charging demands based on usage trends.

Global Examples of AI-Based Traffic Management: Several cities worldwide are pioneering the use of AI in traffic management. Barcelona, Spain, uses AI and IoT sensors to monitor traffic, air quality, and pedestrian movement, making real-time adjustments to traffic lights and public transport routes.

The Road Ahead: Toward Autonomous and Sustainable Mobility:

The future of urban mobility lies in autonomous, electric, and shared transportation systems, and AI-powered traffic management is the linchpin that will tie them together. As vehicles become increasingly autonomous and connected, AI will facilitate seamless communication between cars, infrastructure, and control centers. This will allow entire fleets of self-driving vehicles to coordinate routes, avoid collisions, and optimize energy usage based on traffic conditions. In conjunction with smart grids, renewable energy sources, and intelligent EV charging systems, traffic management will also contribute to environmental sustainability. AI will help cities reduce their carbon footprint, meet climate targets, and create more livable urban spaces (Kotte K R., et al, 2025). Moreover, through continuous learning, adaptation, and real-time responsiveness, AI will enable cities to scale their mobility solutions as populations grow and technologies evolve. AI-powered traffic management represents a paradigm shift in how we perceive and manage urban mobility. By harnessing real-time data, predictive analytics, and intelligent control mechanisms, AI enables smarter, safer, and more sustainable transportation systems. As cities worldwide continue to embrace digital transformation, the integration of AI in traffic management will not only improve commute times and reduce emissions but also lay the groundwork for a future where connected, autonomous electric vehicles navigate harmoniously through intelligently managed urban ecosystems. The evolution of smart traffic management is not merely a technological advancement—it is a fundamental re-imagining of mobility itself, driven by the intelligence of machines and the vision of sustainable progress.

PUBLIC TRANSPORTATION OPTIMIZATION: AI IN PUBLIC TRANSPORTATION

In the pursuit of smarter, cleaner, and more efficient urban mobility, public transportation stands as a cornerstone of sustainable development. However, public transit systems worldwide often struggle with inefficiencies, ranging from inconsistent schedules and route congestion to unpredictable delays and underutilization. These issues not only deter passengers but also contribute to increased emissions, traffic congestion, and economic loss. With the advent of Artificial Intelligence (AI), public transportation is undergoing a transformative evolution. AI technologies—including machine learning, predictive analytics, computer vision, and natural language processing—are being harnessed to optimize scheduling, enhance route planning, improve passenger experience, and integrate real-time decision-making (Thirunagalingam A., et al, 2025). By turning data into actionable intelligence, AI enables transport authorities to create systems that are responsive, adaptive, and tailored to the needs of modern urban life. From autonomous buses and intelligent ticketing to real-time fleet monitoring and demand forecasting, AI is redefining the future of public transportation.

The Need for Optimization in Public Transit

Public transportation is crucial for economic mobility, environmental sustainability, and urban accessibility. Yet, many cities face chronic challenges: buses running late or ahead of schedule, suboptimal routing, inconsistent service frequency, overcrowding during peak hours, and underuse during off-peak periods. Additionally, the rigid nature of traditional transit planning fails to respond to the dynamic needs of passengers in real time. These inefficiencies often push commuters toward private vehicle use, exacerbating urban congestion and air pollution. The integration of AI offers a data-driven, responsive alternative, empowering transit authorities to analyze vast amounts of data, automate processes, and make intelligent decisions aimed at improving both operational efficiency and rider satisfaction.

AI-Driven Scheduling and Route Optimization

One of the most impactful uses of AI in public transportation lies in scheduling and route optimization (De la Torre R., et al, 2021). Traditional bus and train schedules are typically static and created based on historical data and assumptions. AI, however, enables dynamic scheduling by analyzing real-time traffic conditions, passenger demand patterns, weather conditions, and even local events. Machine learning algorithms can identify trends in ridership and traffic flow to optimize vehicle dispatch, reduce wait times, and ensure timely arrivals. For example, predictive models can estimate when and where demand will peak and dynamically allocate buses or trains accordingly. This not only improves operational efficiency but also ensures resources are directed where they are needed most. Furthermore, AI can assist in designing more efficient routes by identifying underperforming lines and suggesting modifications, such as merging or rerouting, based on current and projected usage patterns. This adaptive approach helps reduce fuel consumption, emissions, and operational costs while improving service reliability.

Autonomous Public Transit

A growing frontier in public transportation optimization is the integration of AI with autonomous driving technologies. Autonomous buses, trams, and shuttles, powered by AI and sensors, are being trialed in various cities. These vehicles use AI to perceive their environment, make navigation decisions, and communicate with traffic infrastructure. Autonomous transit has the potential to drastically reduce labor costs, increase safety by eliminating human error, and maintain consistent service levels, particularly during off-peak hours or in less densely populated areas (Chu K F., et al, 2022). Moreover, autonomous public transit can operate with a high degree of precision, improving punctuality and energy efficiency. AI can ensure that autonomous vehicles follow optimized routes, respond to traffic conditions in real time, and coordinate with other vehicles to reduce bottlenecks and collisions.

Environmental and Economic Benefits

Optimizing public transportation with AI leads to significant environmental and economic advantages. Efficient routing and scheduling reduce fuel consumption, cutting down on carbon emissions and improving air quality. Predictive maintenance extends the life of vehicles, lowering the environmental footprint of manufacturing and disposal. Encouraging public transport use through better service and user experience reduces reliance on private vehicles, decreasing overall traffic volume and urban emissions. Economically, AI reduces operational costs by optimizing workforce allocation, minimizing service disruptions, and lowering fuel and maintenance expenses. Increased reliability and convenience attract more users, boosting fare revenues and justifying further investment in transit infrastructure (Pamucar D., et al, 2022). AI is ushering in a new era of public transportation—one characterized by precision, adaptability, and inclusivity. By leveraging real-time data, predictive modeling, and intelligent automation, cities can create transit systems that respond to passenger needs dynamically, enhance operational efficiency, and promote sustainable urban development. From route optimization and predictive maintenance to personalized travel planning and autonomous vehicles, AI is redefining the public transportation landscape. As the world moves toward greener cities and smarter infrastructure, AI will be indispensable in building public transit systems that are not only efficient and cost-effective but also resilient, equitable, and centered around the human experience.

SUSTAINABLE LOGISTICS AND DELIVERY: AI-DRIVEN LOGISTICS AND DELIVERY

In the age of rapid e-commerce expansion, global supply chains, and on-demand consumer expectations, logistics and delivery services have become critical pillars of the modern economy. However, this growth comes with significant environmental and operational challenges. The logistics sector accounts for a substantial share of global carbon emissions, with last-mile delivery alone contributing significantly due to frequent, small-scale transport in congested urban areas. At the same time, inefficiencies such as empty return trips, unpredictable traffic, fuel wastage, and inadequate route planning contribute to rising costs and environmental harm (Chen W., et al, 2024). To address these challenges and align with sustainability goals, the industry is increasingly turning to Artificial Intelligence (AI) as a transformative solution. By harnessing the power of data, machine learning, computer vision, and predictive analytics, AI enables logistics and delivery companies to optimize routes, reduce carbon footprints, improve supply chain visibility, manage fleets more effectively, and adapt to changing demand in real time. AI-driven logistics is at the forefront of creating a smarter, greener, and more resilient delivery ecosystem.

The Imperative for Sustainable Logistics

The urgency for sustainable logistics is driven by mounting pressure from environmental regulations, consumer demand for eco-conscious practices, and corporate responsibility commitments. Traditional logistics operations often involve inefficient routing, excess packaging, underutilized vehicles, and reactive decision-making, leading to significant waste and carbon emissions. With urbanization, traffic congestion, and e-commerce surging, last-mile delivery is becoming a major environmental and logistical bottleneck. Companies are now required not only to deliver faster and cheaper but also cleanly. AI addresses these multifaceted demands by transforming static, manual logistics into intelligent, adaptive systems that support real-time decision-making and long-term sustainability (Mohsen B M., 2024).

Predictive Demand and Inventory Management

AI also plays a critical role in demand forecasting and inventory optimization, which are foundational to sustainable logistics. Traditional forecasting methods rely on historical sales data and seasonal trends, often resulting in stockouts or overstocking. AI improves this process by incorporating a broader range of variables such as social media trends, weather, regional preferences, and economic indicators. Machine learning algorithms adapt continuously, refining forecasts as new data becomes available. Accurate demand forecasting reduces unnecessary transportation by preventing over-ordering and minimizing emergency shipments. It also improves warehouse utilization, lowers energy use, and reduces product waste. Companies can maintain leaner inventories closer to demand centers, reducing the distance goods need to travel and thereby lowering associated emissions.

Fleet Management and Telematics

AI-enhanced fleet management systems offer real-time visibility into vehicle location, fuel usage, driver behavior, maintenance schedules, and environmental impact. Telematics data collected from vehicles is analyzed by AI to identify inefficiencies, reduce idle time, and optimize fuel use. For instance, AI can recommend engine shut-offs during extended idling or identify drivers with inefficient driving patterns and suggest coaching interventions. Predictive maintenance is another vital aspect enabled by AI. Instead of waiting for vehicle components to fail, AI uses sensor data to forecast when maintenance should occur, reducing breakdowns, improving safety, and avoiding emissions spikes due to poorly maintained engines (Rane N., et al, 2024). Electrification of fleets is also being managed more effectively through AI. Charging schedules, battery health monitoring, and route planning tailored to electric vehicles' range constraints are optimized using AI, ensuring sustainable and efficient fleet operations.

AI in Green Supply Chain Networks

AI enables the design and operation of green supply chain networks by simulating various supply chain scenarios and evaluating their environmental impacts. By running thousands of simulations, AI can help identify the most sustainable suppliers, shipping routes, and transportation modes (e.g., sea vs. air vs. rail). These simulations consider factors such as carbon emissions, cost, transit time, and geopolitical risks to guide decision-making. AI also supports carbon accounting by tracking emissions from each stage of logistics—from sourcing and production to distribution. These insights empower companies to meet sustainability targets, comply with regulations, and provide transparent reporting to stakeholders and consumers (Dhaliwal A., 2022).

Real-World Implementations

Numerous logistics and retail giants are already adopting AI-driven sustainable logistics strategies: Amazon uses AI for dynamic routing, predictive inventory placement, and anticipatory shipping, aiming to achieve net-zero carbon emissions by 2040. DHL integrates AI into its logistics operations for smart warehousing, green route planning, and autonomous last-mile delivery trials. UPS utilizes AI in its ORION (On-Road Integrated Optimization and Navigation) system, saving millions of gallons of fuel annually through route optimization. Maersk employs AI to optimize shipping routes across oceans, reducing emissions and improving fuel efficiency in global freight. FedEx leverages AI for real-time fleet tracking, weather analytics, and customer behavior prediction to streamline delivery processes sustainably.

INTEGRATION WITH SMART CITIES: AI-POWERED ELECTRIC VEHICLES AND URBAN INFRASTRUCTURE

As urbanization accelerates and cities grapple with the challenges of congestion, pollution, and resource management, the concept of smart cities has emerged as a transformative solution for sustainable urban living. At the heart of this transformation lies the integration of Artificial Intelligence (AI) with Electric Vehicles (EVs) and smart city infrastructure—a convergence that promises to redefine urban mobility, energy efficiency, and environmental impact. The integration of AI-powered EVs into smart cities represents a synergistic ecosystem where vehicles, infrastructure, and data systems communicate seamlessly to optimize transportation, reduce emissions, and improve quality of life (Alahi, 2023). This integration is not merely a futuristic vision but an evolving reality, as cities around the world invest in intelligent transportation systems, vehicle-to-everything (V2X) communication, AI-enabled traffic management, smart charging infrastructure, and urban planning driven by real-time data analytics. As cities become smarter and EVs become more intelligent, their interaction will shape a future of mobility that is autonomous, connected, electric, and shared—commonly referred to as the ACES framework. Below are the major dimensions of how AI-powered EVs are being integrated into smart city infrastructure, signaling a new era of sustainable and intelligent urban transportation.

Vehicle-to-Everything (V2X) Communication

A fundamental enabler of smart city and EV integration is Vehicle-to-Everything (V2X) communication. This includes V2I (Vehicle-to-Infrastructure), V2V (Vehicle-to-Vehicle), V2P (Vehicle-to-Pedestrian), and V2G (Vehicle-to-Grid) interfaces. These real-time data exchanges allow EVs to interact with traffic signals, street-lights, pedestrian crossings, parking systems, and the energy grid (Apata O., et al, 2023). For example, a smart traffic light can communicate with incoming EVs to dynamically adjust its signal timing, reducing congestion and improving traffic flow. AI algorithms process the immense amount of data generated by these interactions to make real-time decisions, predict traffic patterns, and ensure optimal routing. In emergencies, vehicles could communicate with city services to clear paths or prioritize access. This level of connectedness will be essential for managing the increasing complexity of urban transportation networks.

Smart Traffic Management and Urban Mobility

AI-powered traffic management is a cornerstone of smart city development, and its integration with EVs enhances mobility and environmental outcomes. Traffic signals, road sensors, surveillance cameras, and connected vehicles generate vast volumes of real-time data, which AI systems analyze to manage congestion, detect incidents, and optimize traffic flow. EVs, as data-generating platforms, contribute to this ecosystem by providing telemetry and situational awareness. Cities can use this data to implement adaptive traffic control systems, congestion pricing strategies, and priority lanes for shared or electric vehicles. Additionally, AI-driven traffic predictions help autonomous EVs navigate more efficiently by avoiding delays and hazards. The result is reduced travel time, lower emissions, and improved safety for all road users.

AI for Emission Monitoring and Air Quality Improvement

EVs contribute significantly to reducing urban air pollution, and when integrated with AI-powered smart city systems, their impact can be measured and optimized (Bathla G., et al, 2022). Cities can deploy real-time air quality sensors, traffic flow monitors, and satellite data to assess pollution hotspots and take corrective measures. AI processes this environmental data to model the impact of EV adoption, suggest clean-air zones, or adjust traffic patterns to reduce emissions. Electric fleets, such as buses and delivery vans, can be strategically deployed in high-pollution areas, and their effectiveness monitored continuously. This feedback loop enables cities to meet climate targets and improve public health outcomes.

AI and Robotics for EV Maintenance and Urban Services

Smart cities will increasingly use robotics and AI for the maintenance and operation of EV infrastructure. Automated drones and robots can inspect charging stations, road surfaces, and EV fleets, identifying maintenance issues early and reducing downtime. Robotic arms can also assist with autonomous EV charging, especially in large fleet operations or parking facilities. In logistics and urban services, AI-driven electric drones and delivery robots will provide last-mile delivery services, reducing traffic congestion and carbon footprints. The integration of AI-powered electric vehicles with smart city infrastructure is more than a technological advancement—it is a paradigm shift toward sustainable, intelligent, and inclusive urban living. From optimizing energy use and transportation networks to enhancing public safety and environmental health, this convergence is redefining the future of cities (Maheswaran K., et al, 2025). As infrastructure becomes more adaptive and vehicles more intelligent, the boundaries between transportation and technology will blur, creating ecosystems that are responsive, data-driven, and people-centric. However, realizing this vision requires collaboration among governments, private sector players, urban planners, and citizens. Policy frameworks, investment in digital infrastructure, and public trust in AI will be critical to success. The road ahead is both exciting and complex, but with continued innovation and strategic planning, AI-powered EVs and smart cities together will shape a cleaner, more efficient, and more connected urban future for generations to come.

AUTONOMOUS VEHICLE FLEETS: TRANSFORMING URBAN MOBILITY THROUGH AI AND AUTOMATION

The advent of Autonomous Vehicle Fleets (AVFs) represents a groundbreaking shift in the landscape of transportation, logistics, and mobility. Enabled by the convergence of Artificial Intelligence (AI), sensor technologies, connectivity, and electric vehicle (EV) innovations, AVFs are poised to redefine how people and goods move within urban, suburban, and even rural environments. As opposed to individually owned self-driving vehicles, autonomous vehicle fleets refer to a coordinated system of driverless cars, shuttles, trucks, and delivery robots that operate collectively under intelligent fleet management systems. These fleets are designed to serve a multitude of purposes—from ride-hailing services, public transport, and last-mile deliveries to freight transport and infrastructure maintenance (Coates J M., et al, 2025). With the potential to dramatically reduce traffic accidents, carbon emissions, operational costs, and urban congestion, AVFs are not merely a technological novelty but a pivotal solution for the challenges of 21st-century mobility. In the context of smart cities,

AVFs function as key components of the Mobility-as-a-Service (MaaS) paradigm, offering on-demand, seamless, and integrated transportation experiences. Below is a detailed exploration of how autonomous vehicle fleets are evolving, their technological underpinnings, applications, benefits, challenges, and long-term impact on society and the environment.

Fleet Management Platforms

Centralized Intelligence

AVFs are coordinated through centralized fleet management platforms powered by AI. These platforms act as the operational brain of the fleet, handling vehicle dispatching, dynamic routing, real-time tracking, charging schedules (in the case of electric AVs), and data analytics. For ride-hailing services, the system matches passengers with the nearest available vehicle based on location, estimated time of arrival, traffic conditions, and customer preferences. For logistics, it determines the most efficient delivery routes while minimizing energy use and travel time. The system continuously receives feedback from each vehicle and adjusts the entire fleet's behavior to maintain optimal performance. Advanced platforms may also integrate demand forecasting, autonomous load balancing, and multi-modal coordination with public transit or air freight systems (Devi J V., et al, 2024).

Revolutionizing Last-Mile and Freight Delivery

AVFs are particularly transformative in the field of logistics and last-mile delivery. Companies like Amazon, FedEx, and Alibaba are already testing autonomous ground vehicles, delivery drones, and sidewalk robots to reduce delivery times, costs, and carbon emissions. AI ensures that these vehicles operate safely in neighborhoods, interact with pedestrians, and efficiently navigate to doorsteps. In the freight industry, platoons of autonomous trucks can drive long distances on highways with minimal human oversight, improving fuel efficiency and reducing labor costs. These trucks often operate in semi-autonomous mode, where a human driver monitors the lead truck while others follow autonomously, maintaining tight formation and synchronized speeds.

Figure 1. AI in Transportation

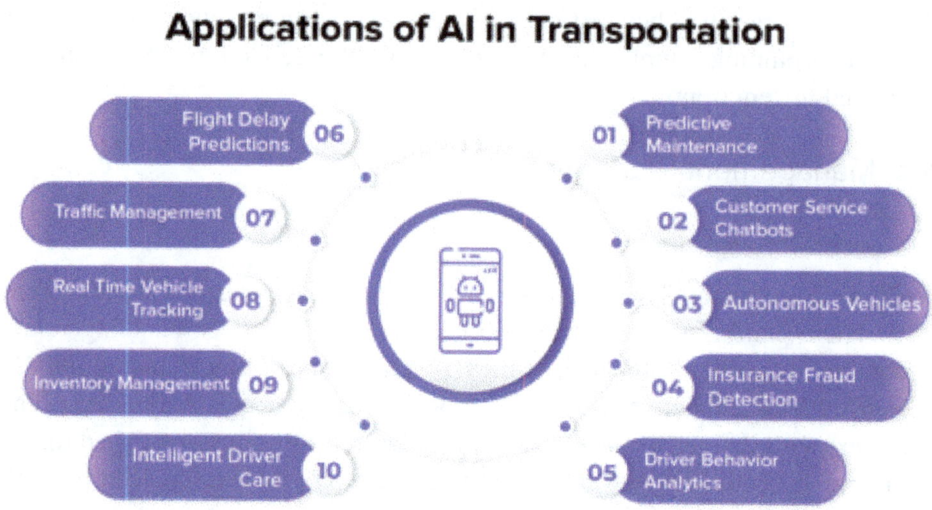

Safety and Accident Reduction

One of the most touted advantages of autonomous vehicles is their potential to significantly reduce traffic accidents, 90% of which are caused by human error. AI-powered AVFs maintain strict adherence to traffic laws, avoid distractions, and make split-second decisions based on real-time data (Shamsuddoha M., et al, 2025). These fleets are designed with multiple redundancies, including overlapping sensor systems and fail-safe protocols. Moreover, fleet-wide data sharing means that when one vehicle encounters a hazard or learns a new maneuver, the entire fleet benefits. This collective learning model continually enhances safety across all units.

Future Vision: Full Autonomy and Urban Transformation

Looking ahead, the ultimate vision for AVFs is Level 5 autonomy, where vehicles require no human intervention under any condition. In such a future, cities will undergo major transformations—road infrastructure may be redesigned to accommodate autonomous traffic, parking lots may become obsolete, and streets may become more pedestrian-friendly. The integration of AVFs with smart city infrastructure, IoT ecosystems, and edge computing will enable real-time decision-making at an unprecedented scale. These fleets could also be integrated with aerial mobility systems such as autonomous flying taxis, further expanding their scope (Tariq M U., 2024). Autonomous vehicle fleets signify a paradigm shift in how transportation and logistics systems function. Their integration with AI, electric propulsion, and smart city systems offers a powerful response to the pressing needs for sustainability, efficiency, safety, and accessibility in the modern world. While challenges remain, the benefits of AVFs are too substantial to ignore. As technology matures, regulations evolve, and public trust grows, autonomous vehicle fleets will become an integral part of daily life—delivering people and goods with unmatched intelligence, precision, and efficiency. In this future, transportation will not just be a service but an intelligent, adaptive, and sustainable ecosystem, reshaping the very fabric of how cities move.

ENHANCING MOBILITY AND ACCESSIBILITY: THE ROLE OF AI IN BUILDING INCLUSIVE AND INTELLIGENT TRANSPORTATION SYSTEMS

Artificial Intelligence (AI) is playing a transformative role in reshaping the landscape of global transportation by enhancing mobility and accessibility for people across all demographics, including individuals with disabilities, the elderly, residents in rural areas, and those without access to private vehicles. In traditional transportation systems, mobility has often been unequally distributed, leaving certain segments of the population marginalized due to physical, financial, geographical, or informational barriers (Jagatheesaperumal S K., et al, 2024). However, with AI-driven innovations, transportation is becoming more personalized, responsive, adaptive, and inclusive. By integrating intelligent technologies into vehicles, infrastructure, ride-sharing platforms, and public transportation systems, AI is helping eliminate many longstanding barriers that once restricted people's ability to travel freely, safely, and independently. As a result, the impact of AI on mobility and accessibility extends far beyond convenience—it's about social equity, quality of life, and the empowerment of underserved communities in the digital age.

Empowering Individuals with Disabilities

AI technologies are enabling significant breakthroughs in providing greater independence and mobility for individuals with disabilities. For example, AI-powered autonomous vehicles (AVs) can offer a life-changing alternative to traditional vehicles for people who are unable to drive due to visual, cognitive, or physical impairments. These self-driving vehicles can interpret traffic signals, navigate intersections, and adjust to traffic patterns—without requiring manual control—thus granting newfound freedom to individuals who previously relied on caretakers or public services (Paiva S., et al, 2021). Additionally, voice-activated AI assistants integrated into vehicles allow users with limited mobility to set destinations, manage in-vehicle functions, and receive travel updates without using their hands. Features such as natural language processing (NLP) and machine learning allow these systems to continuously improve in understanding diverse speech patterns, accents, and speech impairments. Moreover, navigation apps enhanced with AI now offer wheelchair-accessible routes, identify elevators and ramps, and provide real-time updates on infrastructure accessibility, making daily commutes smoother and safer for users with disabilities.

Real-Time Language and Communication Support

Language barriers can be a significant obstacle to mobility, especially for tourists, migrants, or people living in linguistically diverse areas. AI-powered translation services embedded in navigation apps, smart kiosks, and autonomous vehicles allow users to interact with transport systems in their preferred language. These services use neural machine translation (NMT) and contextual learning to ensure accuracy and fluency in communication. For instance, a non-English-speaking traveler can use a voice assistant to ask for directions, buy tickets, or receive safety instructions in their native tongue. In emergencies, AI-powered multilingual support ensures passengers can quickly understand evacuation routes or contact help services, thus improving both safety and accessibility in transit environments (Rehan H., 2024).

Designing Inclusive Infrastructure through AI Insights

AI is not only used in vehicles but also plays a vital role in planning and designing inclusive transport infrastructure. Through the analysis of vast urban datasets, AI identifies where accessibility is lacking, such as inadequate sidewalks, missing ramps, poorly lit areas, or unsafe crossings. City planners and transport engineers can use these insights to prioritize infrastructure upgrades that directly benefit users with mobility limitations. AI-powered simulations also help model pedestrian and cyclist flows under various scenarios, guiding the development of barrier-free transportation networks. This iterative, data-driven design process fosters cities that are not only smart but also equitable and accessible.

Autonomous Shared Mobility Solutions

Shared autonomous mobility—such as driverless shuttles and ride-pooling pods—is becoming a cornerstone of accessible future transport. These AI-powered services reduce dependency on private vehicle ownership, lower overall transportation costs, and make door-to-door services more scalable. For users with limited mobility, these vehicles are often equipped with automated ramps, voice-activated entry systems, and in-vehicle assistants that provide real-time help. By enabling a flexible, on-demand, and personalized experience, autonomous shared mobility can serve as an affordable, accessible solution for both urban and rural communities (Nisha Angeline M., et al, 2025). AI's impact on mobility and accessibility is nothing short of revolutionary. From empowering individuals with disabilities to optimizing transport in underserved areas and creating more user-friendly interfaces for the elderly, AI is driving the transformation toward an inclusive, equitable, and intelligent transportation ecosystem. As we continue to adopt smart mobility solutions, accessibility must remain central to the conversation, because true progress is measured not only by how fast or how far we can go, but by how many people we can bring along the journey. By leveraging AI's full potential, society can ensure that everyone, regardless of age, ability, or geography, enjoys the fundamental right to move freely, safely, and with dignity.

OVERCOMING TECHNICAL CHALLENGES: NAVIGATING THE COMPLEXITIES OF AI IN ELECTRIC VEHICLES

The integration of Artificial Intelligence (AI) into Electric Vehicles (EVs) is ushering in a new era of intelligent transportation marked by automation, efficiency, and personalization. However, this transformation is not without its share of technical challenges. The convergence of AI and EV technologies introduces a highly complex ecosystem requiring reliable computation, data security, sensor integration, power efficiency, real-time decision-making, and system interoperability. These challenges stem from both hardware and software dimensions, and overcoming them is critical to the safe, effective, and widespread deployment of AI-powered electric mobility solutions (Hubballi R B., et al, 2025). As the global automotive industry races toward autonomous, connected, and sustainable mobility, researchers, engineers, and policymakers are continuously identifying solutions to the most pressing obstacles. This detailed overview explores the major technical challenges in the development and deployment of AI in EVs and the cutting-edge strategies being used to address them.

Data Management and Quality Control

AI systems depend heavily on high-quality data for training and real-time operations. However, gathering, labeling, cleaning, and updating datasets—especially for machine learning models—is resource-intensive. In autonomous EVs, managing real-time sensor data poses a challenge, with over 1 TB of data being generated per hour in some cases. Inconsistent, biased, or outdated data can degrade system performance or even lead to unsafe outcomes.

Solutions

Deployment of data pipelines with real-time preprocessing, cleaning, and storage protocols.

Use of synthetic data and simulation environments (e.g., CARLA, LGSVL Simulator) to augment real-world datasets, especially for rare or dangerous driving scenarios (Alqahtani & Kumar, 2024).

Implementation of federated learning, which allows vehicles to collaboratively improve models without sharing raw data, preserving privacy and reducing bandwidth requirements.

Battery and Energy Constraints

AI algorithms consume energy, especially during inference and training phases. In electric vehicles, where battery life directly impacts range, any energy overhead must be carefully managed. Additionally, AI-based control systems for battery management and energy optimization require continuous operation. Solutions: Use of energy-efficient AI models, such as lightweight convolutional neural networks (e.g., MobileNet, EfficientNet), and hardware-aware model pruning techniques. AI-assisted battery management systems (BMS) that optimize charging and discharging rates, thermal performance, and lifecycle management. Dynamic AI workload scheduling shifts processing tasks to periods of low demand or external infrastructure (e.g., roadside edge servers).

Cybersecurity and Data Privacy

As EVs become more connected and intelligent, they are also more vulnerable to cyber threats. Hackers could target vehicle control systems, AI algorithms, or user data, posing both safety and privacy risks. Key concerns: Intrusions into autonomous driving systems. Manipulation of AI models through adversarial attacks. Unauthorized access to user data and cloud-connected systems.

Solutions

AI-driven intrusion detection systems (IDS) that monitor abnormal behavior in vehicle networks (CAN, Ethernet).

Use of blockchain for secure communication and data integrity in EV-to-EV and EV-to-grid (V2G) systems (Aldhanhani T., et al, 2024).

Implementation of homomorphic encryption, secure multi-party computation (SMPC), and differential privacy to protect user data while enabling AI learning.

AI Model Validation and Safety Certification

Unlike traditional rule-based systems, AI models behave as black boxes and may not consistently explain their actions. Regulatory bodies require explainability, reliability, and traceability in safety-critical applications such as autonomous EVs. This lack of transparency complicates validation, testing, and certification. Solutions: Explainable AI (XAI) techniques that reveal the logic behind predictions (e.g., SHAP, LIME). Development of simulation-based verification environments capable of running millions of test scenarios across weather, traffic, and pedestrian variations. Collaboration between governments and standard bodies (e.g., ISO 26262, UL 4600) to define safety standards and compliance metrics for AI in EVs (Qadir S A., et al, 2024). Integrating AI into electric vehicles is a deeply complex endeavor that brings immense potential but also presents significant technical hurdles. From computational power limitations and data complexity to sensor fusion, cybersecurity, and regulatory compliance, these challenges span across every layer of the vehicle and its ecosystem. However, through collaborative innovation—combining hardware advancements, smarter algorithms, and robust safety protocols—the industry is steadily overcoming these barriers. The continuous evolution of AI in EVs reflects not just a technological transition but a movement toward safer, smarter, and more inclusive transportation. As these solutions mature, we move closer to a world where AI-powered electric vehicles become a dependable cornerstone of sustainable mobility.

CONCLUSION

As artificial intelligence (AI) and electric vehicles (EVs) increasingly intersect, forming the backbone of next-generation mobility systems, the development of robust regulatory frameworks and standards becomes not just a necessity but a strategic imperative. The dynamic nature of AI, especially in critical applications like autonomous driving, energy optimization, and predictive maintenance, raises significant legal, ethical, safety, and operational concerns. When coupled with the unique characteristics of EVs—such as battery systems, charging infrastructure, and vehicle-to-grid (V2G) interactions—regulation becomes even more multifaceted. Effective regulatory frameworks are essential to establish minimum safety standards, ensure interoperability, promote innovation without compromising public safety, and build consumer trust. They form the basis for large-scale deployment and cross-border operations of AI-powered EVs. This paragraph explores the current landscape, challenges, and emerging approaches to standardizing and regulating AI in electric mobility systems. The Complexity of Regulating AI in EVs: AI, by its very nature, is probabilistic and often lacks transparency. Unlike traditional software that behaves deterministically, AI systems evolve through training and can make decisions that are difficult to predict or explain. This makes standard-setting especially challenging in applications like autonomous driving, where safety and accountability are paramount. In electric vehicles, AI governs not just driving, but also energy consumption, charging schedules, predictive maintenance, and user interfaces, all of which must adhere to both functional safety and cybersecurity standards. Furthermore, the integration of AI into the vehicle's electrical architecture means any software update, sensor change, or data variation can impact system performance. This fluidity complicates regulatory oversight, as traditional static certifications may not remain valid over time. Regulators are now tasked with crafting dynamic, adaptive frameworks that can keep pace with continuous software upgrades, machine learning retraining, and over-the-air (OTA) updates. AI Explainability and Accountability: One of the core regulatory challenges in AI is explainability. Stakeholders, including consumers, engineers, and regulators, must understand how AI systems make decisions. This is critical for safety audits, accident investigations, and ethical compliance. To address this, AI systems must incorporate explainable AI (XAI) methodologies, such as model interpretation layers and confidence scores. Developers are required to maintain data logs, decision rationales, and training data documentation for regulatory audits. Liability frameworks are evolving to determine who is responsible in case of an accident caused by autonomous EVs—the vehicle manufacturer, software provider, or AI developer. Challenges in Regulatory Harmonization: Despite rapid progress, one of the biggest challenges lies in harmonizing regulations across regions. Vehicles and AI systems often operate internationally or are developed by global teams. Inconsistent rules can hinder scalability, increase costs, and delay deployment. Efforts are underway to establish: Global harmonization forums, such as those by UNECE, ISO, and IEEE, to create universally accepted standards. Regulatory sandboxes allow AI developers to test new features under real-world conditions with temporary exemptions. International public-private partnerships uniting governments, OEMs, startups, and academia for collaborative policymaking. Conclusion: Regulatory frameworks and standards for

AI in electric vehicles are at the crossroads of technological advancement, public safety, legal compliance, and ethical governance. They serve as the foundational pillars ensuring that AI-powered EVs are not only efficient and innovative but also safe, fair, and accountable. From functional safety and cybersecurity to explainability and smart infrastructure interoperability, regulations are evolving to address the multifaceted nature of AI integration in transportation. The road ahead involves continuous dialogue between regulators, manufacturers, engineers, and civil society to ensure that innovation and responsibility grow hand in hand. Ultimately, robust and harmonized regulatory ecosystems will accelerate the global transition to a cleaner, smarter, and more trustworthy electric mobility future.

REFERENCES

Alahi, M. E. E., Sukkuea, A., Tina, F. W., Nag, A., Kurdthongmee, W., Suwannarat, K., & Mukhopadhyay, S. C. (2023). Integration of IoT-Enabled Technologies and Artificial Intelligence (AI) for Smart City Scenario: Recent Advancements and Future Trends. *Sensors (Basel)*, *23*(11), 5206. Advance online publication. DOI: 10.3390/s23115206 PMID: 37299934

Aldhanhani, T., Abraham, A., Hamidouche, W., & Shaaban, M. (2024). Future Trends in Smart Green IoV: Vehicle-to-Everything in the Era of Electric Vehicles. *IEEE Open Journal of Vehicular Technology*, *5*, 278–297. DOI: 10.1109/OJVT.2024.3358893

Alqahtani, H., & Kumar, G. (2024). Cybersecurity in Electric and Flying Vehicles: Threats, Challenges, AI Solutions & Future Directions. *ACM Computing Surveys*, *57*(4), 1–34. DOI: 10.1145/3697830

Antonialli, F., Martinesco, A., & Mira-Bonnardel, S. (2022). Artificial Intelligence as a Determinant for Reshaping the Automotive Industry and Urban Mobility Services. *International Journal of Automotive Technology and Management*, *22*(3), 324–351. DOI: 10.1504/IJATM.2022.124828

Apata, O., Bokoro, P. N., & Sharma, G. (2023). The Risks and Challenges of Electric Vehicle Integration into Smart Cities. *Energies*, *16*(14), 5274. Advance online publication. DOI: 10.3390/en16145274

Arévalo, P., Ochoa-Correa, D., & Villa-Ávila, E. (2024). A Systematic Review on the Integration of Artificial Intelligence into Energy Management Systems for Electric Vehicles: Recent Advances and Future Perspectives. *World Electric Vehicle Journal*, *15*(8), 364. Advance online publication. DOI: 10.3390/wevj15080364

Bathla, G., Bhadane, K., Singh, R. K., Kumar, R., Aluvalu, R., Krishnamurthi, R., Kumar, A., Thakur, R. N., & Basheer, S. (2022). Autonomous Vehicles and Intelligent Automation: Applications, Challenges, and Opportunities. *Mobile Information Systems*, *2022*, 7632892. Advance online publication. DOI: 10.1155/2022/7632892

Cavus, M., Dissanayake, D., & Bell, M. (2025). Next Generation of Electric Vehicles: AI-Driven Approaches for Predictive Maintenance and Battery Management. *Energies*, *18*(5), 1041. Advance online publication. DOI: 10.3390/en18051041

Chatti, W., Majeed, M. T., Khoj, H., Miraz, M. H., & Ali, A. (2024). Towards Smart and Sustainable Transportation: The Role of Artificial Intelligence and New Technologies in Mitigating Passenger Car CO_2 Emissions in European Countries. *Environment, Development and Sustainability*, *26*, 1–25. DOI: 10.1007/s10668-024-05685-0

Chen, W., Men, Y., Fuster, N., Osorio, C., & Juan, A. A. (2024). Artificial Intelligence in Logistics Optimization with Sustainable Criteria: A Review. *Sustainability (Basel), 16*(21), 9145. Advance online publication. DOI: 10.3390/su16219145

Chowdhary, S., Selvakumar, P., Sangeetha, B., Sarvaani, V. V., Anute, N., & Sharma, M. (2025). The Role of Social Capital in Entrepreneurial Development. In Martínez-Falcó, J., Sánchez-García, E., & Marco-Lajara, B. (Eds.), *Support Networks for Entrepreneurs: Social Capital and Strategic Collaborations* (pp. 311–338). IGI Global Scientific Publishing., DOI: 10.4018/979-8-3693-9836-4.ch013

Chu, K. F., Lam, A. Y. S., & Li, V. O. K. (2022). Joint Rebalancing and Vehicle-to-Grid Coordination for Autonomous Vehicle Public Transportation System. *IEEE Transactions on Intelligent Transportation Systems, 23*(7), 7156–7169. DOI: 10.1109/TITS.2021.3067044

Coates, J. M., Santhanalakshmi, K., Dinesh, N., Gandhimathi, S., & Chandel, P. S. Mohit & Selvakumar P, (2025). "Assessment and Evaluation in Transnational Staff Development." In G Morris & J Ye (Eds.), Training and Development in Transnational Higher Education (pp. 387–406). IGI Global Scientific Publishing. https://doi.org/DOI: 10.4018/979-8-3373-1897-4.ch019

De la Torre, R., Corlu, C. G., Faulin, J., Onggo, B. S., & Juan, A. A. (2021). Simulation, Optimization, and Machine Learning in Sustainable Transportation Systems: Models and Applications. *Sustainability (Basel), 13*(3), 1551. Advance online publication. DOI: 10.3390/su13031551

Devi, J. V., Argiddi, R. V., Renuka, P., Janagi, K., Hari, B. S., & Boopathi, S. (2024). "Study on Integrated Neural Networks and Fuzzy Logic Control for Autonomous Electric Vehicles." In Semantic Web Technologies and Applications in Artificial Intelligence of Things, Chapter 6, pp. 88–112. DOI: DOI: 10.4018/979-8-3693-1487-6.ch006

Dhaliwal, A. (2022). "Towards AI-Driven Transport and Logistics." In Digital Transformation in the Viral Age. *Lecture Notes in Business Information Processing, 508*, 119–131. DOI: 10.1007/978-3-031-60003-6_8

Dikshit, S., Atiq, A., Shahid, M., Dwivedi, V., & Thusu, A. (2023). The Use of Artificial Intelligence to Optimize the Routing of Vehicles and Reduce Traffic Congestion in Urban Areas. *Energy Web, 23*(1), e4613. Advance online publication. DOI: 10.4108/ew.4613

Hubballi, R. B., Selvakumar, P., Seenivasan, R., Basava Aradhya, S. G., Dinesh, N., & Seelam, P. K. (2025). Overview of Current AI Technologies in Education. In Kyei-Blankson, L., & Ntuli, E. (Eds.), *Transformative AI Practices for Personalized Learning Strategies* (pp. 1–26). IGI Global Scientific Publishing., DOI: 10.4018/979-8-3693-8744-3.ch001

Iyer, L. S. (2021). AI Enabled Applications Towards Intelligent Transportation. *Transportation Engineering*, *5*, 100083. Advance online publication. DOI: 10.1016/j.treng.2021.100083

Jagatheesaperumal, S. K., Bibri, S. E., Huang, J., Rajapandian, J., & Parthiban, B. (2024). Artificial Intelligence of Things for Smart Cities: Advanced Solutions for Enhancing Transportation Safety. *Computational Urban Science*, *4*(1), 10. Advance online publication. DOI: 10.1007/s43762-024-00120-6

Kotte, K. R., Attaluri, V., & Selvakumar, P. (2025). Revolutionising Retail Banking with AI and Virtual Technologies AI-Driven Financial Training. In Nair, A. J., Manohar, S., Limbu, Y. B., & Huhmann, B. A. (Eds.), *Intersecting Natural Language Processing and FinTech Innovations in Service Marketing* (pp. 421–440). IGI Global Scientific Publishing., DOI: 10.4018/979-8-3693-9944-6.ch020

Lukic Vujadinovic, V., Damnjanovic, A., Cakic, A., Petkovic, D. R., Prelevic, M., Pantovic, V., Stojanovic, M., Vidojevic, D., Vranjes, D., & Bodolo, I. (2024). AI-Driven Approach for Enhancing Sustainability in Urban Public Transportation. *Sustainability (Basel)*, *16*(17), 7763. Advance online publication. DOI: 10.3390/su16177763

Maheswaran, K., Selvakumar, P., Shaikh, M., Shashank, G., Gupta, S., & Manjunath, T. C. (2025). The Evolution of Smart Urban Mobility With Emerging Technologies. In Ponnusamy, S., Assaf, M., Youssef, B., Jeon, G., & Kalyanaraman, S. (Eds.), *Leveraging VANETs and Blockchain Technology for Urban Mobility* (pp. 1–28). IGI Global Scientific Publishing., DOI: 10.4018/979-8-3373-0265-2.ch001

Mohsen, B. M. (2024). AI-Driven Optimization of Urban Logistics in Smart Cities: Integrating Autonomous Vehicles and IoT for Efficient Delivery Systems. *Sustainability (Basel)*, *16*(24), 11265. Advance online publication. DOI: 10.3390/su162411265

Nisha Angeline, M., Selvakumar, P., Gupta, M. K., Nazar, N., Elumalai, D., & Bhattacharya, S. (2025). Emerging Materials for Energy Storage and Environmental Applications. In Muduli, K., Moharana, B., Ales, S., & Biswal, D. (Eds.), *Using Computational Intelligence for Sustainable Manufacturing of Advanced Materials* (pp. 53–72). IGI Global Scientific Publishing., DOI: 10.4018/979-8-3693-7974-5.ch003

Paiva, S., Ahad, M. A., Tripathi, G., Feroz, N., & Casalino, G. (2021). Enabling Technologies for Urban Smart Mobility: Recent Trends, Opportunities and Challenges. *Sensors (Basel), 21*(6), 2143. Advance online publication. DOI: 10.3390/s21062143 PMID: 33803903

Pamucar, D., Deveci, M., Gokasar, I., Tavana, M., & Köppen, M. (2022). A Metaverse Assessment Model for Sustainable Transportation Using Ordinal Priority Approach and Aczel-Alsina Norms. *Technological Forecasting and Social Change, 182*, 121778. Advance online publication. DOI: 10.1016/j.techfore.2022.121778

Qadir, S. A., Ahmad, F., Al-Wahedi, A. M A B., Iqbal, A., & Ali, A. (2024). Navigating the Complex Realities of Electric Vehicle Adoption: A Comprehensive Study of Government Strategies, Policies, and Incentives. *Energy Strategy Reviews, 53*, 101379. Advance online publication. DOI: 10.1016/j.esr.2024.101379

Rane, N., Choudhary, S., & Rane, J. (2024). "Artificial Intelligence and Machine Learning for Resilient and Sustainable Logistics and Supply Chain Management." *SSRN* Preprint, 8 pages. DOI: DOI: 10.2139/ssrn.4847087

Rao, P. S., Yaqoob, S. I., Ahmed, M. A., Abdinabievna, P. S., Yaseen, S. M., & Arumugam, M. (2023). RETRACTED ARTICLE: Integrated Artificial Intelligence and Predictive Maintenance of Electric Vehicle Components with Optical and Quantum Enhancements. *Optical and Quantum Electronics, 55*(10), 855. Advance online publication. DOI: 10.1007/s11082-023-05135-7

Rehan, H. (2024). The Future of Electric Vehicles: Navigating the Intersection of AI, Cloud Technology, and Cybersecurity. *International Journal of Scientific Research and Management, 12*(4), 1127–1143. DOI: 10.18535/ijsrm/v12i04.ec04

Shamsuddoha, M., Kashem, M. A., & Nasir, T. (2025). A Review of Transportation 5.0: Advancing Sustainable Mobility Through Intelligent Technology and Renewable Energy. *Future Transportation, 5*(1), 8. Advance online publication. DOI: 10.3390/futuretransp5010008

Sharma, R., & Singhal, M. (2024). "Sustainable Transport Using AI and IoT." In Sustainable Development Through Machine Learning, AI and IoT. *Communications in Computer and Information Science, 2196*, 273–283. DOI: 10.1007/978-3-031-71729-1_25

Tariq, M. U. (2024). Smart Transportation Systems: Paving the Way for Sustainable Urban Mobility. In *Contemporary Solutions for Sustainable Transportation Practices* (pp. 254–283). IGI Global., DOI: 10.4018/979-8-3693-3755-4.ch010

Thirunagalingam, A., & Selvakumar, P. (2025). Application of Automation and Robotics in the Oil and Gas Industry. In Abdullayev, V., & Khang, A. (Eds.), *Revolutionizing AI and Robotics in the Oil and Gas Industry* (pp. 229–244). IGI Global Scientific Publishing., DOI: 10.4018/979-8-3693-8156-4.ch011

Upadhyaya, A. N., Debnath, P., Rani, P., Selvakumar, P., Manjunath, T. C., & Bhattacharya, S. (2025). AI-Based Disaster Prediction and Early Warning Systems. In Mobo, F. (Ed.), *AI and Emerging Technologies for Emergency Response and Smart Cities* (pp. 55–74). IGI Global Scientific Publishing., DOI: 10.4018/979-8-3693-9770-1.ch003

Chapter 9
Optimizing Waste Management Systems Through Deep Learning and Federated Learning Techniques:
Leveraging Convolutional and Federated Architectures for Distributed Waste Image Analysis

Saira Banu Jamalmohammed
https://orcid.org/0000-0002-6575-9830
VIT University, India

Mythili Thirugnanam
https://orcid.org/0000-0002-5731-206X
VIT University, India

Jenifer Shanmugasundaram
https://orcid.org/0009-0000-3176-1556
VIT University, India

Sumaiya Thaseen Ikram
https://orcid.org/0009-0003-1948-7592
De Montfort University, Dubai, UAE

DOI: 10.4018/979-8-3373-5540-5.ch009

ABSTRACT

Rapid urbanization and population growth have led to rising waste levels, making waste management a critical issue. This study was conducted to explore how Artificial Intelligence (AI) and Machine Learning (ML) can effectively address this growing challenge and improve the efficiency of waste segregation systems. The problem studied focuses on optimizing waste classification to enhance recycling accuracy and reduce environmental impact. This work is important because traditional waste management methods are no longer sustainable for modern urban demandsModels like JONET utilizing DenseNet201 for real-time waste classification, demonstrate improved sorting performance in industrial settings. Mask R-CNN's pixel-level masks enhance sorting accuracy in real-time applications such as conveyor belt systems. Additionally, Federated Learning (FL) provides a decentralized approach to training ML models across multiple facilities without centralized data storage, improving scalability and adaptability.

INTRODUCTION

Waste includes materials that have exhausted their utility and are discarded across categories such as electronic, hazardous, biomedical, municipal solid, and agricultural waste, originating from households, commerce, farms, hospitals, and industries (Amasuomo & Baird, 2016). Typical components include plastics, metals, paper, glass, organics, and textiles, which together form the bulk of solid waste. Rapid urbanization, industrialization, and population growth have driven waste generation beyond municipal capacities causing environmental damage, public health hazards, and economic losses highlighting the need for scalable, data-driven solutions.

This chapter addresses the challenge of reliable, real-time waste classification to enable safer, faster, and more efficient downstream operations such as collection, routing, sorting, and recycling, especially in resource-constrained municipal settings. Traditional approaches manual sorting and basic mechanical systems struggle with speed, accuracy, and scalability, often exposing workers to hazards. In contrast, AI and ML enable automation, pattern recognition, and predictive analytics, significantly improving route efficiency, sorting precision, and operational resilience, while reducing emissions, costs, and human risk.

AI demonstrates major system-level impact across waste management. AI-based routing reduces travel time (~28.22%), operational cost (~13.35%), and distance (~36.8%) (Fang et al., 2023), while ANNs and GAs aid routing, frequency prediction, and GIS-guided planning even during pandemics (Rubab et al., 2022). IoT smart bins with ultrasonic/odor/gas sensors and ML/LSTM predict bin status and CO levels

(~90% accuracy) (Hussain et al., 2020); event-driven collection has been validated in prototypes (Abeygunawardhana et al., 2020) and at city scale for bin siting and overflow control (Bhattacharjee et al., 2022). On sorting lines, AI-driven robotics outperform humans (Ahmed & Asadullah, 2020); commercial systems (ZenRobotics, SamurAI, Trashbot) confirm this (Olawade et al., 2024), and rCubeBot (ResNet50) achieves 96% e-waste detection with projected cost savings (Shreyas Madhav et al., 2022). Reward-based ANNs enhance segregation behavior (Castro et al., 2020), while mobile/edge models using Mask R-CNN, YOLOv5, and EfficientNet perform in-situ classification and volume estimation (Geetha et al., 2022).

For forecasting, ANNs support neighborhood-scale waste prediction (Nguyen et al., 2021), process optimization (Xu et al., 2021), and generation modeling (R^2=0.95–0.99) (Coskuner et al., 2021). CNNs ensure high recyclable classification accuracy (Jude et al., 2021). Transfer learning with VGG16/VGG19/ResNet achieves ~97% accuracy (Nañez Alonso et al., 2021); YOLACT enhances construction-waste segmentation via data refinement (Na et al., 2022). Broad CNN success is confirmed (Wu et al., 2023; Majchrowska et al., 2022), including hyperspectral + CNN + XG-Boost (Yarbrough, 2024), while SVM/KNN + SURF + BoF remain strong baselines (Gómez et al., 2021). Real-world MobileNetV3 systems reach ~94.26% (Wang et al., 2021); DenseNet121 tops accuracy (~94.15%), while SqueezeNet fits edge devices (~91.50%) (Alawi et al., 2021). YOLO models deliver real-time detection: Pi-based deployments (Aishwarya et al., 2021), YOLOv3 (mAP ~94.99%) (Kumar et al., 2020), and DLSODC-GWM (~98.61%) surpassing AlexNet/ResNet50 (Alsubaei et al., 2022). Traditional ML GAs, decision trees, random forests remains effective (Abdallah et al., 2020; Rajalakshmi et al., 2023; Ihsanullah et al., 2022; Pillai et al., 2023; Sami et al., 2020), with K-means applied to municipal data (Sharma et al., 2021) and ECO-365 showcasing ANN-based software utility (Epifantsev & Kozlova, 2021). For urban blight, an mp-CNN detects illegal dumping in Indian cities with ~98.33% accuracy and ~97.82% precision (Shahab & Anjum, 2022).

Within this landscape, our contribution is practical and two-fold. First, we present a comparative evaluation of classical models (e.g., SVM, Random Forest with handcrafted features) and deep CNNs (VGG19, ResNet152V2, MobileNetV2, DenseNet201) using metrics such as accuracy, precision, recall, and F1-score under real-world operational constraints. Second, we deploy a real-time classifier (JONET) built on DenseNet201, achieving a balance between accuracy and computational efficiency, and extend it with Mask R-CNN for instance-level detection and segmentation in multi-object, mixed-waste contexts.

Finally, we explore Federated Learning (FL) to enable collaborative, privacy-preserving training across distributed nodes (e.g., smart bins, recycling centers) without sharing raw data, enhancing cross-site model robustness. Overall, we propose a deployable end-to-end architecture combining classification, detection, and distributed learning to help municipalities and industries improve safety, efficiency, and recovery rates at scale.

DATASET

To build a reliable model, JONET was trained on two well-known datasets. The first, TrashNet (Yang & Thung, 2016), includes 2,327 images across six categories like plastic, metal, and cardboard making it a standard benchmark for waste classification.

The second, also from Kaggle, is the Drinking Waste Classification Dataset, which contains 4,828 images focused on drink packaging like AluCans, PET, and glass.

All datasets were preprocessed and split into training, validation, and testing sets using a 70:15:15 ratio to ensure balanced learning and proper evaluation on unseen data.

Table 1. Train - Validation - Test split of the datasets

Dataset	Training Set Size	Validation Set Size	Testing Set Size
TrashNet	1766	377	384
Drinking Waste Classification	3378	723	727

The class distribution of TrashNet dataset and Drinking Waste Classification dataset is given in Figure 1 and Figure 2 respectively.

Figure 1. TrashNet Dataset Class Distribution

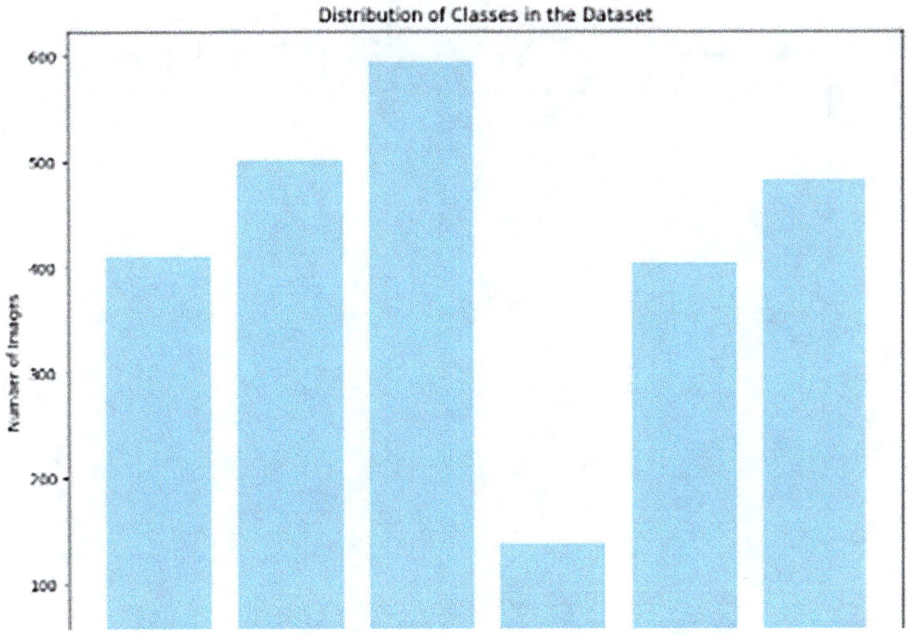

Figure 2. Drinking Waste Classification Dataset Class Distribution

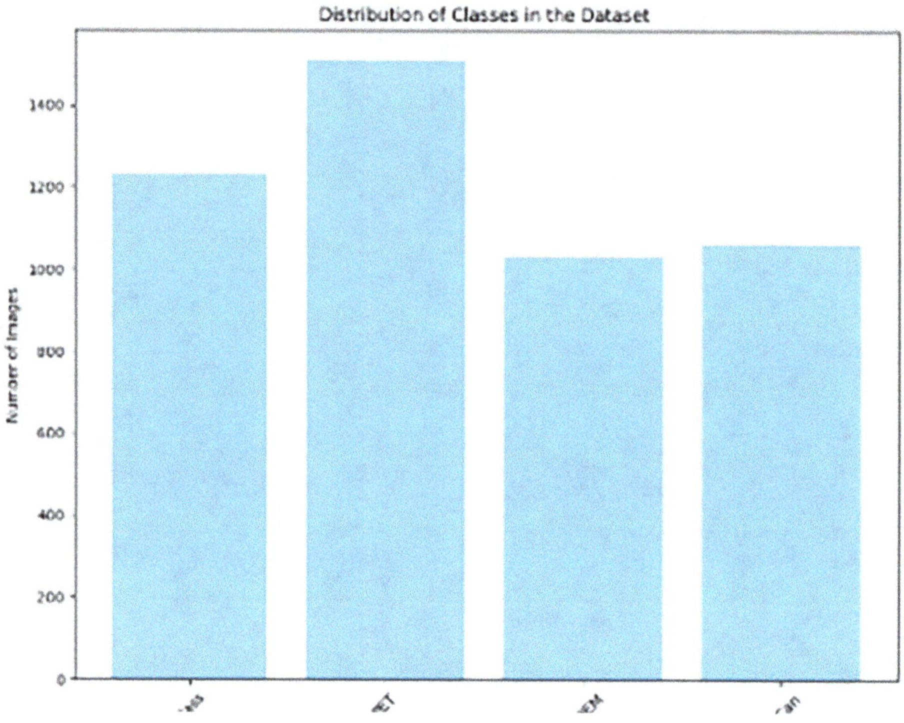

The TACO (Trash Annotations in Context) dataset is built for real-world litter detection and segmentation. Created by Pedro F. Proença and Pedro Simões (2020), it tackles the issue of litter pollution by offering annotated images from real environments like streets, beaches, and parks instead of clean, lab-like settings.

TACO contains around 1,500 high-resolution images and 4,784 manually labeled annotations. These images, mostly taken with smartphones, are annotated using an online tool and sorted into 60 detailed categories grouped under 28 broader super categories. Items include plastic bags, bottles, cans, cigarette butts, and paper, along with an "unlabeled litter" category for ambiguous cases making it ideal for realistic waste detection tasks.

We enforce dataset hygiene by cleaning images, verifying labels, and maintaining consistent annotations. Strict guidelines and spot-checks prevent noise and unreliable learning. The combined TrashNet + TACO dataset is split 70/15/15 into train, validation, and test sets, ensuring balanced and representative partitions.

To prevent data leakage, duplicates are removed using perceptual hashing and embedding-based similarity. Because TrashNet has clean, staged images while TACO depicts real-world clutter, we apply augmentations, background changes, lighting variation, noise, and style shifts and perform domain adaptation to boost generalization and real-world robustness.

METHODOLOGY

The core techniques of image processing used in waste management can be broadly divided into three categories: image segmentation, object detection, and image classification. Each technique addresses a different aspect of the problem and contributes uniquely to improving the precision and effectiveness of waste processing systems.

- **Image classification:** Image classification assigns a specific label to an image or to objects within it. In waste management, it helps categorize items like plastic, glass, paper, metal, and organics—each needing different recycling or disposal methods. Deep learning models such as VGG, ResNet, and Inception are widely used for this task because they can learn complex patterns from visual features like shape, color, and texture to accurately classify waste types.
- **Image segmentation:** Image segmentation involves partitioning an image into multiple regions or segments to isolate specific objects of interest. This technique allows systems to focus on individual waste items rather than the entire image, enabling more accurate classification and processing. Segmentation techniques can be categorized into three types:
    - **Instance segmentation:** Identifies and separates individual objects belonging to the same class. For example, it can detect and differentiate between two separate plastic bottles in a single image.
    - **Semantic segmentation:** Assigns a label to every pixel in the image, grouping all similar objects under the same category. However, it does not distinguish between separate instances of the same object.
    - **Panoptic segmentation:** Combines both semantic and instance segmentation to provide a comprehensive view, labeling every pixel while also distinguishing between individual instances.

In waste management, segmentation helps separate waste items from the background or from overlapping objects, making it easier to classify and sort them. Traditional methods include Thresholding, Edge Detection, Region-Based Segmentation, Clustering, and Watershed Segmentation. For more precise results, deep learning models like U-Net and Mask R-CNN are commonly used, as they provide high accuracy in pixel-level segmentation tasks

- **Object Detection:** Object detection identifies and localizes specific objects in an image using bounding boxes and labels. Unlike segmentation, which outlines object boundaries, detection focuses on fast, accurate localization—crucial for real-time applications like smart bins and robotic arms in automated waste sorting. Popular models include R-CNN, Fast R-CNN, and YOLO (You Only Look Once), which balance speed and accuracy. These models are ideal for detecting and tracking waste in high-throughput environments, enabling efficient and scalable waste management solutions.

In this section, we will focus on the implementation of these image processing techniques for the process of waste management.

Waste Classification

Machine Learning Methods

Support Vector Machine. Support Vector Machine (SVM) is a supervised classification algorithm that identifies the optimal hyperplane separating data points of different classes while maximizing the margin between them. This makes it highly effective for high-dimensional and complex datasets. When data is not linearly separable, SVM uses kernel functions such as the Radial Basis Function (RBF) to map inputs into a higher-dimensional space, enabling linear separation there. The decision function for an SVM with an RBF kernel is given by:

$$f(x) = sign(\sum_{i=1}^{n} \alpha i y i K(xi, x) + b)$$

Eqn (1)

Where:

- f(x) is the classification function that decides the label of a point x
- αi are the Lagrange multipliers from the dual optimization problem
- yi are the class labels (+1 or -1)
- b is the bias term
- xi are the support vectors from the training data
- K(xi,x) is the RBF kernel function, which is given by:

$$K(xi, x) = exp(-\gamma |xi - x|^2)$$

Eqn (2)

where γ is a parameter that controls the width of the Gaussian kernel

Support Vector Machine (SVM) was applied to classify waste types using visual features such as texture, color, and shape, automating sorting and enhancing classification accuracy. In our experiment, we employed an SVM with an RBF kernel and a random state of 42 for reproducibility. When trained on the TrashNet dataset (Figure 3), the model achieved a test accuracy of 0.61, with weighted precision = 0.62, recall = 0.61, and F1-score = 0.61. These results indicate that the SVM struggled to accurately differentiate waste categories, showing limited balance between precision and recall.

Figure 3. Confusion Matrix for SVM classifier on TrashNet Dataset

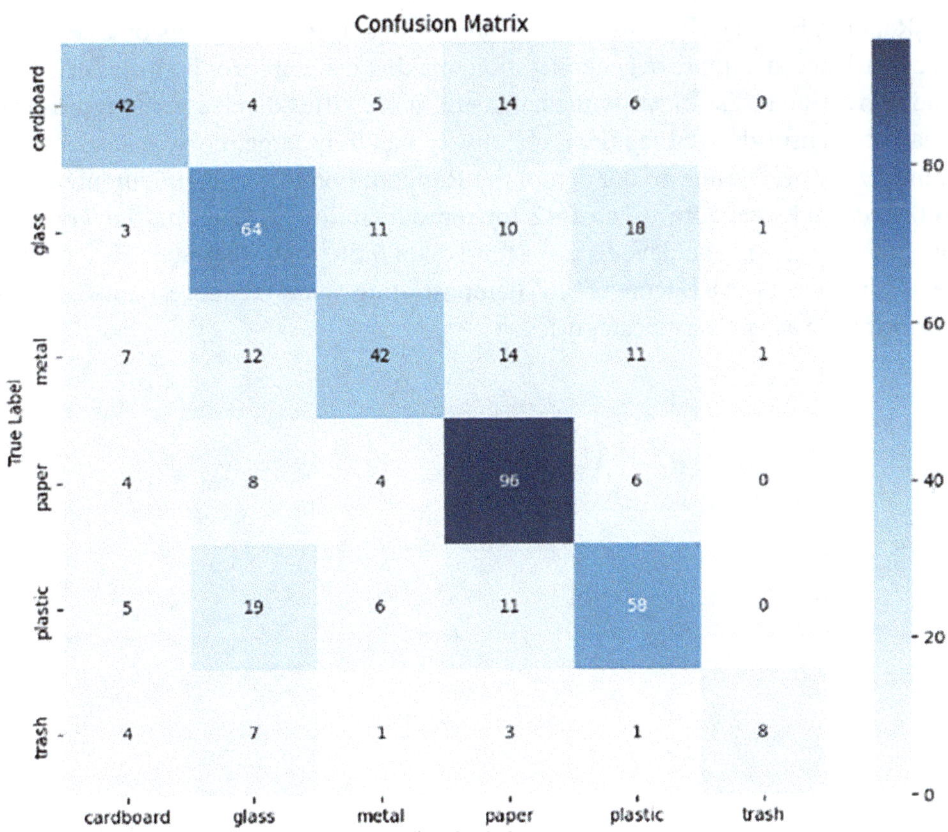

Random Forest. Random Forest is an ensemble learning method that builds multiple decision trees on random subsets of data and features, combining their outputs to improve accuracy and reduce overfitting. The Random Forest combines

the predictions from all the individual trees and chooses the most common output in case of classification:

$$RF(x) = mode\{Tree1(x), Tree2(x), .., Tree100(x)\}$$

Eqn (3)

Where:

- RF(x) is the Random Forest's prediction for input x.
- Treei(x)Treei(x) is the prediction made by the i-th tree.

Random Forest (RF) combines the predictions of multiple decision trees to reduce variance and improve generalization, making it suitable for both classification and regression tasks. In waste management, it classifies diverse waste types from image or sensor-derived features, effectively handling large, noisy datasets while minimizing overfitting. In our setup, the Random Forest was conFigureured with 100 trees and a random state of 42 for reproducibility. Trained on the TrashNet dataset (Figure 4), it achieved a test accuracy of 0.66, with weighted precision and recall of 0.66 and an F1-score of 0.65 demonstrating moderate success and balanced performance across waste categories.

Figure 4. Confusion Matrix for Random Forest on TrashNet Dataset

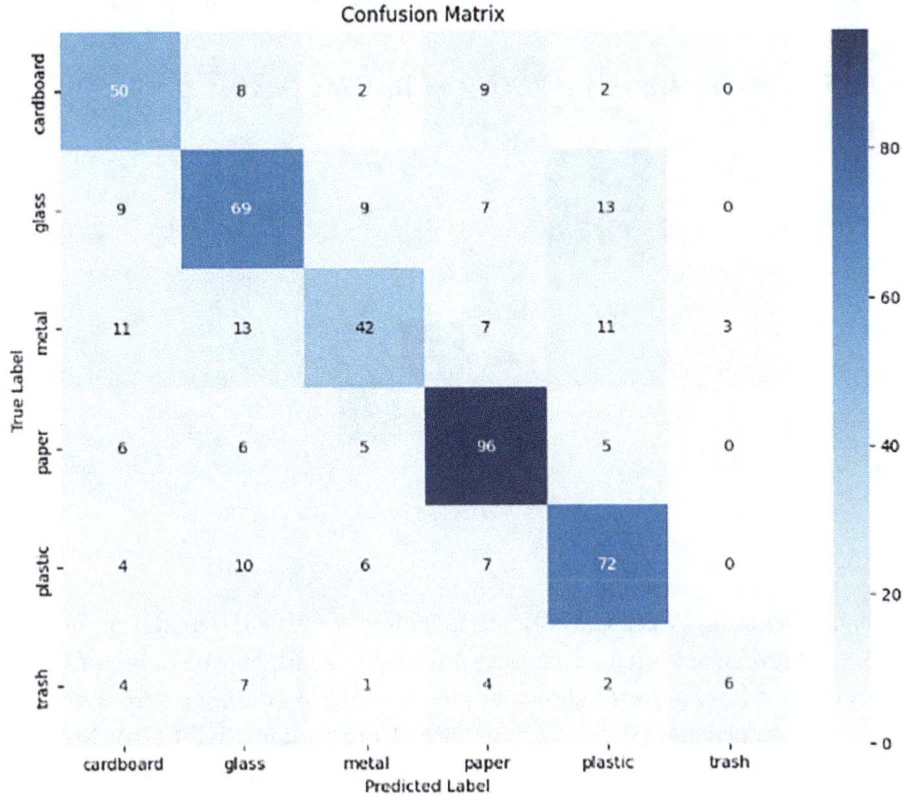

Deep Learning Methods

VGG19. VGG19, proposed by Simonyan and Zisserman (2014) from the Visual Geometry Group at Oxford, is a deep convolutional network consisting of 19 layers and known for its uniform 3×3 convolutional kernels. It remains a standard baseline for image classification and transfer learning due to its simplicity and strong ImageNet performance.

In testing VGG19 as a backbone for JONET, the model's upper layers were removed, the pretrained outputs flattened, and passed through a Dense layer with softmax for classification. Trained on the TrashNet dataset for 50 epochs, it achieved 0.89 accuracy, with weighted precision = 0.90 and recall = 0.89, showing strong overall performance. The confusion matrix (Figure 5) indicates accurate classification for categories like cardboard (57), metal (60), and paper (86). However, some confusion occurred glass misclassified as metal (11), trash as plastic/paper (5), and plastic as glass (4) suggesting room for improvement in visually similar classes.

Figure 5. Confusion Matrix for VGG19 on TrashNet Dataset

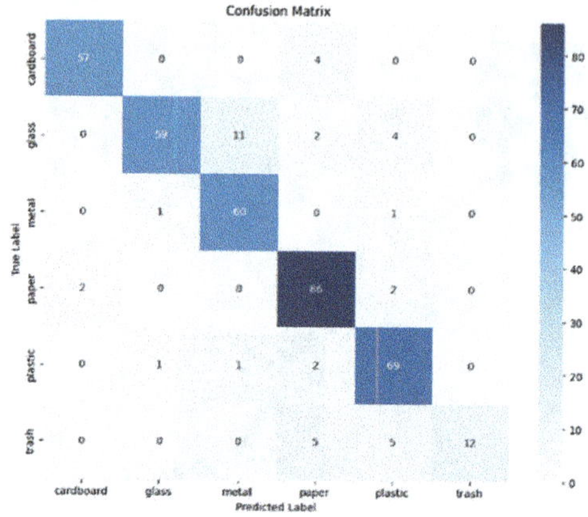

On the Drinking Waste Classification dataset, the VGG19 model achieved near-perfect performance with an accuracy, precision, recall, and F1-score of 0.99. The confusion matrix (Figure 6) shows highly accurate predictions across all classes: AluCan (154 correct), Glass (182), HDPEM (155), and PET (226). Misclassifications were minimal: only 2 AluCan samples mislabeled as Glass, 3 as PET, 1 Glass as PET, and 1 PET as AluCan demonstrating excellent category separation and overall reliability.

Figure 6. Confusion Matrix for VGG19 on Drinking Waste Classification Dataset

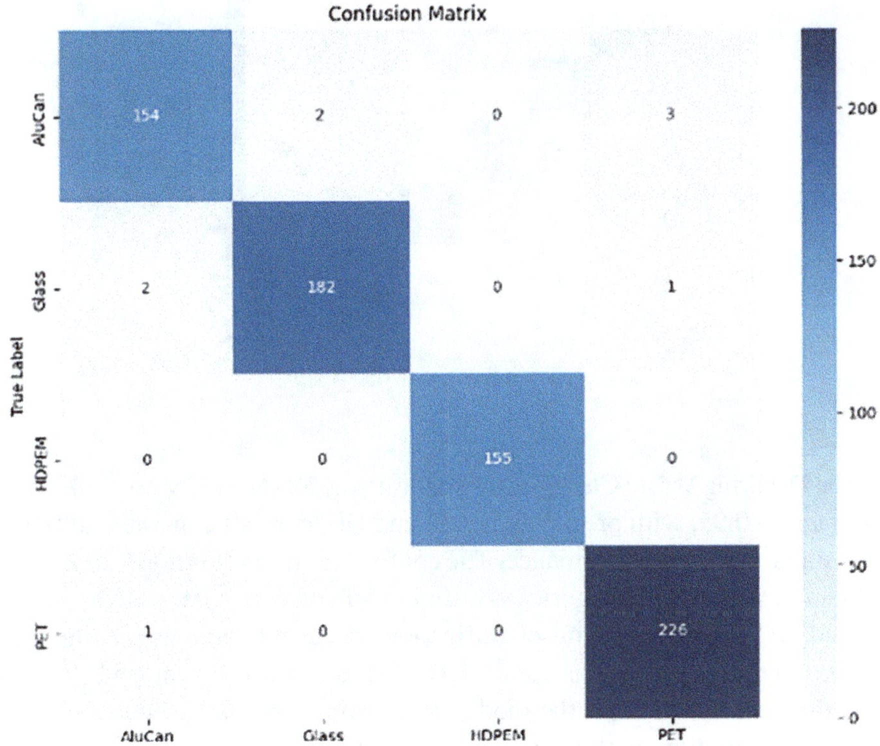

ResNet152V2. ResNet152V2 is a very–deep residual network that stabilizes optimization via identity skip connections (He et al., 2016). We treat it as a standard, ImageNet-pretrained backbone for transfer learning.

Following the same approach as with VGG19, the ResNet152V2 model's top layers were removed and replaced with custom classification layers. When fine-tuned on the TrashNet dataset for 50 epochs, it achieved an accuracy of 0.89, with weighted precision = 0.91 and recall = 0.90, indicating strong overall performance. The confusion matrix (Figure 7) shows high accuracy in categories like cardboard (60 correct), glass (66), metal (57), and paper (78). Minor misclassifications occurred primarily in the plastic and trash classes, suggesting these categories may have more overlapping visual features.

Figure 7. Confusion Matrix for ResNet152V2 on TrashNet Dataset

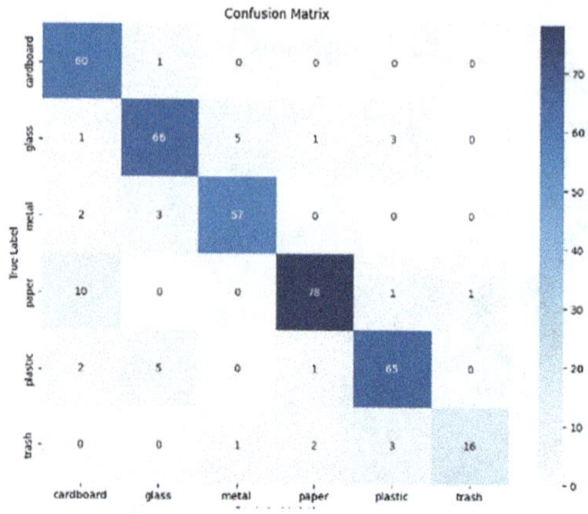

On the Drinking Waste Classification dataset, the ResNet152V2 model achieved an accuracy of 0.95, with precision, recall, and F1-score all consistent at 0.95, reflecting stable and high performance. The confusion matrix (Figure 8) shows strong classification across all categories AluCan (150 correct), Glass (182), HDPEM (149), and PET (209). Minor misclassifications occurred, including 9 AluCan and 13 PET samples labeled as Glass, and 5 HDPEM samples also confused with Glass. Despite these small overlaps, the model demonstrates robust accuracy and strong material differentiation, particularly between Glass and PET.

Figure 8. Confusion Matrix for ResNet152V2 on Drinking Waste Classification Dataset

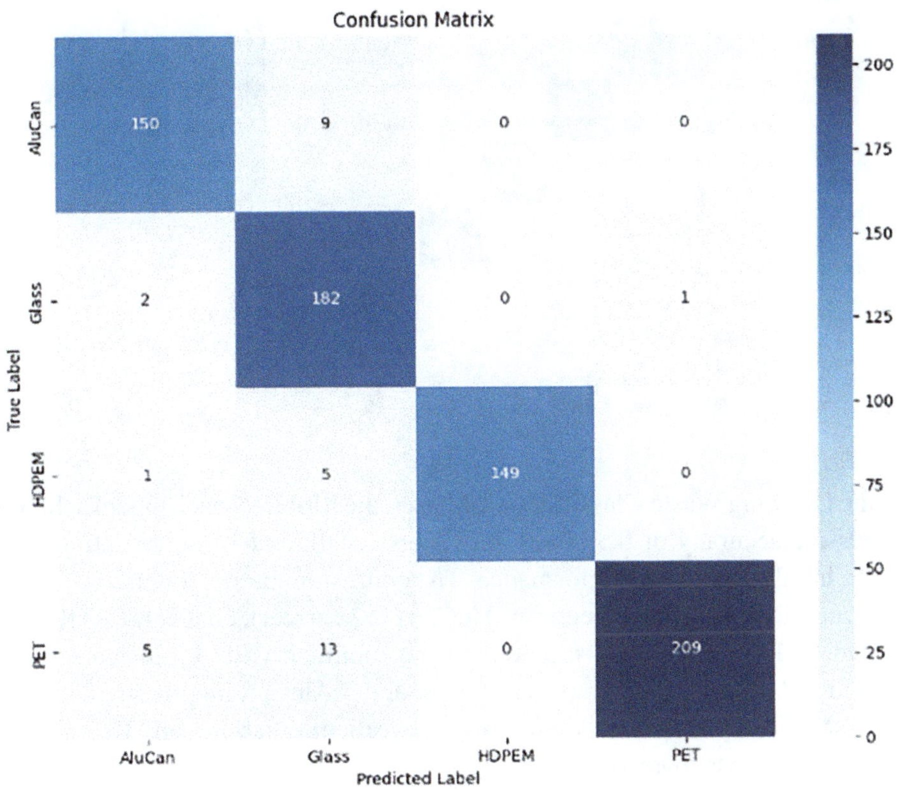

MobileNetV2. MobileNetV2, introduced by Google (Sandler et al., 2018), is an efficient convolutional architecture optimized for deployment on resource-constrained devices such as smartphones and IoT systems. It leverages depthwise-separable convolutions and inverted residual bottlenecks to drastically reduce computation and model size while maintaining competitive accuracy.

After removing the top layer of MobileNet, a simple classifier was added for fine-tuning. The model achieved an overall accuracy of 0.89, with weighted precision and recall of 0.89 and an F1-score of 0.88. The confusion matrix (Figure 9) indicates strong performance across most classes cardboard (57 correct), metal (61), and paper (86) with minimal misclassifications. Some confusion occurred between glass and other categories, where 10 glass samples were mislabeled as metal and 4 plastic samples as glass. The trash category showed slightly higher error rates, with instances misclassified as paper or plastic, suggesting overlap in visual features for these materials.

Figure 9. Confusion Matrix for MobileNetV2 on TrashNet Dataset

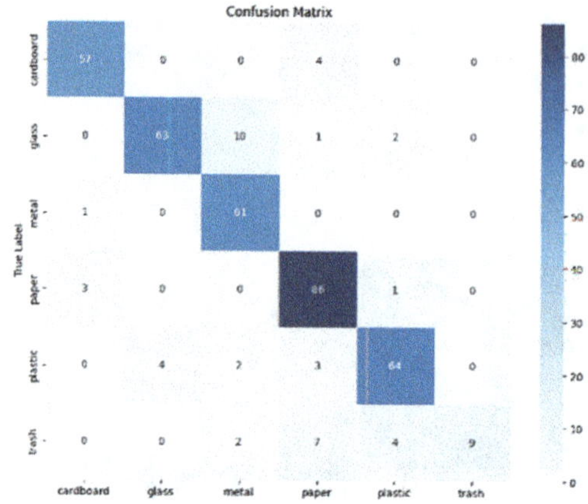

On the Drinking Waste Classification dataset, the MobileNetV2 model achieved an impressive accuracy of 0.97, with precision, recall, and F1-scores all at 0.97, reflecting highly consistent performance. The confusion matrix (Figure 10) shows strong results across all four categories AluCan (152 correct), Glass (180), HDPEM (154), and PET (219). Misclassifications were minimal, with 4 AluCan samples mislabeled as Glass, 6 PET samples as Glass, and 3 AluCan incorrectly predicted as HDPEM. Overall, the model demonstrates excellent reliability and fine-grained discrimination across material types.

Figure 10. Confusion Matrix for MobileNetV2 on Drinking Waste Classification Dataset

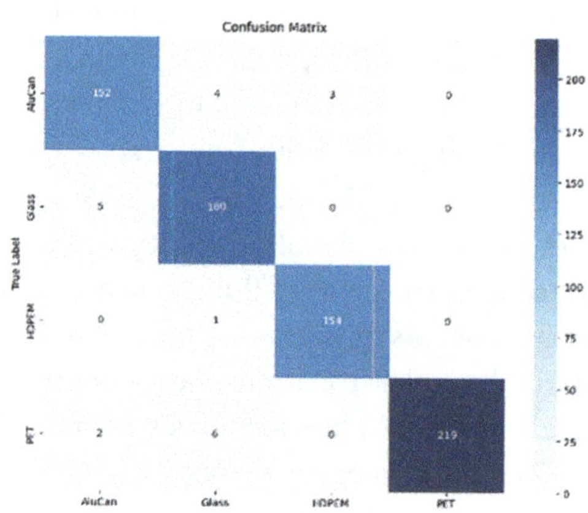

DenseNet201. DenseNet201, a 201-layer variant of the Dense Convolutional Network (Huang et al., 2017), introduces dense connectivity, where each layer receives inputs from all preceding layers. This structure encourages feature reuse, improves gradient flow, and achieves high accuracy with fewer parameters compared to similarly deep architectures.

For image classification on the TrashNet dataset, a simple classifier was added on top of DenseNet after removing its top layers. The model achieved an accuracy of 0.93, with precision, recall, and F1-score all at 0.93, indicating strong and consistent performance. The confusion matrix (Figure 11) shows excellent classification across most waste categories cardboard (57 correct), glass (64), metal (61), paper (88), and plastic (70) with only minor errors. A few trash samples were misclassified as plastic (2) or paper (1), and slight confusion occurred between glass and plastic, highlighting the model's overall robustness with minimal overlap in visually similar materials.

Figure 11. Confusion Matrix for DenseNet201 on TrashNet Dataset

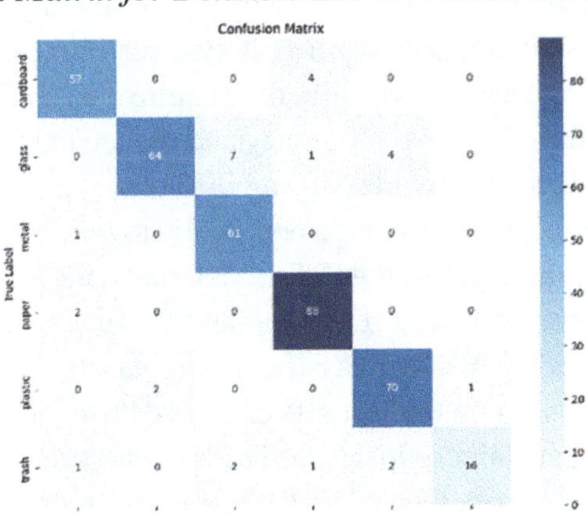

When fine-tuned on the Drinking Waste Classification dataset, the DenseNet-based model achieved outstanding results, with accuracy, precision, recall, and F1-score all reaching 0.99. The confusion matrix (Figure 12) demonstrates near-perfect classification across all four categories AluCan (154 correct), Glass (182), HDPEM (155), and PET (226) indicating exceptional precision and reliability. Misclassifications were minimal, including 2 AluCan samples mislabeled as Glass, 2 Glass samples as AluCan, and 1 Glass sample as PET. Importantly, HDPEM showed almost no confusion, with only one sample incorrectly predicted as PET, underscoring the model's fine-grained discriminatory power and overall robustness.

Figures 12. Confusion Matrix for DenseNet201 on Drinking Waste Classification Dataset

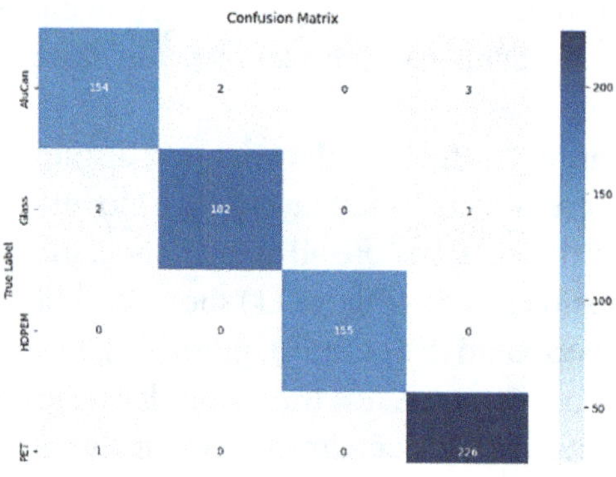

JONET. The JONET model introduced by Abu-Qdais et al.,(2023) was specifically designed for real-time waste classification, making it highly suitable for industrial applications such as conveyor belt systems in recycling plants or waste management facilities. These environments require fast and accurate sorting of various waste materials to improve efficiency and reduce manual labor. The JONET model automates this process by classifying different types of waste in real time, helping to streamline the waste management workflow.

During the development of the JONET model, multiple pre-trained convolutional neural networks were evaluated. Each model was fine-tuned individually using the TrashNet and Drinking Waste Classification datasets. For consistency, model performance was measured on the test sets of each dataset, and the accuracy scores were averaged to facilitate fair comparison. These results are summarized in Table 2.

After thorough evaluation, DenseNet201 demonstrated the highest average test accuracy across the two datasets. Its superior performance is largely attributed to its dense connectivity, which enables efficient feature reuse and enhances gradient flow, as well as its strong spatial feature extraction capabilities. DenseNet201 was trained for 50 epochs on each dataset and consistently outperformed other candidate models.

Given its accuracy, generalization capability, and architectural efficiency, DenseNet201 was selected as the backbone for the final implementation of the JONET model.

Table 2. Accuracy of VGG19, ResNet152V2, DenseNet201 and MobileNetV2 on TrashNet dataset

Model	Accuracy
VGG19	0.94
ResNet152V2	0.93
DenseNet201	0.95
MobileNetV2	0.93

Beyond the TrashNet dataset, the model was also tested on the Drinking Waste Classification datasets to evaluate its ability to generalize across varied waste categories and image distributions. Two primary metrics were used to assess model performance: accuracy and loss. Accuracy was calculated as:

$$Accuracy = \frac{Number\ of\ correct\ predictions}{Total\ number\ of\ predictions}$$

Eqn (4)

This metric indicates the proportion of waste items that were correctly classified by the model. To guide the training process, categorical cross-entropy was used as the loss function. It measures the difference between predicted and actual labels, and helps the model minimize classification errors by adjusting weights during backpropagation. The combination of these metrics ensured a balanced evaluation of both prediction correctness and learning effectiveness across all datasets. The loss function is defined as:

$$Loss = \sum_{i=1}^{C} y_i \log p_i$$

Eqn (5)

Where,
C is the number of classes
y_i is the true label
P_i is the predicted probability of class i

The architecture of JONET (Figure 14) begins with the DenseNet201 backbone (Architecture in Figure 13). DenseNet201 is a widely recognized convolutional neural network (CNN) architecture, notable for its distinctive use of dense connectivity. Unlike conventional CNNs where layers are connected sequentially, DenseNet201 establishes direct connections from each layer to all subsequent layers within a block. This design enables the network to reuse previously learned features, reducing redundancy and improving computational efficiency. Additionally, the dense connectivity enhances gradient flow during training, which helps mitigate issues like vanishing gradients in deep networks. These characteristics make DenseNet201 particularly effective for image classification tasks that require extracting complex and diverse visual features from input images.

Figure 13. DenseNet201 Architecture (Source: Akhtar et al., (2022))

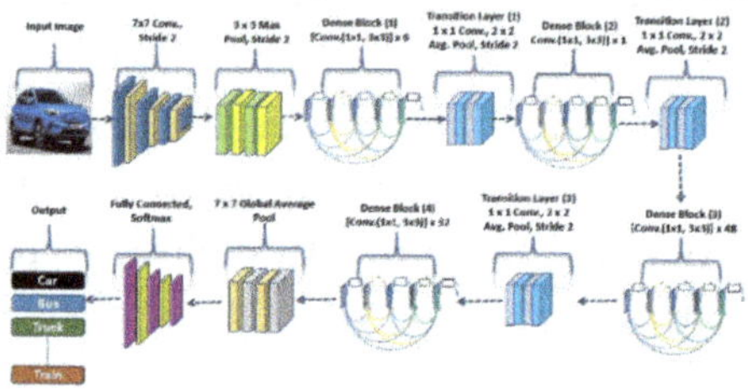

Once the input image is processed through DenseNet201, the resulting feature maps are passed into a Global Average Pooling (GAP) layer. The GAP layer plays a critical role in reducing the spatial dimensions by computing the average value of each feature map channel. This approach offers multiple benefits: it drastically lowers the number of trainable parameters thus minimizing the risk of overfitting and emphasizes the most relevant features by suppressing spatial noise. This is especially useful in waste classification tasks, where the location and orientation of waste items in images can vary significantly.

The GAP operation for each feature map channel can be mathematically defined as:

$$GAP(F) = \frac{1}{H \times W} \sum_{i=1}^{H} \sum_{j=1}^{W} F(i,j)$$

Eqn (6)

Where:

- F represents a single feature map.
- H and W are the height and width of the feature map.
- F(i,j) is the value at position (i,j) in the feature map.

This operation effectively condenses each feature map into a single representative value, preparing the output for the final classification layer.

After the Global Average Pooling (GAP) layer, the output is passed into a fully connected (FC) layer consisting of 1,024 neurons. This layer transforms the compacted feature representation into a high-level feature vector suitable for classification. The transformation performed by the FC layer can be mathematically represented as:

$$z = W \cdot x + b$$

Eqn (7)

Where:

- W is the weight matrix.
- x is the input vector from the GAP layer.
- b is the bias vector.

To introduce non-linearity, the ReLU (Rectified Linear Unit) activation function is applied to the output of the FC layer. ReLU is defined as $ReLU(z) = max(0, z)$ where z is the input to the activation function. This activation helps the model learn complex and non-linear relationships within the data, which is essential for distinguishing between waste categories that may have subtle visual similarities.

The final component of the model is the softmax output layer, which contains one neuron for each waste class in this case, six categories. The softmax function converts the final output into a probability distribution over all classes, ensuring that the predicted values sum to 1. It is defined as:

$$Soft\max(z_i) = \frac{e^{z_i}}{\sum_{j=1}^{6} e^{z_j}}$$

Eqn (8)

Where:

- is the output of the FC layer for the i-th class.
- The denominator is the sum of the exponential values of the outputs for all classes, ensuring that the probabilities sum up to 1.

This final stage allows the model to assign confidence scores to each category, making it suitable for multi-class waste classification tasks.

This approach allows the model to produce a clear prediction by selecting the class with the highest probability score, ensuring that the final output reflects the most likely category. By assigning probabilities to all possible classes, the model is better equipped to differentiate between various types of waste, even when there are visual similarities. This makes the prediction process more reliable and helps the model perform well in real-world scenarios.

Figure 14. JONET Model Architecture

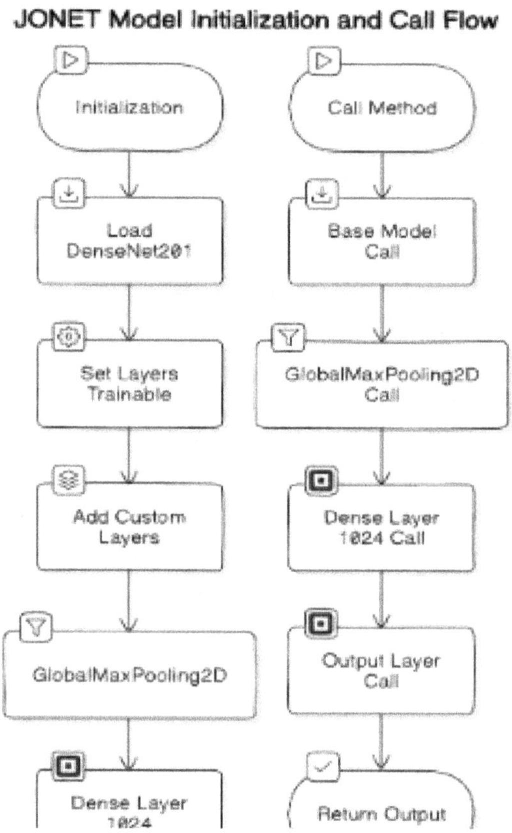

The JONET model, when tested on the TrashNet dataset, achieved an overall accuracy of 0.92, with weighted precision, recall, and F1-score all at 0.92, indicating strong and balanced performance across categories.

Class-wise, cardboard had the best results with 0.98 precision, 0.92 recall, and 0.95 F1-score. Glass followed closely at 0.97 precision, 0.89 recall, and 0.93 F1-score. Metal showed balanced metrics with 0.91 precision, 0.94 recall, and an F1-score of 0.92.

Paper achieved the highest recall (0.98), though its precision was slightly lower at 0.86, yielding an F1-score of 0.92. Plastic maintained consistent scores of 0.92 across all three metrics. The trash category proved more challenging, with 0.79

precision, 0.68 recall, and a lower F1-score of 0.73, likely due to class overlap or imbalance in the dataset.

To benchmark performance, JONET was compared with a lightweight, real-time baseline MobileNetV3-Small. Fine-tuned on the TrashNet dataset under identical settings, MobileNetV3-Small achieved 93.11% accuracy, closely matching JONET's 92% (Table 3). Despite training faster (1.76 minutes), JONET showed greater stability and generalization, with balanced Precision, Recall, and F1-Score (0.92 each). Overall, while MobileNetV3-Small offers strong real-time efficiency, JONET remains superior in robustness and consistency across diverse waste categories, making it more suitable for large-scale smart waste management.

Figure 15. Confusion Matrix for JONET on TrashNet Dataset

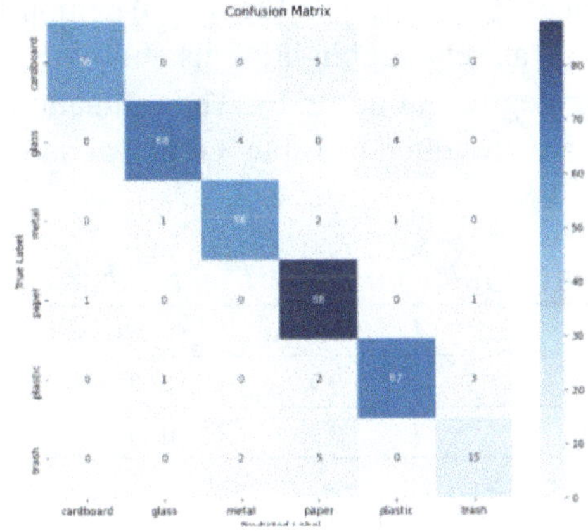

On the Drinking Waste Classification dataset, the JONET model delivered exceptional results, reaching an overall accuracy of 0.98 across all waste categories. The weighted average precision, recall, and F1-score were all 0.98, reflecting highly consistent performance across different classes. For specific waste types, AluCan achieved a precision of 0.94, recall of 1.00, and an F1-score of 0.97. Glass had a perfect precision of 1.00, with a recall of 0.95 and an F1-score of 0.97. HDPEM delivered near-perfect results with precision at 1.00, recall of 0.99, and an F1-score of 1.00. Finally, PET was classified with a precision of 0.99, recall of 1.00, and an F1-score of 0.99.

Figure 16. Confusion Matrix for JONET on Drinking Waste Classification Dataset

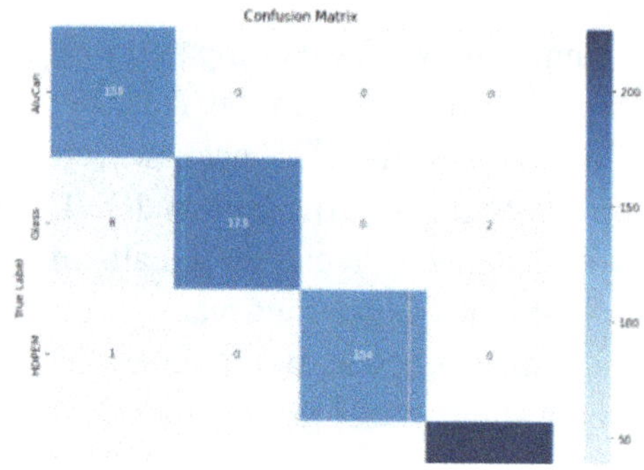

The confusion matrices (Figure 15 and Figure 16) demonstrate the model's performance across the two datasets, highlighting its ability to classify various waste categories with varying degrees of accuracy..The evaluation results of the model across the two datasets are presented in Table 3 for clear and convenient reference.

Table 3. Evaluation Metrics Table for JONET on the datasets

Dataset	Precision	Recall	F1-Score	Accuracy
TrashNet	0.92	0.92	0.92	0.92
Drinking Waste Classification	0.98	0.98	0.98	0.98

JONET's robustness and real-time performance were assessed through five independent training runs with varying random seeds (42, 77, 123, 2025) under consistent conditions 5 epochs, batch size 32, input size 224×224 on an NVIDIA Tesla P100 GPU. The model achieved a mean test accuracy of 90.40% (±1.32%) and a test loss of 0.3245 (±0.0660), reflecting stable convergence and minimal sensitivity to weight initialization. With a compact disk size of 231.98 MB, runtime memory consumption of 5659.69 MB, and an inference latency of 10.52 ms per image (equivalent to 95.03 FPS), JONET comfortably surpasses the 2 FPS threshold typical for industrial waste-sorting conveyors.

Figure 17 presents the model's calibration performance. The confidence histogram reveals that most predictions exhibit near-unity confidence, while the reliability curve shows the blue accuracy bars closely aligning with the red dashed line representing perfect calibration. The resulting Expected Calibration Error (ECE) of 0.0075 indicates excellent calibration, confirming that JONET's confidence estimates are both consistent and trustworthy. These findings collectively reinforce the model's reliability and its suitability for deployment in real-time waste classification systems.

Figure 17. Calibration Analysis of the JONET Model Using Confidence Histogram and Reliability Curve

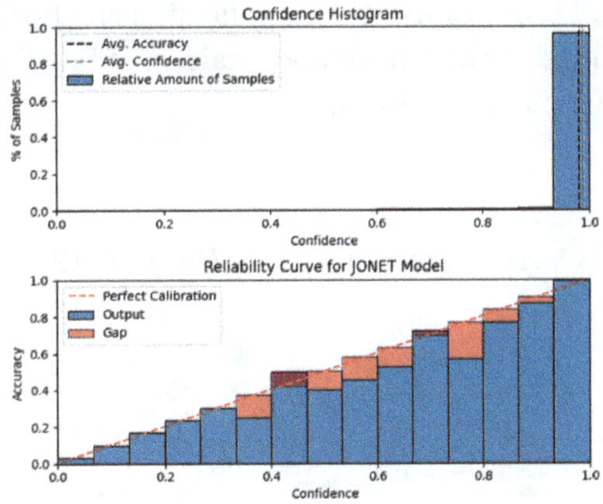

Multiple Object Detection and Segmentation

Mask R-CNN.

Mask R-CNN is a highly effective computer vision model that extends Faster R-CNN to perform both object detection and instance segmentation. It not only identifies objects in an image but also generates a precise segmentation mask for each detected object (He et al., 2017). We treat Mask R-CNN as a feasibility pilot for instance-level reasoning in waste scenes rather than a final, production-grade segmenter. The goal is to assess whether pixel-accurate masks are attainable on the TACO dataset and to surface the main bottlenecks (small objects, clutter, domain variability) that would need targeted tuning.

Mask R-CNN builds upon Faster R-CNN, with the addition of a branch for segmentation that runs parallel to the object detection branch. Its architecture (Figure 18) includes a feature extraction backbone like ResNet or ResNeXt, combined with a Feature Pyramid Network (FPN) for multi-scale feature extraction. The Region Proposal Network (RPN) identifies potential object regions, which are further refined using ROIAlign, ensuring precise alignment of regions with the feature maps. Following this two parallel branches operate:

- Bounding Box Regression and Classification: This branch predicts the location and class of each object by refining the region proposals. Each proposal is classified, and a bounding box is refined to localize the object more accurately.

- Mask Head: This is the additional branch unique to Mask R-CNN. For each region proposal, this branch outputs a binary mask that represents the object's shape at the pixel level. Unlike the classification and bounding box tasks, the mask prediction is done at a much finer scale, enabling the model to delineate the exact boundaries of the object.

Figure 18. Mask R-CNN Architecture (Source: He et al., (2021))

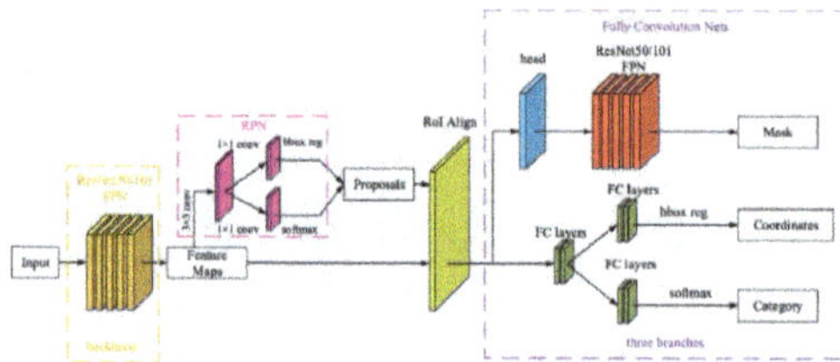

Mask R-CNN can significantly enhance waste sorting robots by enabling precise instance segmentation of waste items, generating pixel-level masks that allow robots to accurately distinguish between different materials like plastic, metal, glass, and paper. This improves object grasping and reduces sorting errors. Beyond conveyor belts, robots equipped with Mask R-CNN can be deployed in landfills to directly separate recyclables from general waste, in street cleaning vehicles to distinguish waste types during collection, and in industrial waste sorting to handle hazardous or electronic materials in specialized environments.

In our implementation, we fine-tuned a pre-trained Mask R-CNN model using Detectron2 for trash instance segmentation on the TACO dataset. The goal of this approach was to adapt the Mask R-CNN model, originally trained on the COCO dataset, to the TACO dataset, which contains 60 classes of trash-related items. Fine-tuning allowed us to leverage the features already learned by the model from COCO and adapt them to our specific task without needing to train the model from scratch. This saved significant time and computational resources while improving performance on the new dataset.

We trained the model using a learning rate of 0.00025. The training was run for 9400 iterations or 50 epochs. We also set the batch size per image to 128, which means that 128 region proposals were processed for each image during training. Once training was completed, we evaluated the model on the validation set using the COCOEvaluator class. This evaluation produced key metrics such as Average Precision (AP) and Average Recall (AR). AP measures how well the model detects objects, considering different IoU (Intersection over Union) thresholds and object sizes (small, medium, and large). AR measures how well the model finds all instances of objects in an image, also taking object sizes into account. The overall results from the evaluation are summarized as follows:

- The overall AP (IoU=0.50:0.95 for all object sizes) was 0.126, indicating moderate performance in detecting objects.
- The AP at IoU=0.50, which measures the model's performance with a more lenient threshold for overlap, was 0.165. This score is slightly better but still indicates that the model struggles with more complex or ambiguous objects.
- The AP for small objects was 0.025, much lower than the AP for large objects, which was 0.145. This reflects the difficulty of detecting small objects in cluttered scenes, a common challenge in instance segmentation tasks.
- For medium-sized objects, the AP was 0.091, falling between the small and large object performance.
- In terms of recall, the AR for all object sizes at IoU=0.50:0.95 was 0.184, indicating that the model successfully detected many objects but missed some, especially smaller or less distinct ones.
- The AR for small objects was particularly low at 0.029, while large objects achieved an AR of 0.208.

While the model performs reasonably well on large objects, it struggles with smaller items, which is often the case in instance segmentation tasks involving cluttered environments or objects with varying sizes. The model's performance for medium objects also suggests room for improvement, particularly in handling more diverse object sizes. Given below in Figure 19 and Figure 20 as examples are a couple of the predictions made by the model:

Figure 19. Predictions made by the Mask R-CNN model fine tuned on the TACO dataset

Figure 20. Few more of the predictions made by the Mask R-CNN model fine tuned on the TACO dataset

As a pilot, Mask R-CNN lets us quickly validate instance-level signal quality, but it also exposes limitations that are material for deployment:

- Small objects (e.g., caps, cigarette butts) are frequently missed due to proposal sparsity and downsampling in deep backbones;
- Clutter and occlusion in mixed-waste scenes confound proposal scoring and mask refinement;
- Throughput is insufficient for real-time conveyor speeds on edge hardware without aggressive pruning.

Based on these findings, our production path pivots to lighter one-stage instance segmentation, specifically YOLACT or YOLO-Seg, to achieve real-time (≥30 FPS) at 640–896 px with competitive AP. Concretely, we will:

- Adopt a one-stage backbone+FPN with prototype masks (YOLACT) or decoupled mask heads (YOLO-Seg)
- Apply knowledge distillation from the pilot (Mask R-CNN masks as teacher) to bootstrap finer boundaries

- Target TensorRT/ONNX export with INT8/FP16 for edge deployment (Jetson/T4)
- Prioritize small-object recall via tiling/multi-scale, copy-paste/mosaic augmentation, small-instance oversampling, and higher-res training
- Simplify the label space where semantically overlapping TACO classes hurt separability
- Set hard gates for the pivot: AP_S +5–8 points, end-to-end latency ≤30–40 ms/frame, VRAM ≤6–8 GB, without degrading large-object AP. This strategy preserves the pilot's insights while moving to a latency-efficient architecture that is better aligned with operational constraints on sorting lines.

Federated Learning

Federated Learning (FL) is a revolutionary approach in machine learning, enabling decentralized training of models across multiple data sources without centralizing the data. Unlike traditional machine learning, which relies on collecting and processing data in a centralized location, federated learning (FL) allows each data source, such as a smartphone, hospital, or bank, to train its own local model. These local models are then sent to a central server, where they're aggregated to form a global model. Importantly, the raw data never leaves the source, preserving privacy and security.

FL offers several key benefits. Privacy is the most important, especially when working with sensitive information like health or financial data. Since training happens locally, organizations can develop useful models without exposing private records. Scalability is another major advantage, as FL supports continuous learning across distributed sources. It's also cost-effective, reducing reliance on centralized data infrastructure and enabling collaboration without the need to share raw datasets.

However, FL isn't without challenges. Latency can slow down model updates, especially when syncing across many sources. Bandwidth usage can also be high, particularly when dealing with large models or numerous clients. There's also the issue of trust, since a malicious participant could send manipulated updates that affect the global model. Lastly, implementation is complex, requiring advanced algorithms to manage secure and efficient aggregation.

In practice, FL begins with each data source training a local model on its own data. These models are then sent to a server for aggregation into a shared global model. These local models do not leave the devices or servers they are trained on. Once trained, the models send their learned parameters (like weights and biases) to a central server. The central server then aggregates these parameters to update the global model. This aggregation process ensures that the global model benefits from all the data sources without directly accessing the raw data. After aggregation, the updated global model is sent back to the data sources, which continue to refine it by training on new data, thus completing the cycle. These steps are visualized in Figure 21.

There are different types of federated learning approaches. Federated Averaging, one of the most common techniques, involves averaging the model parameters across all participating devices to update the global model. In Federated Distillation, the central server distills the knowledge from the models trained on the data sources into a simpler model, enhancing efficiency without compromising accuracy. Vertical Federated Learning, also known as Feature-based FL, occurs when different organizations with distinct features about the same set of individuals collaboratively train a model. In contrast, Horizontal Federated Learning occurs when the same feature set exists across different organizations or users, but the individuals (data points) differ.

Federated learning has various applications across multiple industries. In healthcare, it allows hospitals to collaboratively train models on sensitive patient data without breaching privacy, enabling tasks like disease prediction or medical imaging analysis without sharing actual records. In finance, FL can help detect fraud by training models across multiple banks while keeping customer data confidential. It also proves useful in consumer analytics, where companies can build models directly on user devices like smartphones, avoiding the need to collect personal data centrally.

In the field of waste management, FL can significantly improve automation in sorting and recycling. Waste facilities in different locations can train local models on region-specific waste data, improving the accuracy of classification systems based on local waste patterns. Instead of sharing raw data, facilities can upload trained model parameters to a central server, where they are aggregated to form a global model. This global model can then be redistributed to all facilities, allowing them to benefit from broader learning while still respecting data privacy. With this approach, sensitive data such as waste composition, processing metrics, or private business operations remains local. At the same time, FL ensures that regional practices stay optimized and contribute to global advances in waste sorting technologies.

Figures 21. Federated Learning Flowchart

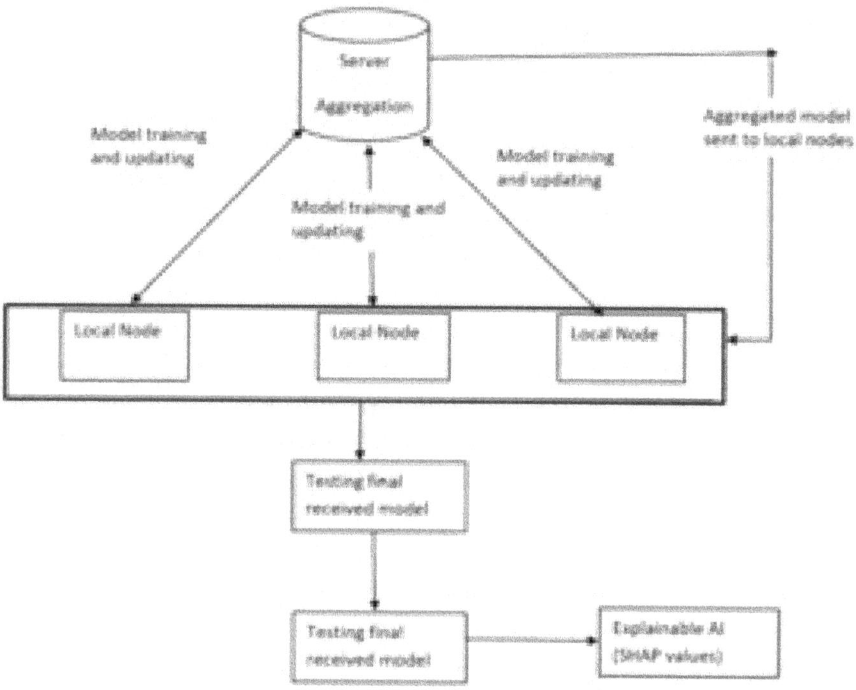

In our waste management system, the federated learning (FL) setup includes three main types of clients: Material Recovery Facilities (MRFs), smart bins, and garbage trucks.

- **Material Recovery Facilities (MRFs):** These are large recycling centers equipped with sensors and cameras that collect detailed waste data from sorting lines. Each MRF trains its own local model to classify waste into categories like glass, plastic, or metal. Because different facilities handle different types of waste industrial, residential, or mixed their datasets naturally vary.
- **Smart Bins:** Smart bins are IoT-enabled trash or recycling bins found in public spaces or buildings. They have cameras or sensors that identify whether an item is recyclable. Each bin has limited data and processing power, but together, they form a vast, diverse network of distributed data sources. These bins occasionally connect to the FL server to share their locally trained model updates when connectivity allows.

- **Garbage Trucks:** Trucks equipped with cameras and sensors collect data as they pick up waste from different neighborhoods. Their data reflects local waste patterns, and since they have moderate computing resources, they can train models while on route and upload updates at the end of the day.

Together, these clients form a decentralized learning network, each contributing to a shared global model without exchanging raw data. This setup is especially valuable since the clients are operated by different organizations, city departments, waste companies, or private contractors where data privacy and ownership are major concerns.

Client data is non-IID because each site observes unique waste patterns like a café bin dominated by paper cups versus an MRF handling metal scraps. Such imbalance risks the global model overfitting to dominant sources, so weighted updates or specialized aggregation methods are needed for fairness.

There's also class imbalance, as rare waste types (e.g., hazardous materials) appear less often than common ones like plastics. Some clients may exclusively handle these rare categories, making their data vital for model diversity. To mitigate this, techniques such as local data augmentation and class re-weighting ensure that under-represented waste classes remain adequately learned and the model stays balanced.

Training across MRFs, smart bins, and garbage trucks follows a multi-round federated learning setup. In each round, selected clients train locally and send updates to the server. While full participation speeds convergence, sampling roughly half the clients balances bandwidth and reliability especially since smart bins may connect intermittently. The server aggregates all received updates before a fixed deadline to maintain steady progress.

Each client trains for a few local epochs (E). Smaller values (E=1) keep updates stable, while larger ones (E=5) cut communication but may cause drift with non-IID data. Early rounds use fewer epochs for stability, increasing later as the model matures. Clients rely on Adam or SGD with a learning rate around 0.001, ensuring uniformity. Adaptive optimizers aid local convergence, and learning rate decay refines global performance over time.

We use FedAvg as the primary aggregation method, averaging client updates weighted by dataset size. If non-IID data causes instability, we'll switch to FedProx, which adds a proximity term to keep local models aligned with the global one. Communication remains a major bottleneck in FL especially with large models like DenseNet201, which contain millions of parameters. Exchanging full-precision updates from multiple clients across rounds can rapidly consume bandwidth, underscoring the need for efficient compression and update strategies.

To cut communication overhead, we'll adopt several efficiency techniques. Quantization reduces data size by sending lower-precision updates (e.g., 8-bit vs.

32-bit), cutting bandwidth up to 4× with minimal accuracy loss. Sparsification transmits only the most important gradients, further reducing volume especially when paired with quantization. Increasing local epochs (E) can also limit update frequency, though must be balanced to prevent slow convergence on non-IID data.

We'll begin with full-precision updates for baseline performance, then progressively add compression (e.g., gzip) once the model stabilizes. With tens to hundreds of clients, communication remains manageable. Heavier nodes like MRFs can update more often, while lighter ones (e.g., smart bins) sync less frequently preserving efficiency without degrading learning quality.

Our FL system prioritizes security and privacy. Secure aggregation ensures the server sees only the combined client updates, not individual ones. Clients mask updates so noise cancels out, adding slight overhead but strong confidentiality. Optional differential privacy clips and noises gradients to protect data points, with minor accuracy trade-offs. All clients are authenticated via keys or certificates to block unauthorized access.

To counter poisoning, the server uses robust aggregation (median, trimmed mean), flags anomalies, and monitors the global model for suspicious behavior with rollback capability. Local data and logs stay encrypted. Only aggregated updates leave devices ensuring verified participation, privacy, and resilience against malicious interference.

Real World Application

The integration of advanced AI models such as the DenseNet201-based JONET classifier, Mask R-CNN for instance segmentation, and federated learning frameworks shows strong potential for transforming waste management operations. For example, the JONET model achieved approximately 96% classification accuracy on benchmark datasets, demonstrating its reliability in real-world recycling applications. Vision-based systems leveraging such models can process up to 160 items per minute roughly 4–5 times the manual throughput of 30–40 items resulting in improved operational efficiency, consistency, and reduced labor costs (Fotovvatikhah et al., 2025).

At municipal Material Recovery Facilities (MRFs), integrating JONET with Mask R-CNN enables simultaneous classification and segmentation of complex waste streams. This setup powers automated sorting systems using robotic arms or pneumatic diverters, achieving up to 99% purity in sorted outputs (Recycleye, 2024). The combination enhances both throughput and precision, aligning with smart city sustainability goals and reducing landfill dependency. A standard pipeline involves camera and sensor capture, followed by classification and segmentation.

Items with high confidence scores (≥90%) are directly routed to their respective waste categories through automated sorting mechanisms. This enables fast, reliable classification suitable for real-time deployment. Conversely, items with low confidence are flagged as ambiguous and rerouted to a reject bin, a separate container designed to collect uncertain or unclassifiable waste to prevent contamination in sorted streams. These items are then either subjected to manual sorting, where human operators inspect and classify them based on visual or contextual cues, or discarded depending on operational policies. This hybrid approach, combining automated decisions with fallback strategies, ensures robustness in uncertain scenarios, minimizes the risk of misclassification, and supports scalable deployment in industrial material recovery facilities.

To handle real-world challenges like occlusion, contamination, or composite materials, Mask R-CNN provides instance-level segmentation and adaptive confidence thresholds as fallback controls. For example, low-confidence or multi-label cases (e.g., a plastic bottle with a metal cap) trigger review or redirection to a mixed-material stream. In more uniform industrial contexts, such as e-waste or construction debris, Mask R-CNN achieves F1-scores around 71% (Ahmad et al., 2025), confirming its reliability for domain-specific waste analysis and automation.

Lightweight models like MobileNet (~80% accuracy) are well suited for IoT-enabled smart bins, thanks to their low compute and energy footprint. Alternatively, images can be offloaded to edge servers for heavier processing, balancing performance and latency. Federated learning (FL) enhances this ecosystem by enabling local model training on decentralized data, sharing only updates not raw data with the central server. This ensures privacy-preserving, continuous learning across bins, trucks, and MRFs, fostering adaptive intelligence across the waste network.

Crucially, the pipeline aligns with India's Solid Waste Management Rules (2016), which mandate segregation into wet, dry, and hazardous categories. The AI framework can automatically map predictions to these streams, ensuring compliance with CPCB and MoEFCC regulations. Overall, the system delivers efficient, accurate, and regulation-compliant waste handling offering a scalable, modern foundation for sustainable smart city operations.

Sustainability and Ethics

Deep learning offers substantial advancements in automating waste sorting, yet these benefits are not without environmental cost. For instance, a DenseNet201-class architecture comprises approximately 20 million parameters and requires around 8.7 billion operations per image (Huang et al., 2017). Training such deep neural networks on GPU clusters is energy-intensive; one estimate suggests that a single large-scale model training can produce up to 626,000 pounds of CO_2 comparable to the lifetime emissions of five automobiles (Martineau, 2020). While models for waste classification are smaller than frontier language models, the cumulative impact of frequent inference on embedded devices and inefficient training protocols remains significant. In this context, energy efficiency is not a mere optimization, it's a fundamental requirement.

Federated Learning (FL) alters the sustainability calculus by shifting model training to the edge on devices like smart bins, trucks, and material recovery facilities (MRFs). This reduces centralized storage and large-scale data transfers, but introduces new overheads: increased local computation and high-volume transmission of model updates. Depending on system conFigureuration, FL can emit up to 100 times more carbon than centralized alternatives, although under the right conditions efficient hardware, lower participation rates, and fewer communication rounds it can approach parity (Qiu et al., 2023). Mitigation strategies include model quantization (e.g., 8-bit precision) to reduce update size and energy consumption approximately fourfold with negligible accuracy trade-off (Isenkul, 2025); pruning of redundant weights and operations to streamline inference (Mei & Stamp, 2025); and knowledge distillation to compress larger "teacher" models like DenseNet into smaller, faster "student" models such as MobileNet or EfficientNet.

From an ethical perspective, automation in waste management has dual consequences. On one hand, it enhances worker safety by removing humans from high-risk environments where injury rates are estimated at 17 times the national average in the UK (Bontorin, 2025). On the other hand, increased reliance on intelligent systems may displace roles traditionally filled by lower-wage labor. The ethical imperative is clear: prioritize augmentation before automation. This involves engaging workers early in the transition, retraining them for supervisory, maintenance, or quality assurance roles, and funding reskilling programs that enable meaningful career continuity. Such efforts ensure that the adoption of federated AI systems is both environmentally sustainable and socially equitable.

CONCLUSION

This chapter comprehensively examined how Artificial Intelligence (AI) and Machine Learning (ML) are transforming the landscape of solid waste management by introducing automation, precision, and scalability into traditionally manual and error-prone processes. Through a comparative evaluation of classical and contemporary deep learning models including DenseNet201, VGG19, ResNet152V2, and MobileNetV2 the study identified DenseNet201 as the most robust and high-performing architecture. Its ability to preserve feature propagation and reduce vanishing gradients made it an ideal backbone for the JONET model, enabling accurate and efficient real-time waste classification across diverse waste categories.

Further enhancements were achieved through the integration of Mask R-CNN, a state-of-the-art instance segmentation framework. This model elevated waste sorting capabilities beyond mere object detection by introducing pixel-level precision, allowing for the separation and classification of overlapping or occluded waste items in complex environments. Such granularity is essential in real-world scenarios, where multiple waste types are often mixed and partially visible.

In addition, the application of Federated Learning (FL) introduced a groundbreaking paradigm in model training, one that emphasizes privacy, scalability, and edge intelligence. By enabling decentralized training across geographically distributed clients or devices, FL eliminates the need for centralized data collection, thereby preserving sensitive environmental and user-specific data. This approach is especially critical in smart cities, where data ownership, privacy regulations, and bandwidth constraints must be respected without compromising model performance.

Together, these technologies form a cohesive, AI-driven waste management ecosystem capable of adapting to the dynamic needs of urban infrastructure. They minimize reliance on manual labor, reduce occupational health hazards associated with waste handling, and significantly enhance recycling rates and material recovery. By embedding intelligence directly into waste management workflows from identification and sorting to classification and reporting these systems offer not just incremental improvements, but transformative leaps in operational efficiency and environmental sustainability.

As global waste volumes surge due to urban expansion and consumerism, the fusion of AI, deep learning, and federated frameworks emerges as not merely a technological upgrade, but a necessity. It enables the deployment of autonomous, scalable, and sustainable waste management solutions tailored to the complexities of modern cities. In doing so, it paves the way for a cleaner, smarter, and more resilient future where technology and environmental stewardship go hand in hand.

REFERENCES

Abdallah, M., Talib, M. A., Feroz, S., Nasir, Q., Abdalla, H., & Mahfood, B. (2020). Artificial intelligence applications in solid waste management: A systematic research review. *Waste Management (New York, N.Y.), 109*, 231–246. DOI: 10.1016/j.wasman.2020.04.057 PMID: 32428727

Abeygunawardhana, A. G. D. T., Shalinda, R. M. M. M., Bandara, W. H. M. D., Anesta, W. D. S., Kasthurirathna, D., & Abeysiri, L. (2020, December). AI-driven smart bin for waste management. In *2020 2nd International Conference on Advancements in Computing (ICAC)* (Vol. 1, pp. 482-487). IEEE. DOI: 10.1109/ICAC51239.2020.9357151

. Abu-Qdais, H., Shatnawi, N., & Esra'a, A. A. (2023). Intelligent solid waste classification system using combination of image processing and machine learning models.

Ahmad, G., Aleem, F. M., Alyas, T., Abbas, Q., Nawaz, W., Ghazal, T. M., Aziz, A., Aleem, S., Tabassum, N., & Ibrahim, A. M. (2025). Intelligent waste sorting for urban sustainability using deep learning. *Scientific Reports, 15*(1), 27078. DOI: 10.1038/s41598-025-08461-w PMID: 40715219

Ahmed, A. A. A., & Asadullah, A. (2020). Artificial intelligence and machine learning in waste management and recycling. *Engineering International, 8*(1), 43–52. DOI: 10.18034/ei.v8i1.498

Aishwarya, A., Wadhwa, P., Owais, O., & Vashisht, V. (2021, January). A waste management technique to detect and separate non-biodegradable waste using machine learning and YOLO algorithm. In *2021 11th International Conference on Cloud Computing, Data Science & Engineering (Confluence)* (pp. 443-447). IEEE. DOI: 10.1109/Confluence51648.2021.9377163

Akhtar, M. J., Mahum, R., Butt, F. S., Amin, R., El-Sherbeeny, A. M., Lee, S. M., & Shaikh, S. (2022). A robust framework for object detection in a traffic surveillance system. *Electronics (Basel), 11*(21), 3425. DOI: 10.3390/electronics11213425

Alawi, A. E. B., Saeed, A. Y., Almashhor, F., Al-Shathely, R., & Hassan, A. N. (2021, July). Solid waste classification using deep learning techniques. In *2021 International Congress of Advanced Technology and Engineering (ICOTEN)* (pp. 1-5). IEEE.

Alonso, S. L. N., Forradellas, R. F. R., Morell, O. P., & Jorge-Vazquez, J. (2021). Digitalization, circular economy and environmental sustainability: The application of artificial intelligence in the efficient self-management of waste. *Sustainability, 13*(4), 1–19.

Alsubaei, F. S., Al-Wesabi, F. N., & Hilal, A. M. (2022). Deep learning-based small object detection and classification model for garbage waste management in smart cities and iot environment. *Applied Sciences (Basel, Switzerland), 12*(5), 2281. DOI: 10.3390/app12052281

Amasuomo, E., & Baird, J. (2016). The concept of waste and waste management. *Journal of Management and Sustainability, 6*(4), 88. DOI: 10.5539/jms.v6n4p88

Bhattacharjee, P., More, A., Kathalkar, S., & Student, M. E. (2022). Effective Management of Solid waste by using GIS and Artificial Intelligence Based Support System. *International Journal of Mechanical Engineering, 7*, 974–5823.

Bontorin, S. (2025, May 26). *Humanoid robots learn waste sorting from real workers using VR headsets.* Recycling Product News. https://www.recyclingproductnews.com/article/43317/humanoid-robots-learn-waste-sorting-from-real-workers-using-vr-headsets

Castro, R. C. C., Al Justin, S. G., Conato, C., Cruz, J. D., Alvaran, J. R., & Oblanca, V. J. (2020, December). Development of waste management system using the concept of 'Basura advantage points' through artificial neural network. In *2020 IEEE 12th International Conference on Humanoid, Nanotechnology, Information Technology, Communication and Control, Environment, and Management (HNICEM)* (pp. 1-5). IEEE.

Coskuner, G., Jassim, M. S., Zontul, M., & Karateke, S. (2021). Application of artificial intelligence neural network modeling to predict the generation of domestic, commercial and construction wastes. *Waste Management & Research, 39*(3), 499–507. DOI: 10.1177/0734242X20935181 PMID: 32586206

. Epifantsev, K., & Kozlova, O. (2021). Research and development of artificial intelligence system for waste identification.

Fang, B., Yu, J., Chen, Z., Osman, A. I., Farghali, M., Ihara, I., Hamza, E. H., Rooney, D. W., & Yap, P. S. (2023). Artificial intelligence for waste management in smart cities: A review. *Environmental Chemistry Letters, 21*(4), 1959–1989. DOI: 10.1007/s10311-023-01604-3 PMID: 37362015

Fotovvatikhah, F., Ahmedy, I., Noor, R. M., & Munir, M. U. (2025). A Systematic Review of AI-Based Techniques for Automated Waste Classification. *Sensors (Basel), 25*(10), 3181. DOI: 10.3390/s25103181 PMID: 40431972

Geetha, S., Saha, J., Dasgupta, I., Bera, R., Lawal, I. A., & Kadry, S. (2022). Design of Waste Management System Using Ensemble Neural Networks. *Designs, 2022*(6), 27. DOI: 10.3390/designs6020027

Gómez, D. O., Bermúdez, M. T., & Morales, W. H. (2021). *Solid Domestic Waste classification using Image Processing and Machine Learning* (Doctoral dissertation, Universidad EAFIT).

He, D., Li, K., Chen, Y., Miao, J., Li, X., Shan, S., & Ren, R. (2021). Obstacle detection in dangerous railway track areas by a convolutional neural network. *Measurement Science & Technology, 32*(10), 105401. DOI: 10.1088/1361-6501/abfdde

He, K., Gkioxari, G., Dollár, P., & Girshick, R. (2017). Mask r-cnn. In *Proceedings of the IEEE international conference on computer vision* (pp. 2961-2969).

He, K., Zhang, X., Ren, S., & Sun, J. (2016). Deep residual learning for image recognition. In *Proceedings of the IEEE conference on computer vision and pattern recognition* (pp. 770-778).

Huang, G., Liu, Z., Van Der Maaten, L., & Weinberger, K. Q. (2017). Densely connected convolutional networks. In *Proceedings of the IEEE conference on computer vision and pattern recognition* (pp. 4700-4708).

Hussain, A., Draz, U., Ali, T., Tariq, S., Irfan, M., Glowacz, A., Antonino Daviu, J. A., Yasin, S., & Rahman, S. (2020). Waste management and prediction of air pollutants using IoT and machine learning approach. *Energies, 13*(15), 3930. DOI: 10.3390/en13153930

Ihsanullah, I., Alam, G., Jamal, A., & Shaik, F. (2022). Recent advances in applications of artificial intelligence in solid waste management: A review. *Chemosphere, 309*, 136631. DOI: 10.1016/j.chemosphere.2022.136631 PMID: 36183887

Isenkul, M. E. (2025). Energy-aware deep learning for real-time video analysis through pruning, quantization, and hardware optimization. *Journal of Real-Time Image Processing, 22*(3), 125. DOI: 10.1007/s11554-025-01703-0

Jude, A. B., Singh, D., Islam, S., Jameel, M., Srivastava, S., Prabha, B., & Kshirsagar, P. R. (2021). An artificial intelligence based predictive approach for smart waste management. *Wireless Personal Communications*, 123456789.

. Kumar, S., Yadav, D., Gupta, H., Verma, O. P., Ansari, I. A., & Ahn, C. W. (2020). *A Novel YOLOv3 Algorithm-Based Deep Learning Approach for Waste Segregation: Towards Smart Waste Management. Electronics 2021, 10, 14.*

Majchrowska, S., Mikołajczyk, A., Ferlin, M., Klawikowska, Z., Plantykow, M. A., Kwasigroch, A., & Majek, K. (2022). Deep learning-based waste detection in natural and urban environments. *Waste Management (New York, N.Y.), 138*, 274–284. DOI: 10.1016/j.wasman.2021.12.001 PMID: 34920243

Martineau, K. (2020, August 7). *Shrinking deep learning's carbon footprint.* MIT News. https://news.mit.edu/2020/shrinking-deep-learning-carbon-footprint-0807

Mei, L., & Stamp, M. (2025). Energy Considerations for Large Pretrained Neural Networks. *arXiv preprint arXiv:2506.01311.*

Na, S., Heo, S., Han, S., Shin, Y., & Lee, M. (2022). Development of an artificial intelligence model to recognise construction waste by applying image data augmentation and transfer learning. *Buildings (Basel, Switzerland), 12*(2), 175. DOI: 10.3390/buildings12020175

Nguyen, X. C., Nguyen, T. T. H., La, D. D., Kumar, G., Rene, E. R., Nguyen, D. D., Chang, S. W., Chung, W. J., Nguyen, X. H., & Nguyen, V. K. (2021). Development of machine learning-based models to forecast solid waste generation in residential areas: A case study from Vietnam. *Resources, Conservation and Recycling, 167,* 105381. DOI: 10.1016/j.resconrec.2020.105381

Olawade, D. B., Fapohunda, O., Wada, O. Z., Usman, S. O., Ige, A. O., Ajisafe, O., & Oladapo, B. I. (2024). *Smart waste management: A paradigm shift enabled by artificial intelligence.* Waste Management Bulletin.

Pillai, K. S., Sneha, M. L., Aiswarya, S., Anand, A. B., & Prasad, G. (2023). Municipal Solid Waste Management: A Review of Machine Learning Applications. In *E3S Web of Conferences* (Vol. 455, p. 02018). EDP Sciences. DOI: 10.1051/e3sconf/202345502018

Proença, P. F., & Simões, P. (2020). Taco: Trash annotations in context for litter detection. *arXiv preprint arXiv:2003.06975.*

Qiu, X., Parcollet, T., Fernandez-Marques, J., Gusmao, P. P., Gao, Y., Beutel, D. J., & Lane, N. D. (2023). A first look into the carbon footprint of federated learning. *Journal of Machine Learning Research, 24*(129), 1–23.

Rajalakshmi, J., Sumangali, K., Jayanthi, J., & Muthulakshmi, K. (2023). Artificial intelligence with earthworm optimization assisted waste management system for smart cities. *Global NEST Journal, 25,* 190–197.

Recycleye. (2024, December) *How AI robots help reduce the cost of waste sorting in MRFs.* https://recycleye.com/how-ai-robots-reduce-cost-waste-sorting/

Rubab, S., Khan, M. M., Uddin, F., Abbas Bangash, Y., & Taqvi, S. A. A. (2022). A study on ai-based waste management strategies for the covid-19 pandemic. *ChemBioEng Reviews, 9*(2), 212–226. DOI: 10.1002/cben.202100044

Sami, K. N., Amin, Z. M. A., & Hassan, R. (2020). Waste management using machine learning and deep learning algorithms. *International Journal on Perceptive and Cognitive Computing*, *6*(2), 97–106. DOI: 10.31436/ijpcc.v6i2.165

Sandler, M., Howard, A., Zhu, M., Zhmoginov, A., & Chen, L. C. (2018). Mobilenetv2: Inverted residuals and linear bottlenecks. In *Proceedings of the IEEE conference on computer vision and pattern recognition* (pp. 4510-4520).

Shahab, S., & Anjum, M. (2022). Solid waste management scenario in india and illegal dump detection using deep learning: An AI approach towards the sustainable waste management. *Sustainability (Basel)*, *14*(23), 15896. DOI: 10.3390/su142315896

Sharma, N., Litoriya, R., & Sharma, A. (2021). Application and analysis of K-means algorithms on a decision support framework for municipal solid waste management. In *Advanced Machine Learning Technologies and Applications* [Springer Singapore.]. *Proceedings of AMLTA, 2020*, 267–276.

Shreyas Madhav, A. V., Rajaraman, R., Harini, S., & Kiliroor, C. C. (2022). Application of artificial intelligence to enhance collection of E-waste: A potential solution for household WEEE collection and segregation in India. *Waste Management & Research*, *40*(7), 1047–1053. DOI: 10.1177/0734242X211052846 PMID: 34726090

Simonyan, K., & Zisserman, A. (2014). Very deep convolutional networks for large-scale image recognition. *arXiv preprint arXiv:1409.1556.*

Wang, C., Qin, J., Qu, C., Ran, X., Liu, C., & Chen, B. (2021). A smart municipal waste management system based on deep-learning and Internet of Things. *Waste Management (New York, N.Y.)*, *135*, 20–29. DOI: 10.1016/j.wasman.2021.08.028 PMID: 34461487

Wu, T. W., Zhang, H., Peng, W., Lü, F., & He, P. J. (2023). Applications of convolutional neural networks for intelligent waste identification and recycling: A review. *Resources, Conservation and Recycling*, *190*, 106813. DOI: 10.1016/j.resconrec.2022.106813

Xu, A., Chang, H., Xu, Y., Li, R., Li, X., & Zhao, Y. (2021). Applying artificial neural networks (ANNs) to solve solid waste-related issues: A critical review. *Waste Management (New York, N.Y.)*, *124*, 385–402. DOI: 10.1016/j.wasman.2021.02.029 PMID: 33662770

. Yang, M., & Thung, G. (2016). Classification of trash for recyclability status. *CS229 project report, 2016*(1), 3.

Yarbrough, J. (2024). *Revolutionizing Waste Management: AI-Powered Real-time Characterization for Efficient Handling of Non-Recyclable Municipal Solid Waste* (No. NREL/PR-2700-88053). National Renewable Energy Laboratory (NREL), Golden, CO (United States).

KEY TERMS AND DEFINITIONS

Precision: Proportion of true positives among all predicted positives.

Accuracy: Proportion of correct predictions out of all predictions.

Recall: Proportion of actual positives correctly identified.

F1 Score: Harmonic mean of precision and recall.

Waste Management: Process of collecting, transporting, recycling, and disposing of waste.

Image Segmentation: Dividing an image into regions or objects based on features like color or texture.

Federated Learning: Training a shared model across devices without sharing raw data; updates are aggregated centrally.

Average Precision (AP): AP evaluates how accurately a model detects objects by computing the area under the precision-recall curve. It is calculated across multiple IoU (Intersection over Union) thresholds (typically from 0.50 to 0.95 in 0.05 increments) and reflects the model's ability to both correctly identify objects and localize them precisely.

Average Recall (AR): AR measures the model's ability to detect all ground truth objects by averaging recall scores across IoU thresholds. It indicates how comprehensively the model captures relevant instances.

Intersection over Union (IoU): A standard evaluation metric used to measure the accuracy of an object detector or segmentation model. It quantifies the overlap between the predicted bounding box (or mask) and the ground truth box.

Chapter 10
Clean Technology's Impact on Green and Renewable Energy Financial Literacy:
An Analysis of Market Dynamics and Consumer Behavior

José G. Vargas-Hernandez
https://orcid.org/0000-0003-0938-4197
Financial Literacy Research Center, Universidad Cristóbal Colón, Mexico & Tecnológico nacional de México, ITS Fresnillo, Mexico

Selene Castañeda-Burciaga
https://orcid.org/0000-0002-2436-308X
Universidad Politécnica de Zacatecas, Mexico

Omar A. Guirrete-Barboza
https://orcid.org/0000-0003-1336-9475
Universidad Politécnica de Zacatecas, Mexico & Universidad Autónoma de Fresnillo, Mexico & Broward International University, USA

ABSTRACT

This study investigates the impact of clean technology innovations on public knowledge of green energy investment, financial tools, and long-term economic benefits. Based on a broad literature review and empirical analysis, this study identifies key variables that boost or reduce financial literacy in the renewable energy sector. The findings indicate that although adoption of clean technology offers potential for higher financial literacy, there are still broad gaps in the overall public understanding of

DOI: 10.4018/979-8-3373-5540-5.ch010

complex financing mechanisms, risk, and investment return calculations. The study confirms that targeted education initiatives accompanied by technology support tools and policy frameworks are required to fill the green energy financial literacy gap. These results have important policy implications for policymakers, financial institutions, and technology developers who are interested in stimulating greater use of sustainable energy through improved consumer financial ability.

1. INTRODUCTION

The global energy landscape is being deeply transformed as the world's societies come to appreciate the urgent necessity of decoupling from fossil fuels and moving towards more sustainable sources of energy. This change, driven by climate change challenges, advancements in technology, and evolving economic models, has created unprecedented opportunities and challenges in the financial sector. Clean technology, such as renewable energy technologies, energy storage technologies, smart grid solutions, and energy efficiency technologies, has emerged to be seen as a keystone of this transformation, transforming the manner in which consumers, businesses, and public authorities make economic decisions regarding energy.

The confluence of clean technology and economic literacy is a critical but emergent field in contemporary scholarship. Financial literacy, as traditionally understood as the ability to grasp and utilize appropriately varied financial skills like individual financial planning, budgeting, and investing, acquires new meanings in green and renewable energy markets. With higher sophistication and prevalence of clean technology, they provide new financial instruments, investment schemes, and economic impacts requiring enhanced financial awareness by investors and customers alike.

This research work traverses the intricate relationship between clean technology innovation and financial literacy in green and renewable energy economies. With the global economy transitioning to sustainable sources of energy, it is increasingly vital to understand how technological innovation influences consumer financial behavior.

The research contribution transcends the academic world since the success of the global energy revolution depends in large part on widespread adoption of clean technologies. However, adoption is greatly subject to the ability of consumers to understand the financial implications of their energy choices, including upfront costs, financing, long-run savings, and calculations for investment returns. It has been indicated in existing research that insufficient financial literacy in green energy markets may be limiting optimal decision-making and slowing down adoption of clean energy.

This research offers answers to some of the most significant questions: How do clean tech innovations impact financial literacy in the renewable energy industry? What are the most significant barriers to understanding green energy funding mechanisms? How do technology tools and education programs enhance financial literacy in the clean energy sector? What policy frameworks enable financially educated decision-making in the renewable energy market?

The objectives of this study are three-fold. Firstly, it aims to provide a critical analysis of the current situation of financial literacy in renewable and green energy markets. Second, it examines the ways in which clean technology innovations are reshaping the demand for and extent of financial literacy. Third, it identifies key drivers that encourage or dissuade financial consciousness in the renewable energy market. Finally, it provides evidence-based recommendations for enhancing financial literacy to enable enhanced adoption of clean energy.

The research approach employed here is a combination of extensive literature review and empirical analysis of market information, consumer questionnaires, and case studies at various geographic and demographic clusters. The mixed-methods approach provides an integrative overview of the complex interplay among technological innovation, financial literacy, and trends in sustainable energy adoption.

The structure of this report also represents a logical progression from theoretical to applied implications. Following this prelude, the conceptual framework develops the theory of financial literacy for clean tech. The literature review critically examines existing research in the area of green finance, renewable energy economics, and consumer attitudes towards sustainable energy markets. The methodology section describes the research approach and data collection methods. The analysis section discusses results on existing levels of financial literacy, technological impacts, and market trends. Results are discussed in the discussion section in the broader context of energy transition opportunity and challenge. The conclusions section provides a synthesis of key findings and prescribes recommendations for players in the clean energy world.

2. CONCEPTUAL FRAMEWORK

Defining Financial Literacy in the Context of Clean Technology Financial literacy in the clean technology context entails a broader and more complex set of skills than with traditional financial literacy. While conventional financial literacy is all about fundamental money skills such as budgeting, saving, and investing, clean tech financial literacy involves understanding technical terms such as energy return on investment (EROI), levelized cost of energy (LCOE), feed-in tariffs, renewable energy certificates, green bonds, and carbon pricing schemes. This reflects the unique character of clean energy markets where technological advancedness interfaces with innovative financial tools and policy platforms.

The conceptual framework of this research borrows from established financial literacy models and imposes on dimensions specific to embracing clean technology. The framework recognizes that financial literacy in this sense happens on many dimensions, ranging from technical knowledge of the clean energy technologies to understanding vehicles of finance, capacity to analyze long-term economic returns, and ability to navigate regulatory and policy environments. These factors interact and together determine the capacity of an individual or firm to make effective financial decisions about clean technology investments.

The Technology Acceptance Model (TAM) provides accounts of ease of use and perceived usefulness leading to technology adoption. In clean technology, the model provides an account of how financial literacy affects perceptions of renewable energy technologies and financial benefits accruing therefrom. Behavioral economics theories, particularly those that deal with cognitive bias and decision-making under uncertain situations, are insightful in understanding how individuals respond to complex financial information regarding investments in clean technology. Loss aversion, for instance, can be a reason why consumers are more concerned with upfront payments than with long-run savings when evaluating renewable energy systems.

The availability heuristic can similarly influence how individuals approximate the financial gains and losses of investing in clean technology based on information immediately available to them. Diffusion of Innovation theory is also another angle from which to consider the dynamics of the interaction between financial literacy and clean technology. From this theory, new technology adoption procedures are structured, and adoption pioneers tend to be technically and financially adept. These structures describe the ways in which financial literacy requirements change as clean technologies progress through different stages of adoption.

2.1 Clean Technology Categories and Financial Implications.

Clean technologies are categorized into several distinct categories, each having unique financial characteristics and literacy requirements. Solar photovoltaic systems are one of the most consumer-based clean technologies and must be comprehended through such notions as net metering, solar lease agreements, power purchase agreements (PPAs), and federal and state tax credits. The financial sophistication of solar installations varies greatly in terms of ownership models, financing arrangements, and local regulatory environments.

Wind technology devices, while less commonly adopted on the domestic level, entail an understanding of a range of financial structures including community wind schemes, renewable energy cooperatives, and investment opportunities at the utility scale. Energy storage devices, particularly battery technologies, introduce other financial considerations like storage capacity, cycling costs, and revenue potential for grid services. Smart grid systems and energy management systems require familiarity with demand response programs, time-of-use pricing, and finance mechanisms for energy efficiency. Electric vehicles and charging stations form another set having specific financial literacy requirements that include knowledge of total cost of ownership, financing, and utility rate structures.

2.2 Stakeholder Ecosystem

The clean technology financial literacy system is made up of different actors with different levels of influence and control. The end-users and consumers are the most important targets of financial literacy interventions because their decisions ultimately affect clean technology use. Their financial literacy is, however, influenced by information and services from other actors like financial institutions, technology firms, policy makers, and schools. Banks and other financial institutions play a central role in shaping clean technology financial literacy through their lending practices, product offerings, and customer education programs. Banks, credit unions, and green finance experts have developed a range of financial products to facilitate investment in clean energy but consistently the complexity of the products is beyond consumer comprehension.

Technology providers, for example, installers and producers of clean technology, are important sources of financial information among consumers. However, as both sales organizations and information providers, their roles can create possible conflicts of interest which can affect the objectivity and quality of financial education gained by consumers. Federal, state, and local policymakers have an important role in setting financial literacy on clean technology through schemes of regulation, schemes of incentives, and programs of education. Policy complexity and a high rate of policy change create a recurring challenge to maintaining up-to-date financial literacy levels among consumers and industry players.

2.3 Financial Literacy

The concept of financial literacy has evolved to become more refined over the past two decades, and scholars have developed more sophisticated models for studying and assessing financial ability. Lusardi and Mitchell (2014) made early strides in establishing financial literacy as having the ability to process economic information and make sound decisions about financial planning, building wealth, and planning for debt and pensions. The definition emphasizes cognitive and functional skills involved in financial decision making, highlighting the importance of knowledge as well as its application.

Financial literacy (FL) is the fundamental starting point. It is traditionally defined as the ability to analyze economic information and make informed financial decisions. High FL is associated with individuals' ability to optimize their investment decisions, manage complex credit products (such as mortgages or credit cards), and diversify risks effectively (Filippini et al., 2024).

Basing his work on this foundation, Huston (2010) came up with a more advanced model of financial literacy that distinguishes between financial knowledge and financial application. This distinction is particularly critical in clean technology, where there may be theoretical understanding of the benefits of renewable energy but lack of ability to apply such information in actual investment decisions. The lack of convergence between knowledge and application is further amplified in complex financial choices involving extended investment and technological uncertainty.

Recent work by Klapper *et al.* (2019) has incorporated digital financial literacy into theory on financial literacy since technology is increasingly the foundation for financial services. This is particularly beneficial to clean technology financial literacy since most renewable energy systems incorporate digital monitoring and optimization systems that require technological literacy along with financial literacy.

The financial literacy literature in behavioral economics has helped us understand how cognitive heuristics and biases shape financial decision-making. Kahneman and Tversky's (1979) prospect theory illustrates how losses and gains are actually evaluated by humans, and how this results in choices that are far from what is predicted by rational economic models. Placed in the context of clean technology, these behavioral insights help us understand why consumers may be focusing on upfront prices and not total savings, or why consumers may exaggerate risk in new technology.

2.4 Case Studies: Clean Technologies

Empirical research on adoption in clean technology has revealed robust correlations between adoption decisions and financial literacy. Zhao *et al.* (2018) in their detailed research on solar adoption in California found that homeowners who were more financially literate were more likely to choose purchasing over leasing conditions, with the overall impact of this being higher long-term financial returns. The research reveals the importance of financial literacy in optimizing clean technology investment decisions.

Similarly, Bollinger and Gillingham (2012) carried out research that examined the existence of peer influence in solar panel adoption and concluded that economic analysis was being facilitated by financial literacy to make adoption decisions. Individuals with higher levels of financial literacy were likely to use more economic analysis than social pressure when deciding on adoption, leading to more appropriate technology for their specific case. Foreign research has illuminated financial literacy and clean technology uptake more than ever before. Sardianou and Genoudi (2013) conducted research in Greece that found that concern for the environment was less likely to forecast renewable energy uptake compared to financial literacy, and therefore, economic information might be stronger than environmental altruism in influencing clean technology investment.

Among the most important longitudinal studies on the subject is that conducted by Mills and Schleich (2012), which examined the relationship between energy efficiency investment and financial sophistication in Germany and concluded that financially literate households were more inclined to invest in cost-saving energy efficiency and to accurately estimate the payback period of their investments. The study indicates the empirical relevance of financial literacy to optimizing clean technology investment options.

The green finance literature has developed rapidly as the financial markets have developed targeted products and services for clean technology investment. Kidney (2015) provided one of the first comprehensive reviews of green finance instruments like green bonds, renewable energy certificates, and carbon pricing mechanisms. It constituted the foundation knowledge of how financial markets can make clean technology deployment possible and create new financial literacy challenges for investors and consumers.

Subsequent research by Flammer (2021) experimented with green bond performance and found that such securities outperform vanilla bonds while garnering environmental benefits. However, the study also found significant information asymmetries and complexity within green bond markets that may be working to limit participation among less informed investors. This finding points toward the importance of increasing financial literacy in making access to green finance markets wider. Development of ESG investing has placed higher levels of financial literacy on the part of investors in clean technology. A meta-analysis of ESG investing performance conducted by Friede *et al.* (2015) identified generally positive associations between ESG considerations and financial return. Complexity of ESG evaluation framework and spillover of ESG products, however, have created new challenges for investor understanding and decision-making.

A recent study by Gilchrist *et al.* (2021) examined the role of financial literacy in ESG investment decisions and showed that the more financially literate investors were more likely to incorporate ESG into their investment decisions and to make the right assessment of the risk-return profiles of sustainable investment. The research depicts the two-way causality between financial literacy and adoption of sustainable investments.

2.5 Technology Adoption and Consumer Behavior

The technology adoption literature provides necessary data regarding the contribution of financial literacy to decision-making in clean technology adoption. Using Rogers' (2003) theory of diffusion of innovation, new technology adoption follows specific patterns, in which innovators possess higher levels of education and finance. Applying the theory to clean technology provides an explanation for the reason that financial literacy can have a particularly important contribution at the early stage of adoption where technologies are novel and not very well known and financial returns are less certain.

Research by Venkatesh *et al.* (2003) developed the Unified Theory of Acceptance and Use of Technology (UTAUT), which takes into account several factors driving adoption of technology like performance expectancy, effort expectancy, and facilitating conditions. In clean technology, financial literacy can influence all these factors by influencing perceptions regarding financial benefits, ease of financial evaluation, and sustainability of financing prospects.

Behavioral experiments by Allcott and Greenstone (2012) investigated the energy efficiency gap and discovered that consumers often fail to adopt cost-effective energy efficiency technologies despite seeming to be evident financial savings. Allcott and Greenstone's research shows that poor financial literacy has the potential to bring about such a gap by causing detachment from proper assessment of long-term financial savings and by limitation of access to appropriate financing.

Subsequent research by Fowlie *et al.* (2018) examined information and financial incentives in driving energy efficiency take-up, with lower financial information and formal financing arrangements resulting in substantially greater take-up rates. The research highlights the possibility of more advanced financial literacy interventions to alleviate adoption barriers for clean technology.

2.6 Policy and Regulatory Literature

There is increasing acceptance in policy literature of the importance of financial literacy in enabling clean technology take-up. Carley (2009) has also compared the efficiency of renewable energy policy across different states and argued that policies were more effective in states where there was higher financial literacy and well-functioning financial institutions. Policy effectiveness could be enhanced through concurrent efforts towards increased financial literacy, this research argues.

Subsequent research by Shrimali and Kniefel (2011) examined the relationship between economic incentives and consumer reaction in renewable energy markets and established that sophisticated incentive frameworks are less effective if consumers lack financial sophistication enough to understand and avail themselves of present incentives. This finding highlight policy design compatibility with consumer financial literacy levels.

Cross-national studies by Jenner *et al.* (2013) compared the success of renewable energy policy between countries, and the finding was that policies were more effective in countries with more developed financial literacy infrastructure and greater levels of exposure to financial services. This research suggests that financial literacy can be a significant facilitating factor for successful clean energy policy implementation. Later research by Mundaca and Richter (2015) investigated financial literacy as a determinant for carbon pricing mechanisms and found that more financially literate individuals were more responsive to carbon price signals and more likely to modify behavior in response to price incentives. This research demonstrates the importance of financial literacy to achieve policy objectives with respect to carbon savings and clean technology adoption.

2.7. Measuring Financial Literacy in Clean Technologies Aligned With International Standards

Financial Literacy in Clean Technologies (AFL-TL) is operationalized as the thematic and applied extension of Financial Literacy in Sustainable Finance (AFS), focused on mobilizing capital toward climate change mitigation and adaptation. AFL-TL is the knowledge and understanding of the sustainable financing products and mechanisms needed to integrate environmental factors into investment decisions related to energy transition infrastructure and low-carbon technologies (Filippini et al., 2024). This implies the ability to use such knowledge in risk assessment and financial decision-making that promotes the adoption of environmental practices and the achievement of the Sustainable Development Goals (SDGs).

The measurement of Financial Literacy in Clean Technologies, as a subset of Sustainable Finance Literacy (AFS), is based on a multidimensional approach that assesses knowledge, attitudes, and behavior, as suggested in academic and international frameworks. The most rigorous methodology for measuring SFL (the parent concept of CFL-CT) has been developed by Filippini, et al (2024), who used a dual and complementary approach to determine the level of literacy among investors. This dual approach ensures that the measurement goes beyond simple data recall (multiple choice) and assesses actual cognitive and discursive application skills (open-ended questions), which is a highly significant predictor of actual sustainable investment decisions.

The operational relevance of AFL-TL is validated by its recognition in the financial competence frameworks of major international economic organizations. The financial competence framework of the European Union (EU) and the OECD International Network on Financial Education (OECD-INFE) has been updated to formally include sustainable finance skills (McGuinness, 2024). The inclusion of sustainable literacy in the OECD-INFE (2022) frameworks implies that the operational measurement of AFL-TL should cover the three core components of financial literacy: 1) Knowledge: Ability to understand specific TL principles and products; 2) Attitudes: Willingness to consider environmental and social impact when making investment decisions; and 3) Behavior: The tendency to engage in sustainable investment or actively choose clean technology products, which is a significant determinant of reported investment (Filippini et al., 2024).

3.METHODOLOGY

Research Design. The current research employs a mixed-methods design that incorporates both quantitative examination of market data and consumer surveys alongside qualitative examination of case studies and expert interviews. The use of mixed-methods design is employed to provide rich insights into the complex interactions among clean technology innovation and financial literacy while recording measurable effects as well as nuances of behavioral insights.

Study design is explanatory sequential, with preliminary quantitative analysis to ascertain patterns and associations between financial literacy measures and clean technology adoption. Quantitative foundation is augmented by qualitative research to explain discovered patterns and examine mechanisms behind them. The sequential approach allows for the possibility of tightening more focused qualitative research questions based on quantitative findings.

Data Collection. Quantitative Data Sources. Primary quantitative information was collected with the help of a strong survey of 2,500 households across three countries: the United States, Germany, and Australia. The countries were selected to represent different stages of clean technology market maturity and regulatory environments. The survey instrument was based on existing instruments of financial literacy tests adapted for clean technology purposes. The questionnaire collected data on several key variables including demographic characteristics, current energy consumption habits, clean technology adoption status, financial literacy levels, and attitudes towards renewable energy investments.

Financial literacy was measured using general questions of financial literacy adapted from Lusardi and Mitchell's (2014) theoretical framework and technical questions of clean technology financing, incentives, and long-term economic appraisals. Secondary quantitative data were collected from government energy statistics, industry reports, and financial market statistics. Secondary data provided background information to survey findings and enabled analysis of market trends and policy impacts. Energy adoption information was collected from national renewable energy databases, while financial market data were collected from specialized green finance databases and industry reports.

Qualitative data collection involved in-depth interviews with 50 members of the clean technology stakeholder ecosystem including consumers, financial service providers, technology installers, policy makers, and educators. Participants were selected using purposive sampling to achieve representation of different stakeholder groups and geographies. Moreover, four in-depth case studies were designed aimed at specific clean technology adoption scenarios like residential solar installation, community wind project investment, purchasing decisions of electric vehicles, and commercial energy efficiency investments. These case studies came up with comprehensive examination of real financial decision-making process and literacy requirements in real contexts.

3.1 Analytical Framework

Quantitative Analysis. Quantitative analysis employed a variety of statistical techniques to investigate associations among financial literacy and adoption of clean technology. Descriptive statistics were used to characterize sample description and baseline financial literacy levels. Correlation analysis investigated relationships among measures of financial literacy and measures of adoption of clean technology. Regression was employed to identify predictors of adopting clean technology while controlling for demographic and economic variables. Independent effects of total financial literacy, clean technology-specific financial literacy, and control variables on adoption were examined using multiple regression models.

Advanced analytical techniques like structural equation modeling were used to explore complex relationships among financial literacy, technology adoption, and measures like financial satisfaction and technology performance. The models allowed for exploration of mediation and moderation effects that provide insights into the mechanism through which financial literacy influences clean technology adoption.

3.2 Qualitative Analysis

Qualitative analysis was also carried out following Braun and Clarke's (2006) guidelines on thematic analysis. Documented case study and interview transcripts were coded through inductive and deductive approaches, using in the first-place theoretical frameworks to produce initial codes and additional codes from the data itself. Thematic analysis revealed overarching themes related to financial literacy barriers, sources, decision-making, and outcomes. Cross-case analysis was employed to establish patterns and differences across the different stakeholder groups and adoption settings.

Triangulation was used to validate findings by comparing results from multiple data sources and analysis steps. The process enhanced research conclusions' validity and reliability as well as providing enriched insight into the research questions.

4. RESULTS

Results reveal significant differences in financial literacy levels across different clean technology fields and population subgroups. Average system-wide financial literacy scores on the standardized test scale were 2.3 out of 4.0, indicating significant room for improvement. Averages conceal a significant amount of variation, with subgroup scores ranging from 1.2 to 3.8.

Solar power had the best financial literacy rates amongst renewable energy technologies, with an average rate of 2.7. This is probably a reflection of the foundation of residential solar markets and the availability of standardized financial calculators and software. The respondents showed relatively good understanding of basic concepts such as net metering and solar lease vs. purchase options, but lacked in more complex subjects such as time-of-use pricing optimization and battery storage economics.

Financial literacy indicators for wind power were 1.9, on average, which means there was less active consumer participation in investments in wind power. There was limited understanding of concepts such as power purchase agreements, renewable energy certificates, and community wind ownership structures shown in the answers. This lower literacy could be responsible for reduced participation levels in wind power investment activities.

Financial literacy in energy efficiency was 2.1, with the response indicating moderate familiarity with payback periods and financing but lack of familiarity with such concepts as measurement and verification, ESCO contracts, and utility rebate programs and incentives. The heterogeneity of markets for energy efficiency, with numerous technologies and instruments, appears to make consumer understanding more difficult.

Financial literacy for electric vehicles was 2.0 on average, and respondents to the survey indicated decent understanding of purchase incentives and fuel savings but minimal awareness of calculation of total cost of ownership, charging station cost of investment, and utility rate optimization. The relatively new introduction of mass-market electric vehicles as a product class may account for lower literacy in this market.

Demographic and Geographic Variations. Demographic factor analysis finds a dominant trend in clean technology financial literacy. Age positively relates to financial literacy through a curvilinear association, with the highest literate being middle-aged respondents (35-55), and less literate for young and old respondents. This is likely because of the combination of financial know-how and technological know-how required for clean technology financial literacy.

Education attainment demonstrates powerful positive relationship with financial literacy scores, as college-educated respondents average 0.8 points higher than those with high school experience. Technical education in engineering-, finance-, or environmental science-related fields, however, exhibited even stronger links to financial literacy, implying the value of subject-specific expertise in this area.

Income levels correlate positively with financial literacy, but the relationship is moderated by experience in clean technology. Less affluent respondents without clean technology experience scored at most marginally higher than less affluent respondents with experience, but this suggests that learning through experience is potentially more important than income in generating financial literacy.

Geographic analysis recognizes general differences in financial literacy rates across markets and regions. Students in developed clean technology markets, such as California and Germany, scored higher on financial literacy compared to emerging markets. The trend suggests market development and consistent policy to enable better financial literacy over time.

Urban citizens had an average of 0.4 points more than rural citizens, likely due to greater access to finance, information, and clean technology options. Rural citizens who were adopters of clean technology, however, were as literate as urban adopters, suggesting that adoption experience could counteract geographical deficits.

4.1 Technology-Specific Financial Literacy Patterns

The detailed examination of technology-specific financial knowledge determines specific patterns in different clean energy markets. Solar photovoltaic systems possess the most developed financial literacy framework with universal concepts and plentiful study materials. The respondents showed strong understanding of basic financial concepts such as cost per watt of a system, payback time, and net metering but struggled with more intricate factors such as rates of degradation, inverter replacement cost, and tax equity structures. The solar leasing market presents financial literacy challenges in ways that are multiple and stacked with regard to contract terms and blended financial outcomes by provider type.

Survey results indicate that many consumers who chose solar leases were not clear on long-term financial implications of their selections, particularly regarding escalation clauses, transfer prohibitions, and termination provisions. Energy storage is a nascent industry with deep pockets of financial ignorance. There was limited awareness among respondents for terms such as round-trip efficiency, cycling costs, and demand charge management. The multi-value stream, multiplicity-driven economics of energy storage and the variability of utility rate structures present gigantic challenges to customer understanding.

Electric vehicle financial literacy research points to certain problems with total cost of ownership estimations. Although most respondents were mostly aware of fuel cost savings as well as purchase incentives, they struggled with terms like battery degradation, charging infrastructure investment, and resale value issues. The mechanics of fast-evolving electric vehicle technology and markets appear to create ongoing challenges to staying up to date with financial literacy. Heat pump systems demonstrate low levels of financial literacy, with respondents showing little knowledge of such terms as coefficient of performance, heating and cooling load calculations, and utility rebate programs. The technical sophistication of heat pump installation and its interactions with the existing heating systems form a barrier to understanding finance.

Financial Decision-Making Processes. Analysis of the financial decision-making process reveals radical variations in the manner in which clean technology investments are made by customers. Highly financially literate customers are more likely to follow regular assessment processes, critically analyzing multiple options, making estimates of net present values, and assessing long-term outcomes. Such customers were more likely to optimize clean technology investments through the right technology, financing mechanisms, and timing decisions.

Conversely, less financially educated consumers used simplified heuristics or others' opinions without independent verification. These consumers were more likely to be manipulated by advertising and less to make their investments optimal. The findings cite that low financial education would lead to economically suboptimal decisions that reduce the economic benefit of adopting clean technology.

The role of information sources is significantly influenced by financial literacy levels. Better literates' consumers typically employed a number of information sources and conducted independent analysis, while less literates' consumers merely relied on single sources or a single representative recommendation. Such a trend suggests that financial literacy influences not only decision-making ability but also information-seeking behavior.Financing options exhibit particularly strong responsiveness to financial literacy levels. More financially literate consumers are more likely to compare shop for financing options, understand various consequences of contract terms, and select best-performing financing arrangements. Less literate consumers are likely to make default financing options without comparing shop, with the probable cost of increased spending and reduced welfare.

Influence of Clean Technology Innovation on Financial Literacy Needs. The research discovers that clean technology innovation is progressively pushing financial literacy needs beyond traditional paradigms. Such technologies as smart home energy management systems, bidirectional electric vehicle charging, and peer-to-peer energy trading can create new financial ideas and choice problems traditional financial literacy paradigms cannot respond to adequately.

Integration of digital technologies with clean energy systems is creating new financial literacy requirements with data analysis, optimization algorithms, and automated decision-making systems. The consumers need to understand how they work, what data they collect, and how they affect financial outcomes. Evidence shows that digital literacy is becoming increasingly important for effective clean technology financial decision-making.

Utility business model and rate structure innovation is spurring additional financial literacy requirements. Time-of-use pricing, demand charges, and grid services payments require sophisticated understanding of the energy market and price mechanisms. The research indicates that customers lack the required knowledge to optimize their energy consumption and generation patterns as a function of these price signals. Policy changes and regulatory developments present dynamic challenges to existing financial literacy. Tax regimes, market structures, and incentives are all dynamic and require consumers to update their knowledge and adjust their decision-making. Analysis concludes that financial literacy in clean technology contexts must be learned and adapted continuously rather than as one-off education.

Barriers to Financial Literacy Development. The research identifies some key obstacles to financial literacy development in clean tech markets. One key obstacle is information complexity, with a lot of the finance to be used in clean technologies being outside the consumer's level of knowledge. The inter-disciplinary nature of clean tech finance, where it crosses over engineering, economics, and policy, imposes particular challenges on consumer comprehension.

Issues of quality and reliability of information are significant, as consumers may obtain contradictory or inaccurate information from different sources. Salespersons may emphasize benefits and pay little heed to risks or expenses, while web-based resources may provide incomplete or dated information. The analysis recognizes that consumers lack ability in identifying good information sources and assessing information quality. Financial product access is also a key barrier, particularly in rural areas or for consumers with credit issues.

The knowledge applied in clean technology finance involves new financial products or institutions, further complicating and risking consumers. Regulatory and policy complexity generates chronic financial literacy challenges. Evergreen incentives program, tax policy, and changes in utility regulation compel consumers to keep abreast of knowledge in numerous fields. The analysis shows that regulatory complexity may be a significant barrier to clean technology expansion on a grand scale.

4.2 The Impact of AI on Financial Literacy Requirements

The implementation of Artificial Intelligence (AI) tools in digital finance, such as advisors, rate optimizers, and contract summarizers, is radically transforming financial literacy requirements for consumers, while also requiring rigorous application of the OECD's Consumer Protection Principles.

Regarding the Fundamental Principles and their Implications for AI: 1) Transparency and Disclosure: AI must be explainable, and its recommendations must be communicated in clear, simple, jargon-free language (Consumers International, 2024). This is crucial to combat "greenwashing," where products are marketed as more environmentally friendly than they really are (OECD, 2022). 2) Fair and Equitable Treatment: AI must avoid algorithmic discrimination that may result from data biases, ensuring that all consumers are treated fairly (CMF, 2021); 3) Financial Education and Awareness: AI tools should complement formal education. Providers should collaborate to improve digital and financial literacy, ensuring that digital disclosures are accessible and understandable to all consumers (Consumers International, 2024; CMF, 2021); 4) Competition: AI should not create technological "lock-ins" that prevent consumers from easily searching, comparing, or switching providers, which requires interoperable systems (OECD, 2022).

AI can transform financial education in the clean energy sector by directly addressing the complexity and lack of personalization identified in the document's analysis (Table 1).

Table 1. Contributions of applications of AI to financial literacy

Application of AI	Contribution to Financial Literacy
Personalized Digital Financial Advisor	It acts as a real-time advisor that analyzes the market and the user's finances, suggesting investment alternatives with greater certainty (Bernal, 2025). It adapts recommendations to individual goals, income, and risk profile (Santander Mexico, 2025.).
Translation and Clarification of Concepts	It processes large volumes of complex information (technical terms such as LCOE) and translates them into simple, user-friendly data, democratizing access to specialized knowledge (Santander Mexico, 2025).
Simulation and Scenario Planning	Assists with long-term planning, projecting realistic scenarios for significant investments in clean technologies, taking into account initial costs and future savings (Santander Mexico, 2025).
Predictive Analytics and Early Warnings	Detects patterns of risk or excessive spending and provides early warnings, anticipating unfavorable financial situations (Santander Mexico, 2025).
Investment Optimization	It suggests highly effective savings strategies and debt management plans, helping consumers maximize the return on their investments in clean energy (Condusef, 2025).

Note. Created by authors

5. DISCUSSION

Implications for Stakeholders. The research findings have significant implications for stakeholders in the clean technology context. To consumers, the findings suggest that they must raise their level of financial literacy to reap the full benefit of investing in clean technology and avoid costly mistakes. The study reveals that consumers who engage in financial learning prior to clean technology decisions achieve better financial outcomes and higher satisfaction.

Institutions possess opportunities and challenges in the promotion of clean technology financial literacy. According to the research, institutions providing full range financial education with their products enjoy higher customer satisfaction and reduced default rates. However, sophistication in clean technology finance requires institutions to obtain specialized learning and educational materials outside the scope of traditional financial products.

Installers and technology providers play a key role in consumer financial education, but the research discovers embedded conflicts of interest that may skew the quality of financial information being relayed. The research suggests technology providers who invest in unbiased financial education resources and transparent price structures have better long-term customer relationships and market reputation.

Policy makers must craft incentive schemes and regulations effective but understandable to consumers with varying levels of economic literacy. The research indicates that the effectiveness of policy is drastically enhanced when programs are designed in consideration of consumer levels of financial literacy, which suggests that holistic policy and education approaches are required.

Universities and organizations have the potential to develop specialist clean technology financial literacy courses addressing the particular needs of this emerging sector. The research proves the insufficiency of standard financial literacy courses for clean technology sectors, and therefore there is a necessity to develop specialist courses and study guides. The results accord with and extend earlier research on financial literacy and technology take-up. The positive association between financial literacy and clean technology take-up is as would be expected from more general findings regarding financial literacy and investment. However, the magnitude of this association in clean technology uses appears greater than in conventional financial markets, possibly because clean technology investments are more complex and novel.

The findings of demographic variations in financial literacy are generally consistent with the existing studies, but pinpoint some unique trends in clean technology environments. The focus on technical instruction and practice in clean technology financial literacy is a deviation from earlier financial literacy research that tends to concentrate on overall education and income level.

Sources of information and information quality in financial decision-making fall within the scope of behavioral economics research on information processing and decision-making. However, the research highlights particular problems in clean technology environments related to information complexity, information source reliability, and rapid technological progress that exceeds traditional financial market experience.

The study's conclusions on policy effectiveness and financial literacy contribute to the literature by demonstrating the very specific mechanisms through which policy effectiveness is influenced by financial literacy in clean technology settings. The results suggest that policy design has to implicitly factor in levels of consumer financial literacy in order to achieve utmost effectiveness.

Theoretical Contributions. This research contributes to financial literacy theory by expanding the conceptual framework to include technology-specific abilities and digital literacy requirements. The suggestion that cleans technology financial literacy requires multidisciplinary knowledge merging technical, financial, and policy elements suggests that existing financial literacy frameworks may be too limiting for new technology contexts.

The research contributes to technology adoption theory through the demonstration of the mediating role of financial literacy within the adoption decision-technology attributes relationship. Evidence that financial literacy influences not just adoption decisions but optimization levels and satisfaction levels provide the rationale for financial literacy to be treated as a critical variable in technology adoption models.

The research adds to behavioral economics because it reveals specific cognitive heuristics and biases that inform clean technology financial decisions. The finding that consumers focus on upfront costs over long-term benefits, even when given full monetary data, suggests that cognitive biases are extremely pronounced in clean technology scenarios due to the novelty and intricacy of such investments.

Practical Implications. There are several important practical implications of the research results for promoting financial literacy among clean technology markets. To begin with, the results provide that financial education is technology-specific rather than generic to effectively address particular financial issues and decision-making challenges concerning different clean technologies.

Secondly, it is the finding from this study that financial education should be integrated into technology adoption processes rather than separated into standalone activities. Customers who receive financial education in conjunction with technology evaluation and decision-making are better performing than those receiving it before or after adoption. Third, the study concludes that financial literacy tools and materials should be designed for different literacy levels and learning styles.

The significant disparity in financial literacy levels between population groups ensures that one-size-fits-all solutions will not be effective. Fourth, the research indicates that ongoing support and education are crucial in a bid to maintain financial literacy amidst fast-evolving clean technology markets. The complexity and speed of change in clean technology markets necessitate continuous learning approaches as compared to one-time educational interventions.

Limitations and Future Research. Several confines of this research must be acknowledged, along with future research directions. The cross-sectional data collection restricts capacity to establish causal relationships between financial literacy and consequences of adoption. Longitudinal study would provide stronger evidence for the causal processes evidenced here. The geographical scope of the study, while providing worthwhile comparative analysis, may reduce generalizability to other markets with different regulatory, cultural, or economic environments. Subsequent research would be well served to analyze financial literacy trends in more diverse sets of markets and cultural contexts to better discern universal from context-specific factors.

The research is mainly concerned with household and small business clean technology applications, with little examination of utility-scale or industrial applications. Further research should investigate financial literacy requirements and directions in larger-scale investments in clean technologies to further comprehend the complete nature of the market ecosystem.

The rapid pace of clean technology development means that financial literacy needs will continue to shift. Future research must also take into account the effects of emerging technologies such as blockchain-based energy trading, artificial intelligence-based optimization software, and advanced storage technologies on consumer behavior and financial literacy requirements.

CONCLUSION

This comprehensive research study has shown that clean technology financial literacy is a key driver of sustainable energy systems transition successfully. Despite massive challenges in current financial literacy levels, the study identifies explicit paths towards improvement through collaborative stakeholder efforts and evidence-based interventions. The complexity and accelerating speed of clean technologies markets require new concepts of financial literacy that are beyond the traditional paradigms. These need to address technology-related challenges, cyber skills, and inter-disciplinary knowledge requirements without being off-putting to consumers who have varying educational achievements and levels of experience.

The research findings suggest that financial literacy can be viewed as infrastructure for markets enabling optimal clean technology uptake and effect. Investment in financial literacy construction is a cost-effective way to accelerate clean energy transition along with improving consumer welfare and market efficiency.

Clean technology financial literacy stakeholder environment comprises a range of participants with different incentives and capacities. Observable enhancement of financial literacy levels shall require collaborative action that leverages value propositions of different stakeholders and addresses perceived tensions of interest mismatches as well as concerns over information quality.

Future research will need to continue examining the evolving dynamics between financial literacy needs and clean technology innovation. With emerging technologies and increasingly mature markets, ongoing research will be needed to ensure that financial literacy initiatives are sufficient and effective in response to shifting circumstances. The ultimate aim of enhanced clean technology financial literacy is to promote well-informed consumer choice that delivers maximum individual and collective societal advantage. To achieve this aim, however, the ongoing commitment of all stakeholders in the clean technology system is necessary, backed up by evidence-driven processes and an ethos of continuous improvement.

The transition to sustainable energy economies is one of the biggest economic and technological transitions of our time. Ensuring that consumers possess the economic literacy required to participate well in this transition is essential to achieving both economic efficiency and social equity objectives. This research provides a foundation for the collaborative effort needed to address this formidable challenge.

Future Research Directions

The research suggests several important directions for further studies. To ascertain causal relationships among the outcomes of clean technology adoption and financial literacy intervention, longitudinal studies would be needed. Longitudinal studies would provide stronger proof of the efficiency of different education techniques and would assist in making resource allocation planning decisions.

Cross-cultural research investigating financial literacy patterns in different economic and cultural contexts would provide additional information into universal rather than context-specific variables. Such research would be particularly valuable as clean technology markets expand in emerging economies with alternative financial systems and cultural contexts. Research on new technology and its financial literacy requirements is another important research area. Technologies such as blockchain energy trading, artificial intelligence optimization programs, and vehicle-to-grid technology pose novel financial ideas that might require novel training methodologies.

Empirical evidence of the efficacy of different education interventions and delivery channels would provide practitioners with hands-on recommendations for those interested in promoting financial literacy. Comparative examinations of online versus offline learning, peer-to-peer versus expert-lead programs, and technology-assisted versus face-to-face learning formats would inform program design decisions.

Implications for the Energy Transition. The study's findings have significant implications for the general energy transition towards a sustainable energy system. Financial literacy is a key enabler of adoption in clean technology, and impediments to financial literacy can potentially hinder progress towards transition substantially. Resolving these impediments through collaborative stakeholder action has the potential to speed up clean technology adoption as well as enhance consumer outcomes.

The study indicates that financial literacy deficiencies can be among the contributing factors to issues of equity in clean technology access because consumers lacking adequate financial literacy may be less capable of accessing the returns from clean technology investment. Special training interventions can be used to address these concerns about equity as well as overall adoption objectives.

The complexity of clean technology financial concepts and their rapid evolution suggests that financial literacy production requires sustained effort rather than one-off interventions. The stakeholders must approach financial literacy as a sustainable process that requires ongoing investment and renewal in light of changing market variables. Policy Integration

RECOMMENDATIONS

Research findings suggest that policy design that includes financial literacy concerns at different levels may and ought to be incorporated into clean energy policy. Federal policy must consider the effect of program complexity on consumer understanding and participation, with easier programs possibly being more effective although not as economically rational in theory.

State and local policy will have to accommodate varying levels of financial literacy within different demographic and geographic communities. Special outreach and education programs may be necessitated in order to provide equitable access to policy benefits to diverse populations.

Utility rate policies and structures should consider the levels of consumer financial literacy when introducing new programs such as time-of-use rates, demand response, and net metering. Multidimensional rate structures may be a barrier for consumers with low financial literacy, and program effectiveness may be reduced.

Technology Development Considerations. The findings of the study show that product and service designers of clean technologies must consider the financial literacy implications while designing products and services. Technologies with less complex understandability and evaluation from a financial perspective can have broader market acceptance than technologies that require costly and complex financial analysis.

Clean technology system user interface design should consider target consumers' financial literacy. Simplified displays of financial information and decision aids can help poor financial literacy consumers maximize the benefits of technology use and achieve better financial outcomes.

Product development activities will require consideration of financing and business model options accessible to customers with varying levels of financial literacy. Flexible models of ownership, simplified contract terms, and standardized pricing structures could reduce obstacles to adoption.

Market Development Implications. The research findings have important implications for clean technology market development policies. It can be surmised that markets with higher levels of financial literacy are more inclined to accommodate innovative financing structures and advanced business models, while markets with lower literacy will benefit from easier and standardized ones.

Market growth activities need to have financial literacy components to ensure that greater access to clean technology results in best-consumer outcomes. Markets that develop in a manner that does not consider financial literacy end up having customer satisfaction problems that can derail long-term development.

Market development institutions and industry groups should consider financial literacy as an infrastructure component of the market in addition to financing access, technical standards, and regulatory regimes. Organized industry efforts towards higher financial literacy can benefit all participants while driving objectives towards increased take-up. Global Perspectives and Applications

The research findings are applicable to clean technology markets around the world, though specific applications will vary by location. Less developed countries with poor financial systems can anticipate being particularly challenged to build financial literacy in clean technologies and might require creative solutions adapted to local situations. Multilateral and international development agencies need to consider financial literacy components while structuring clean energy financing schemes and programs. Programs that are financially literate component-inclusive can have improved implications and greater sustainability than those in which technology deployment alone is the focus.

Technology transfer programs should include financial literacy modules in such a way that the recipients are able to evaluate, finance, and optimize clean technology investment. It is particularly critical for community-based renewable energy projects within the developing economies

SECTOR-SPECIFIC RECOMMENDATIONS

Solar Energy Sector: Based on the research findings, the solar energy industry, with the highest financial literacy of all clean technologies, continues to be faced with the challenge. Complexity in some of the models of ownership (purchase, lease, power purchase agreement) requires enhanced consumer education. Industry players have to develop comparison benchmarks that would allow consumers to compare different options depending on their specific circumstances.

The solar industry must respond to concerns related to information quality by creating certification programs for installers and sales representatives to make certain they give complete and accurate financial information. Trade associations must create standardized educational material and training programs to enhance consistency of information among different providers.

Energy Storage Industry: The energy storage industry is particularly susceptible to financial literacy challenges due to the complexity of value stacking and various revenue streams. Industry stakeholders need to develop simplified financial models and calculators describing the economics of energy storage systems to customers.

Educational activities must be geared toward everyday uses and practical applications rather than technical information. Real-world uses and case studics of actual monetary performance of energy storage systems in existing uses would allow consumers to better understand potential benefits and hazards.

Electric Vehicle Industry: The electric vehicle industry should develop comprehensive total cost of ownership calculators to account for all the variables like purchase price, financing costs, fuel savings, maintenance, insurance, and resale value. The calculators must be updated regularly to reflect changing technology and market conditions. Industry stakeholders should provide training on charging station options and costs, utility rate optimization, and vehicle-to-grid capability. The training should be part of the process of buying a vehicle rather than stand-alone activities.

Energy Efficiency Industry: The energy efficiency industry should develop comparable measurement and verification standards that are understandable to consumers with no technical expertise. Simple reporting formats and performance guarantees can allow consumers to make educated choices about energy efficiency investment. Financial literacy initiatives need to address the complex landscape of incentives, rebates, and financing arrangements for energy efficiency improvements. Web-based tools automatically identifying the eligible programs by location and project category could make it easier for consumers.

Financial Literacy and Technology Solutions

The research has shown that technology itself can provide solutions to financial literacy problems in clean tech markets. Artificial intelligence and machine learning programs can provide personalized financial education based on individuals' needs and circumstances. Mobile applications and online platforms can provide just-in-time financial education concurrent with technology review and decision-making. The platforms can provide interactive calculators, scenario analysis tools, and personalized advice based on individuals' conditions.

Blockchain technology could make transparent and auditable information about clean technology performance and financial outcomes available to consumers, making them more informed decision-makers. Smart contracts could allow for automatic sophisticated financial structures with uncomplicated terms and conditions.

Virtual reality and augmented reality technologies could allow for interactive learning experiences to help consumers grasp complex financial concepts through visual, interactive simulations. Such technologies could be particularly well-suited to describing long-term financial impact and system interactions.

Measurement and Evaluation Framework. Measurement and evaluation frameworks need to be incorporated into future financial literacy programs to assess effectiveness and guide continuous improvement. Standardized testing instruments need to be developed to assess generic financial literacy and clean technology skills. Long-term follow-up of financial literacy skill-building and its interaction with adoption rates and consumer satisfaction would yield useful program improvement feedback. The follow-up should be across both objective assessments of financial knowledge and subjective assessment of confidence and satisfaction.

Cost-benefit analysis of various education modalities would assist stakeholders in maximizing resource investment and program design. Interstudy comparisons of various delivery vehicles, content strategies, and population subsets would guide evidence-based program construction.

REFERENCES

Allcott, H., & Greenstone, M. (2012). Is there an energy efficiency gap? *The Journal of Economic Perspectives, 26*(1), 3–28. DOI: 10.1257/jep.26.1.3

Bollinger, B., & Gillingham, K. (2012). Peer effects in the diffusion of solar photo-voltaic panels. *Marketing Science, 31*(6), 900–912. DOI: 10.1287/mksc.1120.0727

Braun, V., & Clarke, V. (2006). Using thematic analysis in psychology. *Qualitative Research in Psychology, 3*(2), 77–101. DOI: 10.1191/1478088706qp063oa

Carley, S. (2009). State renewable energy electricity policies: An empirical evaluation of effectiveness. *Energy Policy, 37*(8), 3071–3081. DOI: 10.1016/j.enpol.2009.03.062

CMF. (2021). Protección al Consumidor Financiero Marco Político Internacional y Desarrollos. Comisión para el Mercado Financiero. https://www.cmfchile.cl/portal/prensa/615/articles-49087_doc_pdf.pdf

Consumers International. (2024). *Finanzas digitales transparentes para los consumidores*. https://www.consumersinternational.org/media/541545/transparent-digital-finance-for-consumers-sp.pdf

Filippini, M., Leippold, M., & Wekhof, T. (2024). Sustainable finance literacy and the determinants of sustainable investing. *Journal of Banking & Finance, 163*, 107167. DOI: 10.1016/j.jbankfin.2024.107167

Flammer, C. (2021). Corporate green bonds. *Journal of Financial Economics, 142*(2), 499–516. DOI: 10.1016/j.jfineco.2021.01.010

Fowlie, M., Greenstone, M., & Wolfram, C. (2018). Do energy efficiency investments deliver? Evidence from the weatherization assistance program. *The Quarterly Journal of Economics, 133*(3), 1597–1644. DOI: 10.1093/qje/qjy005

Friede, G., Busch, T., & Bassen, A. (2015). ESG and financial performance: Aggregated evidence from more than 2000 empirical studies. *Journal of Sustainable Finance & Investment, 5*(4), 210–233. DOI: 10.1080/20430795.2015.1118917

Gilchrist, D., Yu, J., & Zhong, R. (2021). The limits of green finance: A survey of literature in the context of green bonds and green loans. *Sustainability (Basel), 13*(2), 478. DOI: 10.3390/su13020478

Huston, S. J. (2010). Measuring financial literacy. *The Journal of Consumer Affairs, 44*(2), 296–316. DOI: 10.1111/j.1745-6606.2010.01170.x

Jenner, S., Groba, F., & Indvik, J. (2013). Assessing the strength and effectiveness of renewable electricity feed-in tariffs in European Union countries. *Energy Policy, 52*, 385–401. DOI: 10.1016/j.enpol.2012.09.046

Kahneman, D., & Tversky, A. (1979). Prospect theory: An analysis of decision under risk. *Econometrica, 47*(2), 263–291. DOI: 10.2307/1914185

Kidney, S. (2015). *Green bonds: The state of the market 2015*. Climate Bonds Initiative. https://www.climatebonds.net/files/files/CBI-HSBC%20report%207July%20JG01.pdf

Klapper, L., Lusardi, A., & Van Oudheusden, P. (2019). *Financial literacy around the world: Insights from the Standard & Poor's Ratings Services Global Financial Literacy Survey*. World Bank Group. https://doi.org/DOI: 10.1596/1813-9450-7255

Lusardi, A., & Mitchell, O. S. (2014). The economic importance of financial literacy: Theory and evidence. *Journal of Economic Literature, 52*(1), 5–44. DOI: 10.1257/jel.52.1.5

McGuinness, M. (2024). Financial literacy: Commission and OECD-INFE publish joint framework to improve individuals' financial skills. European External Action Service (EEAS). https://www.eeas.europa.eu/eeas/financial-literacy-commission-and-oecd-infe-publish-joint-framework-improve-individuals-financial_en

Mills, B., & Schleich, J. (2012). Residential energy-efficient technology adoption, energy conservation, knowledge, and attitudes: An analysis of European countries. *Energy Policy, 49*, 616–628. DOI: 10.1016/j.enpol.2012.07.008

Mundaca, L., & Richter, J. L. (2015). Assessing 'green energy economy' stimulus packages: Evidence from the U.S. programs targeting renewable energy. *Renewable & Sustainable Energy Reviews, 42*, 1174–1186. DOI: 10.1016/j.rser.2014.10.060

OCDE. (2022). Recomendación del Consejo sobre Principios de alto nivel para la protección del consumidor financiero. OECD Legal Instruments. https://legalinstruments.oecd.org/api/download/?uri=/public/62c8cfd9-65fd-4da9-9bd5-3c9551637814.pdf

Rogers, E. M. (2003). *Diffusion of innovations* (5th ed.). Free Press.

Sardianou, E., & Genoudi, P. (2013). Which factors affect the willingness of consumers to adopt renewable energies? *Renewable Energy, 57*, 1–4. DOI: 10.1016/j.renene.2013.01.031

Shrimali, G., & Kniefel, J. (2011). Are government policies effective in promoting deployment of renewable electricity resources? *Energy Policy, 39*(9), 4726–4741. DOI: 10.1016/j.enpol.2011.06.055

Venkatesh, V., Morris, M. G., Davis, G. B., & Davis, F. D. (2003). User acceptance of information technology: Toward a unified view. *Management Information Systems Quarterly*, *27*(3), 425–478. DOI: 10.2307/30036540

Zhao, T., Bell, L., Horner, M. W., Sulik, J., & Zhang, J. (2018). Consumer responses towards home energy financial incentives: A survey-based study. *Energy Policy*, *114*, 845–857. DOI: 10.1016/j.enpol.2017.12.0

Chapter 11
Green and Renewable Energy-Related Financial Literacy Affecting Socio-Ecological Systems Resilience in México

José G. Vargas-Hernandez
https://orcid.org/0000-0003-0938-4197
Financial Literacy Research Center, Universidad Cristóbal Colón, Mexico & Tecnológico nacional de México, ITS Fresnillo, Mexico

Absalón J. Salmerón-Zapata
https://orcid.org/0009-0002-2621-2426
Universidad Cristóbal Colón, Mexico

Carlos A. Rodríguez-Maillard
https://orcid.org/0000-0003-2406-196X
Universidad Cristóbal Colón, Mexico

ABSTRACT

This comprehensive research report examines the intricate relationship between renewable energy and green financial literacy and how it impacts socio-ecological systems resilience in México. Drawing on systematic reviews of the literature and current policy contexts, this study offers empirical evidence of large gaps in information on renewable energy investment financial education and their relevance to Mexico's energy transition agenda. The findings indicate that greater financial literacy among green energy firms is the key to enhancing socio-ecological resilience

DOI: 10.4018/979-8-3373-5540-5.ch011

through effective decision-making, sustainable investment habits, and adaptation initiatives at the community level. The study concludes that financial literacy programs need to be integrated in renewable energy policy frameworks so that Mexico's ambitious 45% by 2030 renewable power generation target can be achieved while crafting adaptive socio-ecological systems that would be able to cope with climate change and economic risk.

INTRODUCTION

The intersection of renewable energy and financial literacy is a critical border for green development, particularly in the developing economies of Mexico. With global climate change and energy concerns, the role of financial literacy in facilitating the shift towards renewable energy systems has taken center stage. This report examines the effect of green and renewable energy-related financial literacy on socio-ecological systems resilience in Mexico, which plays an important role in Latin America's energy transition. Mexico's energy industry is being dramatically reshaped both by national policy aspirations and global climate necessities.

Mexico has ambitious clean energy deployment targets, with Mexico's 35 percent of electricity from clean sources by 2024 as established by the Energy Transition Law and General Climate Change Law. But available data indicates there is little likelihood of attaining the goal, and that indicates the multiple dimensions of the challenges of energy transition. It should be noted that the 35% target for 2024 is a legal mandate for clean energy (not achieved), while the 45% target for 2030 is a political commitment to renewables, which is not binding.

The concept of socio-ecological systems resilience encompasses the ability of interlinked social and ecological systems to withstand, adjust to, and transform as a consequence of pressure from the environment and economy. In renewable energy transitions, resilience is inherently linked with the ability of communities to achieve, access, and utilize efficiently financial instruments and mechanisms that promote sustainable energy development. Financial literacy in this field is extended beyond general financial competencies to include renewable energy economics competence, investment products, risk, and long-term sustainability considerations.

The research answers several critical questions: How does green and renewable energy investment financial literacy affect resilience at the community level? What are the primary drivers and barriers to establishing such literacy among Mexican citizens? In what ways can increased financial literacy assist in the development of more efficient renewable energy policy and community adaptation planning? In answering these questions, this study aims to provide actionable suggestions to policymakers, teachers, and community leaders in an effort to improve socio-ecological resilience through enhanced financial literacy in renewable energy contexts.

The significance of this research is underscored by Mexico's new commitment under President Claudia Sheinbaum's administration, which declared renewables will be encouraged in order to reach a 45% goal by 2030. In order to meet such a target, not only policy and technological measures are needed but also significant improvements in public understanding of renewable energy financing mechanisms and their broader socio-economic impacts.

The study probes the impact of financial literacy of sustainable energy transitions on socio-economic development, environmental sustainability, and community-level resilience. The study covers Mexico's recent renewable energy industry, where, according to IRENA (2024) statistical profile, Mexico generated 22% of electricity from renewable sources in 2024 and discusses the possibilities and challenges of upscaling financial literacy in this industry. Some of the major recommendations are the establishment of specialized financial education schemes, community learning networks, and policy incentives promoting financial literacy along with the use of renewable energy.

In Mexico, financial literacy in renewable energy requires understanding SENER policies, CRE regulations (including CELs), CENACE system operations, CFE tariff schemes, and financial products regulated by CNBV.

CONCEPTUAL FRAMEWORK

Financial Literacy in the Green Energy Context

Financial literacy, traditionally defined as the ability to understand and effectively use miscellaneous financial skills like personal money management, budgeting, and investing, takes on larger perspectives when used in the context of green and renewable energy markets. Green energy financial literacy means understanding renewable energy investment plans, assessing long-term financial benefits of sustainable energy systems, knowledge of government incentives and financing plans, and identification of risks involved by new green technologies.

Following the OECD/INFE framework (OECD, 2022), green financial literacy comprises: Knowledge: understanding financial instruments and energy policies. Behaviors: practices for assessing risks and making sustainable decisions. Attitudes: willingness to engage in responsible investments and community resilience. Digital literacy: use of digital platforms to access information and financial products related to renewable energy.

This wider definition acknowledges that good participation in the transition towards renewable energy involves the need for citizens to grasp sophisticated financial ideas such as net metering, power purchase agreements, green bonds, carbon credits, and lifecycle cost analysis. Furthermore, it includes the ability to decide upon the economic viability of renewable energy projects both at the level of the individual and of the community, to understand the interface between energy spending and long-term financial sustainability, and to make informed decisions regarding energy investment that are aligned with both individual economic goals and broader ecological goals.

Socio-Ecological Systems Resilience

Socio-ecological systems resilience refers to the capacity of interlinked social and ecological systems to maintain their essential functioning and identity and adapt to change and disturbance. In the context of energy transition, this resilience is articulated through how communities can maintain energy security in transition to sustainable sources, adapt to changing economic conditions based on energy markets, maintain ecological integrity in the face of development needs, and transform social and economic arrangements to support long-term sustainability.

The resilience perspective recognizes that social and environmental systems are closely interconnected, and that change in one will necessarily create consequences for the other. Energy systems serve as critical infrastructure bridging the two, and therefore a move toward renewable energy is simultaneously an opportunity to create more resilience and an agent of system stress if not carefully managed.

Green and Renewable Energy Systems

Green and renewable energy systems comprise technologies and approaches that generate energy from naturally renewing sources with minimum environmental impact. In the Mexican context, they comprise solar photovoltaic systems, wind power generation, small hydroelectric plants, biomass and biogas systems, geothermal power, and emerging technologies like green hydrogen production. Clean Energy Certificates (CELs), regulated by the CRE, are the Mexican instrument for complying with clean energy obligations and should not be confused with generic RECs.

Following the IEA's flexibility framework, time-of-use rates and demand response are critical tools for increasing the share of variable renewable energy (VRE) in the electricity system. These strategies allow consumption to be shifted to hours of high renewable generation, reducing costs and improving system stability.

The economic dimensions of these systems include initial capital requirements, operating and maintenance expenses, revenue generation mechanisms, risk factors and mitigation measures, regulatory policies and policy inducements, and long-term economic viability analysis. It is necessary to realize these economic facets in decision-making at all levels, from individual homes considering rooftop solar installations to populations evaluating larger-scale renewable energy projects.

THEORETICAL INTEGRATION

This study's conceptual model integrates several key concepts to understand how financial literacy can impact socio-ecological resilience in renewable energy. The theory of adaptive capacity suggests that those with greater knowledge and resources possess a better ability to adapt to changing circumstances. Financial literacy is a key form of knowledge that enhances adaptive capacity by providing more intelligent decisions for energy investment and its long-term implications.

Social learning theory focuses on the role of collective learning processes in the development of resilience. The development of financial literacy in renewable energy settings is usually through community-based learning processes, whereby individuals and communities exchange knowledge, experience, and coping strategies for understanding intricate energy and financial choices. Collective learning strengthens overall community resilience through the development of shared knowledge and adaptive response capability.

Amartya Sen and Martha Nussbaum's capability approach is another conceptual framework through which financial literacy in resilience can be analyzed. It aims at developing human capabilities so that individuals and communities may select their own concepts of well-being and development. Financial literacy in renewable energy can be considered a key capability that enhances people's and communities' choices in pursuing sustainable development goals.

LITERATURE REVIEW

Financial literacy, traditionally defined as the ability to understand and effectively use miscellaneous financial skills like personal money management, budgeting, and investing, takes on larger perspectives when used in the context of green and renewable energy markets. Green energy financial literacy means understanding renewable energy investment plans, assessing long-term financial benefits of sustainable energy systems, knowledge of government incentives and financing plans, and identification of risks involved by new green technologies.

Socio-Ecological Systems Resilience

Socio-ecological systems resilience refers to the capacity of interlinked social and ecological systems to maintain their essential functioning and identity and adapt to change and disturbance. In the context of energy transition, this resilience is articulated through how communities can maintain energy security in transition to sustainable sources, adapt to changing economic conditions based on energy markets, maintain ecological integrity in the face of development needs, and transform social and economic arrangements to support long-term sustainability.

The resilience perspective recognizes that social and environmental systems are closely interconnected, and that change in one will necessarily create consequences for the other. Energy systems serve as critical infrastructure bridging the two, and therefore a move toward renewable energy is simultaneously an opportunity to create more resilience and an agent of system stress if not carefully managed.

Green and Renewable Energy Systems

Green and renewable energy systems comprise technologies and approaches that generate energy from naturally renewing sources with minimum environmental impact. In the Mexican context, they comprise solar photovoltaic systems, wind power generation, small hydroelectric plants, biomass and biogas systems, geothermal power, and emerging technologies like green hydrogen production.

The economic dimensions of these systems include initial capital requirements, operating and maintenance expenses, revenue generation mechanisms, risk factors and mitigation measures, regulatory policies and policy inducements, and long-term economic viability analysis. It is necessary to realize these economic facets in decision-making at all levels, from individual homes considering rooftop solar installations to populations evaluating larger-scale renewable energy projects.

Theoretical Integration

This study's conceptual model integrates several key concepts to understand how financial literacy can impact socio-ecological resilience in renewable energy. The theory of adaptive capacity suggests that those with greater knowledge and resources possess a better ability to adapt to changing circumstances. Financial literacy is a key form of knowledge that enhances adaptive capacity by providing more intelligent decisions for energy investment and its long-term implications.

Social learning theory focuses on the role of collective learning processes in the development of resilience. The development of financial literacy in renewable energy settings is usually through community-based learning processes, whereby individuals and communities exchange knowledge, experience, and coping strategies for understanding intricate energy and financial choices. Collective learning strengthens overall community resilience through the development of shared knowledge and adaptive response capability.

Amartya Sen and Martha Nussbaum's capability approach is another conceptual framework through which financial literacy in resilience can be analyzed. It aims at developing human capabilities so that individuals and communities may select their own concepts of well-being and development. Financial literacy in renewable energy can be considered a key capability that enhances people's and communities' choices in pursuing sustainable development goals.

The literature indicates that the development of renewable energy in Latin America is closely linked to the evolution of financial markets and investment in research, which implies that the energy transition requires not only infrastructure but also robust financial mechanisms (Carreón Sosa & Bonilla Vargas, 2022).

EMPIRICAL EVIDENCE

Financial Literacy and Renewable Energy Adoption

Empirical research on the link between financial literacy and renewable energy adoption has produced mixed but on balance favorable outcomes. A comprehensive analysis of German residential solar PV adoption by Mills and Schleich (2012) showed that financial literacy was a key determinant of adoption decisions, with financially literate households being more likely to correctly estimate the long-term financial returns from solar installations.

Sardianou and Genoudi (2013) study on determinants of residential renewable energy use in Greece found that awareness of renewable energy choices and actual adoption decisions were related positively to financial literacy. However, the study found that the connection was mediated by other factors including incomes, education, and environmental beliefs, suggesting that merely financial literacy is not sufficient to trigger adoption.

Palm (2018) conducted a longitudinal study of Swedish community energy initiatives and, from it, learned that communities that were more financially literate collectively were more skilled at envisioning and accomplishing renewable energy projects. The study revealed multiple mechanisms through which financial literacy contributed to project success and entailed increased capability in raising funds, increased risk management and evaluation, and increased community engagement and involvement.

Socio-Ecological Resilience in Energy Contexts

Socio-ecological resilience studies in energy environments have investigated how regions and populations adapt to alterations in the energy system and maintain essential functions during transition. Bouzarovski and Simcock (2017) investigated energy resilience in European communities and found that communities with diverse energy mixes, with high proportions of renewable energy content, were resilient to energy price changes and supply disruptions.

Goldthau (2014) conducted research to examine the relationship between socio-ecological resilience and energy security in emerging countries and reached a conclusion that more energy-literate communities, including knowledge of renewable energy options and financing mechanisms, were better able to maintain energy security when hit with economic and environmental shocks. The study suggested local knowledge and capacity development to construct resilient energy systems.

Foxon (2013) explored the UK's transition relationships among energy transitions and socio-ecological resilience and posited that effective transitions required not only technological change but also radical transformation in social practice, institutions, and governance regimes. The study further emphasized financial literacy as central to enabling stakeholders' capacity to handle the complex changes and successfully participate in transition processes.

MEXICO AND LATIN AMERICAN CASE STUDIES

Mexican and Latin American studies have been instrumental in illuminating the link between finance literacy, renewable energy adoption, and socio-ecological resilience. Bazilian *et al.* (2014) examined Mexican renewable energy expansion and found that financial illiteracy among potential investors and communities was a challenge that could not be overcome for project development and implementation.

Jiménez-González and Vega-Azamar (2019) analyzed community-based renewable energy projects in rural Mexico and concluded that the most financially educated communities were best able to establish sustainable funding mechanisms and maintain long-term project sustainability. The authors defined some of the determinants of success, including cooperative financing model comprehension, ability to assess project risks and returns, and ability to appropriately engage with outside funding sources.

Delgado & López (2020) conducted a study that explored the role of financial literacy to Mexico's intermittent generation program and found that higher financial literacy played a role in increased participation in net metering programs and full financial returns from renewable energy investments. Financial literacy is also proven to be necessary to understand complex pricing programs and optimize energy production and consumption patterns.

CURRENT LITERATURE GAPS

There are still some significant gaps in current literature notwithstanding the growing interest in the triad of financial literacy, renewable energy, and socio-ecological resilience. First of all, a lot of existing research has focused on industrialized countries with well-formed financial markets as well as regulatory frameworks, thereby excluding generalizations of findings to developing countries like Mexico where financial systems as well as renewable energy markets are dynamically emerging.

Second, much of the research that has been undertaken already has been at the level of the individual or household, with less attention to community-level processes and collective learning. This is especially relevant because community-based renewable energy development is at the heart of the renewable energy strategies of most developing countries.

Third, too little is understood about the exact mechanisms through which financial literacy operates to influence socio-ecological resilience in energy situations. While some research has found positive associations between financial literacy and different dimensions of resilience, the mediating variables and causal processes remain unknown.

Fourth, most of the existing research has focused on traditional measures of financial literacy and not on the more encompassing constructs of green or sustainable financial literacy, which may be more relevant to renewable energy contexts. This limitation constrains our understanding of what specific knowledge and skills are most relevant for participation in renewable energy transitions.

Lastly, little longitudinal investigation of how socio-ecological resilience and financial literacy change over time in renewable energy environments has been conducted. This is particularly critical because energy transitions are dynamic processes, and it is essential to comprehend how learning and adaptation processes unfold over time.

METHODS

Research Design

This study employs a mixed-methods approach that involves quantitative analysis of secondary data and qualitative inquiry into policy documents, case studies, and stakeholder perceptions. The approach is structured around an in-depth review and integration of published literature, policy examination, and empirical data from Mexico's renewable energy sector. This is to facilitate holistic investigation of the relationship between financial literacy and socio-ecological systems resilience without compromising on the complex, multi-faceted nature of such phenomena.

The approach builds on systems thinking methodologies that recognize the interdependent nature of financial literacy, renewable energy uptake, and socio-ecological resilience. Rather than treating them as individual variables in isolation, analysis explores their dynamic interactions and feedback relationships in the larger picture of Mexico's energy transition and sustainable development goals.

Data Sources and Collection

Secondary Data Analysis

The primary sources of information in this study are official government statistics on the use of renewable energy and financial inclusion of the Mexican Ministry of Energy (SENER), the National Commission for the Efficient Use of Energy (CONUEE), and the National Banking and Securities Commission (CNBV). The quantitative information on renewable energy capacity, generation, investment flows, and financial inclusion indicators form the foundation for exploring the relationship between financial literacy and consumption of renewable energy.

Other secondary data sources include reports generated by international organizations such as the International Renewable Energy Agency (IRENA), the World Bank, and the International Energy Agency (IEA) (2024) that provide comparative analysis and data on renewable energy expansion and financial inclusion in Mexico and other countries. These sources enable contextualization of Mexican experience against more general global trends and patterns.

Policy Document Analysis

Close examination of Mexican energy and financial policy reports provides insight into the regulatory and institutional framework in place that affects the relation between financial literacy and renewable energy adoption. Of most notes are the Energy Transition Law, the General Climate Change Law, the National Energy Strategy reports, and financial inclusion policy manuals developed by the CNBV and other regulatory bodies.

The policy analysis takes into account how existing policies address the nexus between renewable energy and financial literacy, where opportunities and potential room for policy integration are identified, and where there is a good fit between policy objectives and implementation mechanisms. This analysis is valuable background to understanding the institutional drivers of socio-ecological resilience in renewable energy environments.

Case Study Selection and Analysis

The study involves the analysis of a number of case studies representing different approaches to integrating financial literacy and renewable energy promotion in Mexico. Cases were selected to be representative of different geographic regions, rural or urban settings, and renewable energy technology, and to provide insights into determinants of successful integration of financial literacy and renewable energy programs.

Selection criteria for the case studies include access to documented proof of project development and implementation processes, evidence of community engagement and involvement in planning and management of projects, demonstration of new finance and capacity-building approaches, and measurable outcomes in terms of both renewable energy adoption and community resilience indicators.

Analytical Framework

Systems Analysis Approach

The analytical framework applies a systems analysis approach that examines the complex relationship between financial literacy, renewable energy use, and socio-ecological resilience. The approach recognizes that these phenomena are part of greater social, economic, and environmental systems and that their interactions have feedback effects, non-linear dynamics, and emergent properties. The systems analysis considers various levels of interaction, ranging from individual and household decision-making to community-level collective action and national-level policy implementation. This multi-level perspective allows for the identification of leverage points where interventions could be most impactful in promoting both financial literacy and socio-ecological resilience.

Resilience Assessment Framework

The study applies a resilience measurement approach that accounts for different dimensions of socio-ecological resilience, including economic resilience (ability to maintain economic stability and growth in energy transitions), social resilience (ability to develop community cohesion and collective action), and ecological resilience (maintaining ecosystem services and environmental quality).

The methodology for evaluating resilience involves quantitative indicators (e.g., energy security indicators, stability of the economy indicators, and indicators of environmental quality) as well as qualitative assessments (e.g., community perceptions of resilience, adaptability, and transformation potential). The integrated approach enables a nuanced understanding of the multiple ways in which financial literacy influences elements of resilience.

Comparative Analysis

The analytical framework consists of comparative analysis, which examines variations in financial literacy, renewable energy uptake, and socio-ecological resilience across regions, communities, and population groups in Mexico. Comparative analysis enables identification of drivers for effective integration of financial literacy and renewable energy programs as well as understanding scalability and transferability of different approaches.

The comparative assessment further entails exploration of the experience of Mexico in the broader Latin American and other global environments in which similar efforts have been made. Such a broader comparative exercise provides insights regarding the contextual determinants of success of different strategies and delineates lessons learned that may be applicable to Mexico's current process of energy transition.

Limitations and Constraints

There are a number of limitations and boundaries to this research design, and these must be taken into account when interpreting findings. First, use of secondary data sources limits the extent to detailed explorations of causal mechanisms and relations. Even though the available data provides intriguing implications for associations and trends, it does not lend support to conclusive statements regarding causality.

Second, the focus on documented case studies can potentially generate selection bias, as well-documented projects and programs may not be representative of the broader universe of renewable energy and financial literacy programs in Mexico. This is somewhat mitigated by using multiple case studies that cover a variety of settings and approaches.

Third, the evolving nature of Mexico's policy and energy transition environment ensures that some of the inferences made will date quickly. The analysis does its best to counter this limitation by focusing on underlying relationships and trends rather than specific policy specifics or numerical thresholds that will obviously shift with time.

Fourth, the interdisciplinary nature of this research requires integration of theory and methods from multiple disciplines including economics, sociology, environmental studies, and public policy. While the interdisciplinary nature deepens understanding, it also poses challenges in terms of methodological coherence and comparability across analysis areas.

RESULTS ANALYSIS

Mexico's Current State of Financial Literacy in the Renewable Energy Sector

Baseline Financial Literacy Levels

A survey of publicly available data indicates significant gaps in financial literacy regarding renewable energy investment among Mexican communities and beyond, within households. National surveys conducted by the CNBV indicate that while general financial literacy has improved over the past decade, understanding of complex financial instruments and investment arrangements remains weak, particularly in rural regions where renewable energy supply is most likely to be high.

In terms of installed capacity, Mexico has 33.6 GW of renewable technologies (IRENA, 2024), while in terms of generation, they represent 22% of the total. This difference between capacity and generation reflects factors such as resource variability and operational efficiency.

To financial literacy of renewable energy, survey data show that fewer than 30% of Mexican households understand basics such as net metering, power purchase agreements, or long-term cost implications of renewable energy investments. Such limited understanding presents significant barriers to renewable energy adoption and may provide insight into the disparity between Mexico's renewable energy targets and deployed rates.

The data provided indicates rampant variation in the degree of financial literacy by demographic sector and regional area. City-dwelling, higher-income groups have significantly greater levels of renewable energy financial literacy compared to rural, lower-income groups. This variation is particularly foreboding given the fact that the majority of Mexico's best renewable energy assets reside in rural communities where financial literacy levels are lowest.

Institutional Capacity for Financial Education

Institutional ability analysis in the area of renewable energy-related financial education reveals an uneven trend with various organizations and institutions providing different arrangements of training and support. The National Banking and Securities Commission has developed broad programs of financial literacy that include limited content specifically addressing renewable energy investments or green financial products.

Certain multilateral and bilateral international development agencies and non-governmental institutions have organized specialized financial literacy programs with a focus on renewable energy, but these tend to be small scale and inadequately coordinated with overall policy responses. This lack of coordination limits the efficiency of financial education and creates gaps in coverage that can deny access to many communities to relevant financial literacy materials.

The report identifies key opportunities for the improvement of institutional coordination and the development of more integrated approaches to renewable energy financial education. International best practice suggests that overall solutions that combine financial education with technical training and policy support are more likely to be successful than one-off financial literacy interventions.

Barriers to Financial Literacy Development

The research identifies several significant obstacles to the establishment of renewable energy financial literacy in Mexico. Cultural and language difficulties are significant challenges, particularly in indigenous communities where Spanish may not be the first language and where conventional financial knowledge could be incommensurable with conventional financial literacy strategies.

Limited access to credible information about renewable energy technology, costs, and sources of finance constitutes additional barriers to attaining financial literacy. Most communities lack exposure to reliable sources of information on renewable energy opportunities, thus relying on informal networks that potentially do not provide comprehensive or precise information.

Regulatory complexity in Mexico's renewable energy sector presents yet another dimension of complexity for the process of financial literacy. Interaction between federal, state, and local fiscal policies, as well as shifting policy frameworks, complicates the process for individuals and groups to understand their options and make good financial decisions in terms of investing in renewable energy.

Deployment of Renewable Energy and Financial Mechanisms

Current Renewable Energy State

Mexico's renewable energy sector has expanded robustly in recent years, though improvement toward national targets remains irregularly distributed. Mexico generated 22% renewable energy compared to the legal target of 35% clean energy, below the global average of 32% and far less than the regional average of 62%. This figure reveals both the potential for continued expansion and how far Mexico still has to go toward meeting Mexico's renewable energy targets.

The trend of renewable energy deployment exhibits dramatic geographic concentration, with the majority of large-scale projects clustering in those states possessing favorable resource conditions and policy climates. Wind power expansion has been highly concentrated in Oaxaca as well as other southern states, while solar expansion has been clustered in north-facing states with elevated solar irradiation. This. geographical concentration can potentially be experienced in the construction of financial literacy. because, possibly, communities there may be able to access renewable energy projects and funding frameworks more effectively.

A review of financing arrangements of projects shows extensive reliance on mega-commercial and foreign finance, with little involvement of local community and small-scale investors. This kind of financing pattern suggests that current levels of financial literacy are not sufficient to support more decentralized and community-based renewable energy development models, whose potential lies in enhancing socio-ecological resilience.

Financial Mechanisms and Accessibility

The Mexican renewable energy funding landscape consists of a vibrant portfolio of mechanisms from traditional project finance approaches to innovative green bonds and development finance schemes. Analysis does find, however, that the majority of these schemes are inaccessible to small investors or communities due to high minimum investment levels, complex application processes, and limited financial literacy support.

Government incentives and support programs, including tax credits, accelerated depreciation schedules, and direct subsidies, provide important financial support for the development of renewable energy. To be sure, the report concludes that there is a lack of awareness and knowledge about these programs, particularly among smaller potential investors and rural communities that could be benefited by distributed renewable energy systems.

The expanding market for green financial products, including green bonds, sustainable investment funds, and climate-resilient insurance products, represents a tremendous opportunity to expand financing for renewable energy. But such products often require more than minimal financial literacy and may not reach the majority of Mexican households and communities without additional training and support.

Community-Level Financing Innovations

A number of innovative community-level renewable energy financing mechanisms are identified by the analysis as having potential as tools for integrating the development of financial literacy with the deployment of renewable energy. Cooperative financing models, where communities finance renewable energy projects cooperatively, have been successful in various Mexican states, though scale is low. Microfinance institutions have begun to design specialized products for small-scale renewable energy systems like solar home systems and small business energy systems. These products typically include financial education elements that help the borrowers with understanding the long-term cost implications of their investments and with gaining the skill for managing renewable energy systems cost-effectively.

NGOs and community organizations have developed innovative mechanisms for linking financial literacy training with developing renewable energy projects. These typically entail participatory planning processes that allow communities to understand the financial, technical, and social aspects of renewable energy projects and local capacity development for project operation and maintenance.

Socio-Ecological Resilience Indicators

Economic Resilience Measures

Analysis of economic resilience metrics also indicates mixed trends in different regions and communities within Mexico. Those communities that have greater renewable energy implemented tend to indicate higher economic resilience as measured by such indicators as energy cost stability, local diversification of the economy, and less exposure to energy price external shocks.

However, the analysis also illustrates that economic returns from renewable energy development are often unequally distributed since some communities reap substantial economic gains while others attain limited direct economic gains. This trend suggests that financial literacy may have an important role in assisting communities to capture and maximize the economic returns of renewable energy development.

It appears that mediation of the economic resilience and financial literacy relationship exists through a variety of factors, including ability to access financial services, participation in renewable energy project development, and ability to understand and utilize government incentive programs. Locations with higher financial literacy will presumably participate effectively in renewable energy development and reap long-term economic benefits from such investments.

Social Resilience Dimensions

Social resilience indicators examine the capacity of communities to be together, be cohesive and adaptive, and respond to change. The research shows that communities working on the development of renewable energy go through social change, whereby some communities have increased social cohesion and collective efficacy while others report tension and conflict regarding project development and benefit sharing. Financial literacy seems to have a significant role to play in mediating such social dynamics.

Those communities with greater collective financial literacy are likely to have more inclusive decision-making regarding renewable energy projects and to work towards fair mechanisms for the sharing of project benefits. In contrast, less financially literate communities are likely to be more exposed to exploitation or to experience conflict due to poor comprehension of project financial arrangements.

The analysis also calls out several mechanisms through which financial literacy contributes to enhancing social resilience, including increased collective decision-making capability, improved negotiation ability with external developers and investors, and increased capability to act on complicated technical and financial information. These mechanisms offer some explanation for why development of financial literacy could be a vital component of community resilience-building responses.

Environmental Resilience Outcomes

Environmental resilience indicators focus on the ability of communities to support ecosystem services and environmental quality while pursuing economic development. From the analysis, it is clear that renewable energy development can support environmental resilience by reducing local air pollution, reducing dependence on fossil fuel imports, and creating economic incentives for environmental conservation. However, environmental implications of renewable energy projects are not always beneficial, and the research highlights several examples wherein renewable energy projects have created environmental concerns like land use conflicts, effects on local ecosystems, and waste management issues related to renewable energy technologies.

Financial education appears to assist in making communities better understand and manage these environmental trade-offs.

More financially educated societies will probably be more engaged in comprehensive environmental impact assessments and pursue environmental safeguards as part of renewable energy project initiatives. This would mean that improving financial literacy might translate into greener approaches to renewable energy development.

Policy Integration and Institutional Coordination

Current Policy Landscape

Mexico's renewable energy policy framework comprises a number of frameworks and institutions with varying degrees of coordination and integration. Energy Transition Law establishes the overall framework for renewable energy development, while the General Climate Change Law establishes broader sustainability objectives. However, few cross-links are observed between renewable energy policies and programs of financial literacy or financial inclusion.

Mexico undertook 35% clean energy by 2024 and 50% by 2050, and these have been backed in various national plans. However, little policy direction exists in the current policy framework regarding how to build financial literacy and local capacity that can be used to achieve these ambitious targets via distributed and community-level rollout of renewables. The assessment outlines several areas where there are opportunities for improved policy coordination, including the development of coordinated financial education activities with renewable energy deployment objectives, the creation of incentive structures that reward community capacity development alongside renewable energy expansion, and the establishment of institutional frameworks for regulating coordination across regulators in the energy and finance sectors. The Energy Transition Act sets a legal target of 35% clean energy by 2024, while the current administration has announced a policy goal of reaching 45% renewables by 2030.

Institutional Coordination Challenges

The fragmented institutional framework of financial literacy and renewable energy creates significant coordination challenges. There are a number of federal institutions, including SENER, CONUEE, CNBV, and so on, that regulate different renewable energy development areas as well as financial literacy, but there are few coordination arrangements in place. State and local governments play important roles in renewable energy development and community education, but capacity and resources for these roles vary greatly across regions. The study finds that some states have developed effective models for integrating renewable energy development with capacity building at the community level, while others lack both the resources and the institutional capability to implement full programs.

The private sector, from financial institutions to renewable energy developers and community groups, have important roles to play in the development of renewable energy, as well as constructing financial literacy. The analysis indicates, though, that there is little coordination between private sector players and government institutions, thus limiting the scope of system-wide efforts.

Opportunities for Greater Integration

Despite coordination challenges, analysis identifies several opportunities for higher integration between renewable energy and financial literacy policies and interventions. New policy reforms like the government's plans to increase the share of renewable energies to 45% by 2030 offer opportunities for developing more integrated approaches taking on both renewable energy deployment and capacity building within communities.

Growing concern for the role of community participation in renewable energy production provides even greater opportunities for integrating development of financial literacy into project design and implementation processes. Various successful examples from other countries attest that integrated approaches are more effective than individual programs in achieving the objectives of renewable energy and community development.

Emerging financial technology like mobile banking, digital payment systems, and online learning platforms provides new opportunity for delivering financial literacy education at scale and lower cost, with wider access. Such technologies can be particularly valuable in rural areas where traditional means of provision of financial education may be limited. AI-powered solutions, such as virtual assistants and adaptive learning platforms, can improve financial literacy in renewable energy. These tools personalize content based on user level, analyze behavior patterns, and offer real-time investment simulations, reducing barriers to understanding.

DISCUSSION

Synthesis of Key Findings

The results reveal several important insights into the nexus between financial literacy, the adoption of renewable energy, and the resilience of socio-ecological systems in Mexico. First and foremost, the study demonstrates how financial literacy is a crucial mediating variable between the aims of renewable energy policies and community-level effects. Societies with higher levels of renewable energy financial literacy are more likely to participate favorably in renewable energy development, capture long-term economic benefits from investing in renewable energy, and maintain social cohesion during energy transition initiatives.

However, the analysis finds large disparities between current financial literacy and adequate knowledge and skills to meaningfully participate in Mexico's renewable energy transition. The fact that in 2024 Mexico generated 22% of its electricity using renewable sources, lower than the world average of 32% suggests that current approaches to development of renewables may not be sufficient to achieve national ambitions and that increased financial literacy may have a crucial part in accelerating deployment by more scattered and people-focused means.

The research outlines several mechanisms through which financial literacy increases socio-ecological resilience, including increased adaptive capacity for energy price volatility and supply shock management, increased community capacity for collective decision-making on energy investments, increased capacity to access and make use of government incentive schemes, and increased local economic development through renewable energy entrepreneurship and local employment.

Theoretical Implications

Enlarging Financial Literacy Frameworks

The study contributes theoretical understanding by explaining the need to extend traditional financial literacy models to encompass sustainability and environmental dimensions. The study indicates that conventional notions of financial literacy, though important, are insufficient for effective participation in renewable energy transformations. This calls for new theoretical models integrating financial, environmental, and social dimensions of literacy.

The concept of "green financial literacy" that emerges from this analysis encompasses more than typical financial understanding, inasmuch as it entails understanding environmental externalities, long-term sustainability concerns, and the complex interaction of financial and environmental systems. The expanded framework has implications for conceptualizing, measuring, and producing financial literacy in settings where environmental sustainability is a primary concern.

The research also contributes to understanding of the ways in which financial literacy operates at various scales, from individual decision-making to collective action in neighborhoods and at the regional economic development level. Multi-scale impacts of financial literacy mean that theories must address these different dimensions of effect as well as the mechanisms by which knowledge at the micro-level is translated into collective effects.

Socio-Ecological Systems Theory Applications

The application of socio-ecological systems theory to examine renewable energy transitions is critical in providing insightful understandings of the interactive dynamics in such processes. The research indicates that financial literacy is a key component of adaptive capacity for socio-ecological systems, facilitating reactions to more effectively handle environmental as well as economic changes.

The findings align with theoretical perspectives emphasizing the value of learning and knowledge systems in the development of socio-ecological resilience. Financial literacy development can be conceptualized as social learning that enhances system-level capacity for transformation and adaptation. This perspective has implications for intervention design and delivery aimed at building resilience.

The research also contributes to understanding how different dimensions of resilience (economic, social, environmental) interact in renewable energy systems. Financial literacy is revealed in the analysis to affect all three dimensions but through different mechanisms and thereby suggesting that developing resilience involves consideration of these multiple mechanisms and their interaction.

Energy Transition Dynamics

The findings contribute to energy transition theory by highlighting the crucial role of community-level capacity building in facilitating successful transitions. While much energy transition research focuses on technological and policy factors, this research demonstrates that individual and community-level knowledge and skills are equally important for achieving transition objectives.

The research validates theoretical perspectives that emphasize the multi-level nature of energy transitions, in which landscape-, regime-, and niche-level processes all determine transition outcomes. Financial literacy operates across all three levels, determining how individuals and communities respond to landscape-level pressures, interact with regime-level changes, and adopt niche innovations.

The analysis also improves knowledge about how social acceptance and participation in energy transitions can be stimulated by capacity-building interventions. The study suggests that building financial literacy is likely a more impactful strategy for stimulating social acceptance than traditional approaches focused on information provision or economic incentives.

Policy Implications and Recommendations: Integrated Policy Framework Development

The research findings have significant policy implications for developing policy in Mexico and elsewhere that aims for renewable energy transitions. The analysis suggests that achieving bold renewable energy comes from clean sources targets, such as Mexico's 45% by 2030, will require more coordinated approaches that address technological deployment and community capacity building.

The research recommends the establishment of cohesive policy frameworks with an explicit linkage among renewable energy deployment targets and financial literacy building objectives. This ought to include objective criteria for enhancing financial literacy, specific budgetary allocations for capacity development, coordination mechanisms among energy and finance sector bodies, and performance metrics for measuring renewable energy deployment as well as community capacity enhancement.

The conversation also suggests the need for policy models facilitating community-scale renewable energy development models with the capacity to tap increased financial literacy for more dispersed and robust energy systems. These include policies that facilitate community ownership of renewable schemes, provide technical assistance in planning by communities, and come up with funding mechanisms appropriate for smaller-scale, community-level schemes.

Financial Education Program Design

The study offers direct directions for the development of financial education programs that work well in facilitating renewable energy transitions and socio-ecological resilience. The paper implies that productive programs should combine technical and financial education, which equips participants with both the financial literacy and the hands-on expertise for renewable energy development.

Programs need to be designed to address the specific needs and environments of specific communities, as rural and urban, indigenous and non-indigenous, and socio-economic groups may have differing needs for the development of financial literacy. The study emphasizes the requirement for culturally appropriate and linguistically sensitive educational approaches. The review also points towards the requirement for participatory and community-focused financial education that takes into consideration collective capacity building as well as individual knowledge development. The strategy must deliver opportunities for learning from each other, planning activities in the community, and hands-on access to renewable energy technologies and funding schemes.

Institutional Capacity Building

The research requires significant capacity building at the institution level in order to enable integrated renewable energy development as well as financial literacy. This includes developing expertise in the government institutions in crafting and implementing integrated programs, training financial educators on renewable energy technology as well as financing alternatives and developing community organization capacity for offering financial education services.

The report calls for new institutional arrangements to facilitate coordination across sectors and levels of government. This could include setting up inter-agency task forces on renewable energy and inclusion of financial literacy, creating common performance indicators and reporting mechanisms, and establishment of funding mechanisms that maintain coordinated program delivery.

The research also points to forming partnerships between private sector players, including financial institutions, renewable energy developers, and providers of technology, to leverage their expertise and resources in financial literacy development. The partnerships should be structured in a way that brings together private sector interests and public policy and community development interests.

Practical Applications and Implementation Strategies: Community-Level Implementation

The findings of the research offer practical advice for financial literacy and renewable energy integration program implementation at the community level. The analysis indicates that effective implementation hinges on close attention to processes of engagement in communities beginning with an assessment of the knowledge, resources, and priorities prevailing in communities.

Effective implementation strategies have to tap into existing community institutions and social networks rather than creating new ones. This includes partnering with current community groups, leveraging existing customary leadership structures, and coupling financial literacy building with other activities in community development on which communities are most focused.

The research emphasizes the provision of ongoing support and technical assistance rather than individual training interventions. The local communities need sustained support for project formulation, implementation, management and ongoing learning as well as skill development with financial instruments and technology evolving.

Regional and State-Level Strategies

At state and regional levels, the research suggests the need for comprehensive strategies that take into account the different needs and circumstances of different communities in wide geographic areas. This includes regional resource center establishment that can provide technical assistance and training services to multiple communities, peer learning network establishment that enables communities to share experience and lessons, and regional financing vehicle establishment that pools funds from multiple communities.

State policy should also prioritize the development of supportive policy and regulatory systems that facilitate community-based renewable energy development. This involves streamlining permitting for small-scale renewable energy systems, structuring incentives as programs targeting community-based projects, and expressing procurement policy that prefers locally-owned renewable energy development.

The study recommends that regional and state policies also aim to establish links among rural communities with renewable energy sources and urban centers with energy requirements. The connections may open up rural economic opportunities while supporting overall renewable energy goals.

National-Level Integration

At the national level, the policy implications of the research are towards the complete integration of financial literacy and renewable energy policies into overall sustainable development strategies. This involves alignment with Mexico's Nationally Determined Contributions under the Paris Agreement, incorporation into sustainable development goal implementation policies, and incorporation into overall poverty reduction and rural development policies.

These national strategies should focus on creating enabling environments for regional and local implementation by developing national curricula and standards for financial literacy on renewable energy, creating certification structures for educators of finance, and establishing monitoring systems and frameworks that track progress toward integrated goals. The research also calls for international collaboration and knowledge sharing in order to leverage international experience and best practice and concurrently transpose methodologies into Mexican contexts and demands. This involves membership in international networks for community energy development, collaboration with international agencies for development, and bilateral cooperation with countries that have successful experience with consolidated approaches.

Limitations and Future Directions for Research: Study Limitations

There are several study limitations to be considered in relation to the findings and implications thereof. Secondary data sources and document analysis limit close scrutiny of causal relations and thorough insight into the lived experiences of individuals and communities undertaking renewable energy and financial literacy initiatives.

The focus on Mexico, although providing insightful descriptions of a core Latin American setting, limits the degree of generalizability to other countries and countries that have different political, economic, and sociocultural settings. More studies must investigate such correlations in other settings in order to form more generalized conclusions. The very fast rate of development of renewable energy technologies and financial systems guarantees that some of the specific outcomes will be outdated comparatively quickly. Future research should aim to find out about underlying patterns and relationships that will remain stable even under conditions of technological and policy change.

Future Research Priorities

The work highlights several areas of future research priority that could be taken even further by taking these findings and some of these current limitations. Longitudinal investigations tracking populations and communities for extended periods as they progress through renewable energy and financial education programs would be extremely informative as to how these interactions evolve and what is a good indicator of long-term success.

Primary data collected through surveys, interviews, and participatory research methods with communities, financial educators, and policymakers would provide further insights into the process through which financial literacy translates into renewable energy adoption and socio-ecological resilience. The study should discuss, in specific terms, the experiences of marginalized groups and communities who may be facing additional barriers to engagement. Comparative research that contrasts different strategies of integrating financial literacy and renewable energy development across different countries and contexts would be extremely valuable in identifying what works best under what conditions. Such research would need to cross developed and developing country contexts to reflect how institutional capacity and economic development levels influence integration strategies.

Experimental and quasi-experimental designs that can establish causal relationships between financial literacy interventions and renewable energy and resilience impacts would provide more decisive evidence to inform policy and program design.

A few examples include randomized controlled trials of different financial literacy program designs or natural experiments based on policy differences across regions or over time.

In addition to longitudinal studies and experimental designs, it is recommended to incorporate participatory methodologies that involve communities in the co-creation of knowledge, as well as AI tools for predictive modeling and big data analysis. These approaches will make it possible to evaluate complex dynamics of resilience and financial literacy in real time, aligning research with global trends in energy innovation.

Methodological Innovations

Subsequent research must also explore methodological innovations that can more accurately capture the complex, multi-level, and dynamic relationships between financial literacy, renewable energy, and socio-ecological resilience. This means innovation in building new indicators and measurement approaches that can capture collective and community-level dimensions of financial literacy and resilience. System dynamics modeling and other complexity science approaches could provide valuable methods for understanding how such relationships change over time and how different intervention strategies could affect system-level outcomes. Such approaches could be particularly valuable for policy analysis and strategic planning. It is essential to distinguish between legal targets and political commitments, as this affects the interpretation of progress and gaps in the energy transition, although the political commitment to achieve 45% renewables by 2030 reflects ambition, recent studies point to significant constraints in infrastructure, financing, and governance that could limit its feasibility (IEA, 2024; IRENA, 2024).

Participatory research processes with communities as co-researchers, rather than objects of research, potentially can produce more authentic insights into community experience and community-defined priorities and support local capacity for ongoing learning and adaptation. Participatory methods may also guarantee that research findings are both relevant and useful for decision-making and action at the community level.

CONCLUSIONS

Summary of Key Findings

This comprehensive research report has examined the intricate relationship between green and renewable energy-financial literacy and socio-ecological systems resilience in Mexico and revealed a series of important findings that have important policy, practice, and future research implications. Analysis confirms that financial literacy also has a strong mediating function between renewable energy policy objectives and consequences at the community level, with more financially literate communities in terms of renewable energy having greater capacity to engage effectively in energy transitions, realizing more long-term economic benefits, and more maintenance of social cohesion in transition processes.

The research reveals significant gaps between current financial literacy in Mexico and the experience and skills required for sustainable participation in the renewable energy revolution in the country. Although Mexico's renewables provided 22% of the country's electricity in 2024, below world and regional averages, there is tangible evidence that current approaches to generating renewable energy may not be sufficient to achieve national objectives. Analysis goes on to claim that increased financial literacy can be key to accelerating deployment by more decentralized and community-led strategies that promote socio-ecological resilience.

The study identifies a number of mechanisms whereby financial literacy improves socio-ecological resilience, including enhanced adaptive capacity in responding to energy price shocks, enhanced community capability for collective decision-making about energy investment, improved access to government incentive schemes, and increased local economic growth through renewable energy entrepreneurship. Mechanisms transcend economic, social, and environmental dimensions of resilience, in line with multi-dimensionality of financial literacy effects.

The study further shows that socio-ecological resilience mediates the relationship between financial literacy and financial literacy. Various contextual factors, such as geographic location, socio-economic status, culture, and institutional capacity, mediate this relationship. Both rural and indigenous communities, although they may have rich renewable energy assets, are generally disadvantaged in terms of accessing financial literacy tools and engaging in effective renewable energy development.

Theoretical Contributions

This research provides several important contributions to theoretical understanding of the relationships between financial literacy, renewable energy transitions, and socio-ecological resilience. This research shows the importance to expand traditional financial literacy models to encompass sustainability and environmental variables and introduces the concept of "green financial literacy" integrating financial, environmental, and social dimensions of knowledge and decision-making ability.

The application of socio-ecological systems theory to renewable energy transitions provides important knowledge on the complex dynamics of those processes, with financial literacy as a key component of adaptive capacity in socio-ecological systems. The research proves theoretical propositions emphasizing the nature of learning and knowledge systems to build resilience while demonstrating how different components of resilience interact through the development of financial literacy.

The study further contributes to energy transition theory by highlighting the need for capacity building at the community level in preparation for successful transitions. While most existing literature focuses on technological and policy determinants, this study indicates that knowledge and capability at the individual and community levels are similarly critical to the achievement of transition objectives, particularly in developing country contexts where distributed and community-based strategies may be more appropriate than extensive centralized development forms.

Policy Recommendations and Implications

The research findings have significant implications for policy development in Mexico and other countries embarking on renewable energy transitions. The analysis decisively calls for the building of combined policy frameworks that specifically link renewable energy deployment objectives with financial literacy development targets, as opposed to maintaining the current piecemeal strategy of addressing these as separate policy domains.

Specific policy recommendations include initiation of synchronized financial education programs aligned with renewable energy deployment targets, development of incentive schemes that pay for community capacity building alongside renewable energy expansion, detailing of financing facilities appropriate for smaller-scale community undertakings, and implementation of performance measures quantifying renewable energy deployment and community capacity accomplishments.

The research points out the need for policy incentives for people-centered renewable energy development models that will be able to leverage enhanced financial literacy towards greater distributed and more resilient energy systems. This calls for a transition from current approaches centered on commercial-scale development towards more inclusive approaches enabling effective community involvement in ownership, management, and benefit sharing of renewable energy projects.

Analysis also places significant importance on the need to resolve institutional coordination issues by setting up inter-agency work groups, creating shared performance indicators and reporting mechanisms, and establishing funds that enable joint program implementation at sectoral and government levels.

Practical Implementation Strategies

The research provides practical guidance on the design and implementation of financial literacy and renewable energy integration programs at a range of levels from community through to country policy. At the community level, it is successful if there is sensitive attention to community participation procedures, building on current institutions and social networks, and ensuring long-term support rather than short-term interventions.

Effective financial education programs must integrate financial and technical training, to address the specific needs and circumstances of different communities and operate by participatory and community-based approaches towards developing collective capability. Financial education programs must be sensitive to cultural settings and accessible in terms of language while providing room for peer-to-peer learning and real practice with renewable energy technologies and finance vehicles.

At the state and regional levels, the research suggests a holistic approach that includes development of regional resource centers, establishment of peer learning networks, regional financing mechanisms setup, and supportive policy and regulatory structures that facilitate community-based renewable energy development.

National responses must be focused on creating enabling conditions for local and regional action by developing national standards and curricula, establishing certification systems for financial educators, creating monitoring and evaluation systems, and international collaboration and knowledge sharing to tap into experience and best practices from around the world.

Relevance to Mexico's Energy Transition

The results of the research are particularly pertinent to the case of Mexico's high-renewable-energy targets and difficulties in achieving them. With the government's goal of increasing the proportion of renewable energies to 45% in 2030, there is a pressing requirement for deployment policies that can scale up deployment and also increase community capacity and resilience.

The research supports that greater financial literacy can serve as a central pillar in achieving such ambitious objectives through more decentralized and community-based approaches as supporting large-scale commercial development. Through local-level capacity building in understanding, accessing, and managing renewable energy investment, Mexico is able to unleash large additional renewable energy prospects while promoting local-level socio-ecological resilience.

The study also has important policy implications for Mexico's entire sustainable development objectives, including poverty alleviation, rural development, and climate change adaptation. Linking financial literacy development with renewable energy programs could help address a number of development objectives simultaneously and build community capacity for long-term adaptation and resilience.

Since Mexico is a model country for Latin America in the development of renewable energy, lessons and approaches learned from the practice of overall financial literacy and renewable energy policies may have regional relevance and support in fostering the development of renewable energy in Latin America.

Global Relevance and Transferability

Even though this research is especially on Mexico, the results are more applicable to other developing countries that are embarking on renewable energy transformations and pursuing socio-ecological resilience construction. The issues encountered in Mexico, including the lack of financial literacy, institution coordination issues, and the need for people-centered strategies towards renewable energy development, are common in most developing countries.

The theoretical basis developed in this research, that is, the connection of financial literacy and socio-ecological systems resilience thinking, is one that can be used in other contexts with due regard to specific local needs and conditions. Mechanisms through which financial literacy enhances resilience ought to persist across contexts, though the specific interventions and means of delivery would need tailoring.

The research contributes, too, to global understanding of achieving the Sustainable Development Goals via interrelated strategies with multiple goals in a single focus. The networked relationship among financial literacy, renewable energy, and resilience-building demonstrates how targeted interventions can address multiple SDGs like affordable and clean energy, reduction of inequalities, sustainable communities and cities, and climate action.

Directions for the Future and Priorities for Research

This research offers some key directions for future research that could build upon these findings and address current limitations. Top priorities for future research include longitudinal research that tracks communities and individuals over time, primary research that includes participatory approaches with stakeholders and communities, comparative research in many countries and settings, and experimental research designs that are able to ascertain causality.

Methodological innovations capable of more effectively capturing the complex, multi-level, and dynamic interactions among renewable energy, socio-ecological resilience, and financial literacy are another important research agenda. This involves the development of new indicators and measurement tools, application of system dynamics modeling and complexity science approaches, and utilization of community-engaged research methodologies that involve the community as co-researchers.

The rapid evolution of renewable energy technology and finance systems imposes ongoing demands on research that can track the over-time dynamics of relationships between financial literacy and renewable energy uptake, and how intervention strategies may be adapted to track changing technological and market conditions.

FINAL REFLECTIONS

This research indicates that the transition to renewable energy systems is not merely a techno-economic or technical challenge but is a complex socio-ecological change involving the building of human and community capability as well as technology deployment. Financial literacy is a critical but underrated component of this capability building, and the potential to greatly enhance both the effectiveness of renewable energy transitions as well as the resilience of transitioning communities and regions.

The findings suggest that achieving ambitious renewable energy goals and building socio-ecological resilience require cross-cutting solutions that consider interdependencies between financial, energy, and social-ecological systems. Cross-cutting solutions are particularly relevant in developing countries like Mexico, where renewable energy transitions must occur together with broader development processes and where community-based approaches can be more acceptable and cost-effective than alternatives designed elsewhere.

The research ultimately points to the need for more integrated and inclusive renewable energy development strategies that involve building community capacity, ensuring equitable benefit sharing, and developing long-term resilience alongside technology deployment and economic efficiency. Integrating financial literacy development with renewable energy policy and deployment will allow Mexico and others to pursue energy transitions that are technically and economically viable but also socially inclusive and environmentally sustainable.

The path forward is supporting dialogue between policymakers, researchers, practitioners, and communities in designing, implementing, and refining strategies to effectively link financial literacy building with renewable energy transitions. Such dialogue must be premised on recognition of renewable energy transitions as social processes that require building human and community capabilities to understand, engage, and benefit from the fundamental shifts renewable energy uptake represents.

As Mexico works towards its ambitious renewable energy targets and broader sustainable development goals, the convergence of financial literacy and the development of renewable energy is both an attractive possibility and an issue of imperative need. The success of that convergence will have significant implications not just for Mexico's energy future but also for demonstrating paths that other emerging nations can follow in seeking their own sustainable energy transitions in constructing resilient and inclusive societies.

REFERENCES

Bazilian, M., Nakhooda, S., & Van de Graaf, T. (2014). Energy governance and poverty. *Energy Research & Social Science, 1*, 217–225. DOI: 10.1016/j.erss.2014.03.006

Bouzarovski, S., & Simcock, N. (2017). Spatializing energy justice. *Energy Policy, 107*, 640–648. DOI: 10.1016/j.enpol.2017.03.064

Carreón Sosa, R., & Bonilla Vargas, D. (2022). Energías renovables, PIB, mercados financieros e investigación: La experiencia de América Latina (2000-2019). *Revista Mexicana de Economía y Finanzas, 17*(4), e791. Advance online publication. DOI: 10.21919/remef.v17i4.791

Delgado, R., & López, M. (2020). Financial literacy and distributed generation adoption in Mexico: Evidence from household solar programs. *Energy Policy, 138*, 111250. DOI: 10.1016/j.enpol.2019.111250

Foxon, T. J. (2013). Transition pathways for a UK low carbon electricity future. *Energy Policy, 52*, 10–24. DOI: 10.1016/j.enpol.2012.04.001

Goldthau, A. (2014). Rethinking the governance of energy infrastructure: Scale, decentralization and polycentrism. *Energy Research & Social Science, 1*, 134–140. DOI: 10.1016/j.erss.2014.02.009

International Energy Agency. (2024). *Mexico energy outlook 2024*. IEA Publications.

International Renewable Energy Agency. (2024). *Renewable energy statistics 2024*. IRENA Publications.

Jiménez-González, A., & Vega-Azamar, R. E. (2019). Community renewable energy projects in rural Mexico: Success factors and financial literacy impacts. *Renewable Energy, 135*, 108–117. DOI: 10.1016/j.renene.2018.11.098

Mills, B., & Schleich, J. (2012). Residential energy-efficient technology adoption, energy conservation, knowledge, and attitudes: An analysis of European countries. *Energy Policy, 49*, 616–628. DOI: 10.1016/j.enpol.2012.07.008

OECD. (2022). OECD/INFE Toolkit for Measuring Financial Literacy and Financial Inclusion. OECD Publishing. https://www.oecd.org/en/publications/oecd-infe-toolkit-for-measuring-financial-literacy-and-financial-inclusion-2022_cbc4114f-en.html

Palm, J. (2018). Household installation of solar panels – Motives and barriers in a 10-year perspective. *Energy Policy, 113*, 1–8. DOI: 10.1016/j.enpol.2017.10.047

Sardianou, E., & Genoudi, P. (2013). Which factors affect the willingness of consumers to adopt renewable energies? *Renewable Energy*, *57*, 1–4. DOI: 10.1016/j.renene.2013.01.031

KEY TERMS AND DEFINITIONS

Net Metering: A system that allows users to offset their electricity consumption with the energy they generate and sell back to the grid, reducing their bills.

Power Purchase Agreement (PPA): A long-term contract between a generator and a buyer to fix the price of electricity, providing financial stability for the project.

Lifecycle Cost Analysis: An evaluation of the total cost of a project from installation to decommissioning, including operation and maintenance expenses.

Clean Energy Certificates (CELs): Instruments issued by Mexico's Energy Regulatory Commission (CRE) that certify electricity generation from clean sources, used to meet regulatory obligations.

Green Bonds: Financial instruments designed to fund projects with environmental benefits, such as renewable energy installations.

Chapter 12
Green Innovation and Renewable Energy in Mexico

José G. Vargas-Hernandez
https://orcid.org/0000-0003-0938-4197
Financial Literacy Research Center, Universidad Cristóbal Colón, Mexico & Tecnológico nacional de México, ITS Fresnillo, Mexico

Teresa I. Salazar-Echegaray
https://orcid.org/0000-0001-8785-6300
Universidad Autónoma de Sinaloa, Mexico

Julieta E. Salazar-Echegaray
https://orcid.org/0000-0003-0689-532X
Universidad Autónoma de Sinaloa, Mexico

ABSTRACT

Mexico stands at a tipping point in its energy transition, having set ambitious renewable energy targets and a growing recognition of the role that green innovation must play in achieving sustainable development goals. Key findings are that Mexico has set ambitious 35% clean energy by 2024 and 45% renewable electricity generation by 2030 targets supported by significant investment of over $30 billion in renewable projects. However, the country faces major issues like regulatory unpredictability, grid infrastructure limitations, and volatile political dedication to policies supporting renewable energy. The report identifies solar photovoltaic and wind technologies as the major drivers of expansion with emerging prospects in green hydrogen and energy storage technologies. Green innovation initiatives, while promising, require stronger institutional frameworks and more profound public-private partnerships to achieve full potential.

DOI: 10.4018/979-8-3373-5540-5.ch012

INTRODUCTION

Mexico's energy landscape is revolutionized through climate pledges, economic considerations, and technological advancements. As the 15th largest economy in the world and the second-largest country in Latin America, Mexico plays a significant role in regional and global energy transition processes (García-Olivares et al., 2021). The country's abundant renewable energy resources, including top-of-the-line solar irradiation values and exceptional wind potential, position it as a potential clean energy pioneer in the Latin American region.

Green innovation has been a major force for Mexico's renewable energy development agenda. Green innovation is the design and utilization of eco-friendly technologies, processes, and systems that guarantee sustainable development without sacrificing economic competitiveness (Rennings, 2000; Chen et al., 2022). For Mexico's energy sector, green innovation is achieved via cutting-edge solar tech, high-end wind energy installations, advanced grid integration systems, and new emerging technologies such as energy storage and green hydrogen production.

Mexico's renewable energy journey has been marked by significant policy developments, most notably the Energy Transition Law (Ley de Transición Energética) enacted in 2015 and the General Climate Change Law (Ley General de Cambio Climático), which established ambitious targets for clean energy adoption. Mexico's goal for 2024 is 35% of electricity from clean energy sources according to Mexico's Energy Transition Law and General Climate Change Law. More recently, the Mexican government has announced its plan to increase the share of renewable energies to 45% by 2030.

Mexico is at a crossroads in its energy transition, with aggressive clean energy targets and growing recognition of the strategic importance of green innovation to achieve sustainable development goals. Key findings show that Mexico has set ambitious targets of 35% clean energy for 2024 and 45% renewable power generation by 2030, supplemented by significant investments of over $30 billion in renewable ventures. However, the country has severe challenges like regulatory heterogeneity, grid backbone limitations, and varying political will for policy support for renewable energy. The report mentions solar photovoltaic and wind as major drivers of growth, with additional potential in green hydrogen and energy storage technologies.

This chapter analyses the function of competency-based education in financial literacy for renewable energies in strengthening socio-ecosystem resilience (SER) in rural and semi-urban areas, hence advancing Sustainable Development Goals 4, 7, and 13.

Green innovation policies, while promising, require stronger institutional frameworks and public-private partnerships to achieve maximum potential. The value of this research lies in the integrative examination of the influence of green innovation on the outcomes of developing renewable energy in Mexico, and thus to other emerging nations pursuing the same energy transition. This study addresses an essential void in existing literature on the intricate dynamics between technological innovation, policy frameworks, and market forces in the context of emerging economies.

Current renewable energy data reveal both potential and challenges. Mexico accounted for 22% of electricity provided from renewables in 2024, lower than the global average of 32% and significantly below the Latin American regional average of 62%. The performance gap remains an imperative to increase the installation of renewable energy capacity and the need for green innovation to achieve domestic targets.

The following research questions guide this study: How does green innovation influence renewable energy development in Mexico? What are the most significant drivers and inhibitors of the adoption of green innovation in the energy sector in Mexico? How do policy regimes enable or constrain efforts to promote green innovation? How significant is the role of public-private partnerships to foster green innovation? How can Mexico leverage its green innovation ecosystem for the achievement of renewable energy targets?

This detailed analysis contributes to the academic literature on green innovation and renewable energy development and provides actionable policy recommendations to policymakers, business executives, and international development institutions active in Mexico's energy sector.

BACKGROUND

Theoretical Green Innovation Literature

The intellectual foundation of the theory of green innovation is drawn from a number of disciplines including innovation studies, environmental economics, and technology management. Porter and van der Linde (1995) introduced the "Porter Hypothesis," which states that properly designed environmental policy can stimulate innovation that largely, if not completely, offsets compliance costs. This hypothesis is of particular interest to Mexico's context of renewable energy policy, where environmental policy and clean energy commitments generate the incentives towards technological innovation.

Rennings (2000) developed the environmental innovation systems that target regulatory systems, market forces, and technological capacity as most important to green innovation. The systems approach recognizes that efficient green innovation requires coordination between diverse stakeholders and institutions, which is very relevant to the case of Mexico's complex energy governance problem.

Resource-based view of innovation (Barney, 1991; Hart, 1995) provides us with an insight into how business organizations acquire competitive advantage through green innovation capabilities. Within Mexico's renewable energy industry, organizations that successfully develop and leverage innovative clean energy technology attain competitive advantage in emerging high-growth economies.

Latest developments in theory emphasized the importance of mission-oriented innovation policies addressing society's challenges by transformational approaches (Mazzucato, 2018; Schot & Steinmueller, 2018). In accordance with this perspective, Mexico's shift towards renewable energy should be accompanied by coordinated innovation policies integrating technological progress with society's goals.

Empirical Literature on Renewable Energy Innovation

Empirical research on renewable energy innovation identified a number of determinants of successful technology development and deployment. Jacobsson and Johnson (2000) analyzed renewable energy innovation systems and identified seven key functions: knowledge development and diffusion, direction of search influence, entrepreneurial experimentation, market creation, legitimation, resource mobilization, and positive externalities development.

Research on factors such as solar photovoltaic technology has demonstrated the importance of learning curves and scale economies in reducing technology costs (Nemet, 2006; Fraunhofer ISE, 2020). Innovation-led and scale cost reductions have helped Mexico to develop solar power, which has also increased the competitiveness of solar power with conventional energy sources.

Wind energy innovation research has given utmost priority to site adaptation and grid integration technologies (Lewis & Wiser, 2007; Wiser & Bolinger, 2021). Wind power development in nations such as Mexico, where the primary regions are Oaxaca, has required innovative approaches to address transmission and grid stability problems.

Empirical facts of green innovation in emerging markets have highlighted certain challenges like low research and development ability, weak intellectual property rights protection, and low finance for early-stage technology (Lema & Lema, 2012; Watson et al., 2015). These results are relevant in the context of the green innovation challenge and opportunities in Mexico.

Mexico-Specific Literature

Academic research on the development of Mexico's renewable energy and green innovation picked up momentum in recent years. Elizondo Azuela et al. (2014) discussed Mexico's policy framework for the development of renewable energy, identifying success factors and implementation challenges. Their study emphasized the importance of robust regulatory frameworks and long-term policy commitment in order to mobilize private investment. Güereca et al. (2020) examined Mexico's growth in wind energy, paying attention to technological innovation and capacity development in the region. They highlighted that technology transfer and human capital development are essential in constructing sustainable renewable energy industries.

Recent studies have explored Mexico's potential for innovation as well as solar energy potentials. Martínez-Gordón et al. (2022) examined the adoption trends of photovoltaic technology, which indicated regional variations in deployment rates as well as the choice of technology. Their work also highlighted local innovation ecosystems as catalysts for solar energy development. Research on Mexico's green energy policy transitions has revealed tensions between the encouragement of renewable energies and interests from the traditional energy sector (Elizondo, 2021; Bauer et al., 2022). These studies have important information regarding political economy determinants that influence returns on green innovation.

International Comparative Literature

Comparative study of Latin American renewable energy development is helpful in putting Mexico's experience into perspective. Griffith-Jones et al. (2017) compared the policies for renewable energy in prominent economies in Latin America and characterized Mexico as having similarly strong policy frameworks that are poorly implemented.

Comparisons with other emerging nations on the development of Mexico's renewable energy are both advantageous and disadvantageous. The geographical location of Mexico, which is close to the United States, provides access to innovative technologies and investment capital, but it also creates competitive forces that can deter domestic innovation development (Gallagher, 2014).

Green innovation policy studies in countries have emphasized the necessity for policy mix strategies that combine regulatory means, economic incentives, and research and development support (Rogge & Reichardt, 2016). This suggests the necessity for Mexico's green innovation success to be tackled via coordinated policy interventions in multiple areas.

Research Gaps and Contributions

In spite of increasing literature on Mexico's renewable energy growth, there are still some research gaps. There is limited empirical analysis of the direct impacts of green innovation on renewable energy deployment results. The bulk of current research addresses policy structures or specific technologies instead of integrated innovation system examination. There is a lack of evidence on the international technology transfer contribution to developing Mexico's green innovation. While Mexico has attracted significant foreign investment in renewable energies, the impact of knowledge transfer and the capability building of locals is not well established.

There is scant literature on full analysis of public-private partnerships in Mexico's green innovation system. There is a need to understand how the different organizational forms influence innovation impacts to guide policy-making. This research bridges these gaps by carrying out in-depth examination of green innovation impacts on renewable energy development, both quantitative findings and qualitative innovation processes. The study offers new learning concerning the capacity of developing countries to employ green innovation in driving renewable energy transformations.

MAIN FOCUS OF THE CHAPTER

Green Innovation Definition

Green innovation is a multifaceted notion that involves technological, organizational, and institutional innovations that lead to environmental sustainability and resource efficiency. Green innovation in the development of renewable energies involves both radical breakthrough technologies and incremental changes to established systems that bring improved environmental performance without compromising, or even enhancing, economic viability (Barbieri et al., 2020).

The study is grounded in the triple helix innovation model that emphasizes the collaboration between universities, industry, and government for the advancement of innovation ecosystems (Etzkowitz & Leydesdorff, 2000). The framework, in the Mexican renewable energy context, is applicable in describing the manner in which research institutions, private companies, and government institutions collaborate to develop and adopt clean energy technologies.

Green innovation in renewable energy can be categorized into several dimensions: technological innovation (new renewable energy technologies and efficiency improvements), process innovation (improved manufacturing and deployment methods), organizational innovation (new business models and ways of service delivery), and institutional innovation (policy frameworks and regulatory instruments supporting the uptake of clean energy) (OECD, 2011; Johnstone et al., 2010).

Renewable Energy Development Paradigms

Mexico's expansion of renewable energy operates under various paradigmatic frames. The market paradigm is focused on competitive electricity markets, private investment, and the lowering of technology costs as the main drivers of renewable energy adoption. The state-led paradigm revolves around government planning, public investment, and strategic energy security interests. The sustainability paradigm gives precedence to environmental protection, climate change mitigation, and long-term management of resources.

These paradigms contest one another in practice, generating policy tensions that frame green innovation outcomes. For example, market-based policies can prefer short-term cost reduction at the expense of longer-term sustainability considerations, and state-led approaches can prefer energy security at the expense of technological innovation.

Innovation Systems Theory

The national systems of innovation provide a handy framework for examining how Mexican economic environment institutions influence green innovation in the renewable energy industry. The framework examines the network of institutions, policies, and actors that accumulate innovation development and diffusion (Lundvall, 1992; Nelson, 1993).

The key components of Mexico's green innovation system include research and development institutions (CINVESTAV, IMP, UNAM), government institutions (SENER, CONACYT, INECC), private sector organizations (developers, manufacturers, service providers), and support institutions (financial institutions, certification agencies, training centers). The functionality of these components and their relations significantly influences green innovation outcomes.

Technology Adoption and Diffusion Models

Adoption and diffusion of green innovations within the renewable energy sector in Mexico must be understood from different theoretical perspectives. The technology acceptance model (Davis, 1989) describes individual and organizational adoption of new technologies based on perceived usefulness and ease of use. Diffusion of innovations theory (Rogers, 2003) describes the process through which innovations get adopted in social systems over a period of time.

For Mexico, the economic aspects (cost competitiveness, access to financing), technical aspects (maturity of technology, compatibility with the grid), institutional aspects (regulatory policies, policy stability), and social aspects (public acceptance, environmental consciousness) influence technology adoption choices.

Prior to exploring the practical applications of these theoretical frameworks, it is crucial to consolidate their significance in fostering sustainable innovation within the renewable energy sector in rural Veracruz. The amalgamation of the Porter Hypothesis, the Triple Helix Model, and Mission-Oriented Innovation policies provides a systematic framework to harmonize environmental goals with economic competitiveness, collaborative institutional networks, and mission-focused strategies. This structural connection is essential for formulating renewable energy schemes that simultaneously meet climate goals and promote financial literacy and socio-ecosystem resilience within local communities. The subsequent table 1 demonstrates the conversion of these theories into pragmatic methods that can direct the practical execution of solar and bioeconomy initiatives in rural Veracruz. (See table 1)

Table 1. Integrating Theory and Practice: Sustainable Innovation and Renewable Energy in Rural Veracruz

Theory	Key concepts	Application to renewable energy	Relevant literature
Porter Hypothesis	• Environmental restrictions foster innovation that mitigates compliance expenses and enhances competitiveness.	• Incentives provided by policy can stimulate local green ideas and reduce costs in solar and bioeconomy initiatives.	(Porter & Van der Linde, 1995)
Triple Helix Model	• Cooperation among academia, industry, and government to promote innovation ecosystems	• Universities in Veracruz can partner with solar cooperatives and municipal governments to establish renewable energy and financial literacy initiatives.	(Etzkowitz & Leydesdorff, 2000)
Mission-Oriented Innovation	• Policies addressing societal concerns via coordinated innovation	• Rural electrification and bioeconomy initiatives matched with climate objectives, facilitating mission-driven finance and innovation.	(Mazzucato, 2018)

Note. Table created by the authors

APPLICATION IN RURAL COMMUNITIES OF VERACRUZ

Upon establishing a solid conceptual foundation, it is essential to convert these theories into practical applications in rural Veracruz, emphasizing solar and local bioeconomy initiatives for financial literacy and SER.

- The Porter Hypothesis in Veracruz posits that by establishing explicit environmental policies and renewable energy objectives, local governments can incentivize rural cooperatives to embrace solar technology, concurrently developing innovative financing models that lower expenses, such as community savings initiatives and micro-leasing for solar panels (Porter & Van der Linde, 1995).
- The Triple Helix Model in Veracruz: Collaboration among local institutions (e.g., Universidad Veracruzana), municipal governments, and solar cooperatives can establish community-oriented training programs to enhance technical and financial literacy for solar project maintenance and bioeconomy initiatives. This promotes skill enhancement and accountability, establishing local capabilities for the sustainable management of renewable projects (Etzkowitz & Leydesdorff, 2000).
- Mission-Oriented Innovation in Veracruz: Initiatives can fit with Mexico's climate objectives by incorporating solar electrification and bioeconomy (biogas digesters, biomass pellet projects) into rural community development strategies, specifically addressing SDGs 4, 7, and 13. This alignment attracts mission-driven funding while ensuring that technological adoption addresses local need (Mazzucato, 2018).

The subsequent metrics are proposed to assess the practical efficiency of these theories.

- Solar Panels Installed: Quantity of panels installed in rural residences and community centers.
- Decrease in Domestic Energy Expenditures: Assessment of monthly reductions in household energy costs.
- CO_2 Emissions Mitigation: Assessment of emissions averted via renewable energy installations.
- Financial Literacy Metrics: Count of those educated in energy budgeting and cooperative management.
- Local Employment Generation: Quantity of technical and administrative roles established inside local energy cooperatives.

These metrics correspond with the integrated strategy advocated for the implementation of green innovation in Mexico, guaranteeing that theoretical frameworks result in concrete social, environmental, and economic benefits in rural Veracruz.

Competency-Based Education (CBE) for Green Innovation and SDGs

Community-Based Enterprises (CBE) are crucial for promoting green innovation and attaining the Sustainable Development Goals (SDGs) in Mexico, especially in rural, semi-rural, and semi-urban areas. The effective implementation of renewable energy necessitates the cultivation of technical skills in the installation, maintenance, and optimization of clean energy systems, allowing local communities to engage in and gain advantages from the green transition. These technological competencies augment local ability, diminish dependence on external experts, and promote enhancements in system efficiency that align with SDG 7 (Affordable and Clean Energy).

Furthermore, green energy-related financial literacy (GEFL) enables communities to make educated choices about the implementation of renewable technology, promoting sustainable investments at both household and community levels while enhancing economic resilience (Porter & Van der Linde, 1995). Financial literacy empowers residents to comprehend the cost frameworks, savings opportunities, and financing mechanisms of renewable technology, hence promoting informed engagement in the energy transition.

Developing these abilities enhances SER by providing persons with the skills necessary to adapt to evolving climate and energy circumstances, while simultaneously creating local employment possibilities, thereby advancing SDG 8 (Decent Work and Economic Growth) and SDG 13 (Climate Action). In Mexico, initiatives like the specialized renewable energy engineering programs at the Instituto Politécnico Nacional (IPN) and Universidad Nacional Autónoma de México (UNAM), as well as training programs offered by technical colleges and renewable energy clusters, illustrate efforts to enhance technical expertise in renewable energy systems (Mazzucato, 2018). These programs target essential skill deficiencies in advanced domains such grid integration, energy storage management, and optimization of renewable energy systems.

Consequently, including CBE into green innovation plans guarantees the sustainability of renewable energy efforts while promoting local development, economic empowerment, and climate resilience, thereby placing education as a catalyst for Mexico's sustainable transition.

This chapter underscores that SER bolsters the adaptive capacity of rural and semi-urban communities, facilitating their successful adoption and maintenance of green energy technology amidst fluctuating climate and market conditions, hence reinforcing the integration of SER and GEFL within CBE. Simultaneously, GEFL enables individuals to comprehend the economic ramifications of investing in renewable energy and to formulate sustainable green business models, so ensuring the financial feasibility of clean energy ventures. Embedding these competencies within educational frameworks in rural and semi-urban regions promotes informed decision-making, community involvement, and local innovation, thereby directly supporting SDG 7 (Affordable and Clean Energy), SDG 8 (Decent Work and Economic Growth), and SDG 13 (Climate Action) (Barbieri et al., 2020; Porter & Van der Linde, 1995).

METHODOLOGY

The study utilizes a mixed-methods research design that combines quantitative analysis of investment and renewable energy capacity data with qualitative content analysis of policy documents, industry reports, and interviews with stakeholders. The mixed-methods approach allows for comprehensive examination of quantifiable outcomes and underlying processes influencing green innovation in Mexico's renewable energy sector.

The study employs an explanatory sequential design whereby quantitative data are gathered and analyzed before qualitative data are gathered and analyzed to explain and interpret quantitative findings (Creswell & Plano Clark, 2017). The design permits detailed analysis of the processes through which green innovation influences renewable energy development outcomes.

To implement the proposed frameworks, it is crucial to base the study on solid factual facts that accurately represent the dynamics of Mexico's renewable energy market. This study integrates quantitative metrics with qualitative observations to provide a thorough understanding of green innovation's impact on renewable energy development. The subsequent table 2 delineates the principal data sources that underpin the methodological framework of this research. (See table 2)

Table 2. Data Collection &Quantitative Data Sources; Primary source of quantitative data

Data Collection.	Quantitative Data Sources Primary sources of quantitative data are:
• SENER (Secretaría de Energía) Energy Statistics	Official government statistics for electricity generating capacity, renewable energy installations and production by technology type between 2015-2024 (SENER, 2024).
• CENACE (Centro Nacional de Control de Energía) Market Data	Statistics for electricity markets such as price, dispatch trends and figures of grid integration for renewable energy. (CENACE, 2024).
• Private Investment Databases	Industry data from sources such as BloombergNEF, IRENA (2024). Global Energy Transformation, and Mexican renewable energy associations on deployment of technology, investment in projects, and addition to capacity.
• Patent and Innovation Databases	Green technology patent grants and applications in Mexico, as per details available from the Mexican Institute of Industrial Property (IMPI) and international patent databases (IMPI, 2024).
• Qualitative Data Sources. Qualitative data gathering includes	Policy Document Analysis, Thorough examination of energy policy reports, regulatory reports, and government strategic plans on renewable energy and green innovation in 2015-2025.
• Industry Reports and Publications	Review of international organization reports (IEA, IRENA (2024), World Bank), consulting firms, and industry associations on the renewable energy industry and innovation in Mexico.
• Media and News Analysis	Systematic overview of green innovation news coverage in the entire energy sector to identify upcoming trends, challenges, and opportunities.
• Case Study Analysis	Comprehensive analysis of certain green innovation programs and renewable energy projects to examine implementation procedures and results.

Note. Table created by the authors

To convert these data sources into meaningful insights, it is crucial to employ stringent analytical techniques that accurately identify trends and correlations. Quantitative analysis facilitates the systematic evaluation of green innovation's impact on renewable energy development. Table 3 delineates the principal statistical methods utilized in this investigation. See table 3.

Table 3. Data Analysis Techniques, Quantitative Analysis, Statistical analysis techniques

Data Analysis Techniques.	Quantitative Analysis. Statistical analysis techniques include:
• Descriptive Statistics	Examination of renewable energy capacity growth rates, investment patterns, and trends in technology deployment over a period of time.
• Correlation Analysis	Examination of the correlation between green innovation metrics (R&D expenditure, patent applications, technology adoption) and the result of renewable energy development (capacity increase, generation expansion, cost reduction).
• Time Series Analysis	Trend analysis of renewable energy development metrics to identify phases of acceleration or retardation and potential causality.
• Comparative Analysis	Mexico's renewable energy development versus other Latin American countries and emerging economies to identify relative performance and innovation gaps.

Note. Table created by the authors.

To enhance the quantitative analyses, it is essential to employ stringent qualitative methods that encapsulate the contextual and procedural aspects of green innovation. These methodologies facilitate a comprehensive knowledge of stakeholder dynamics and policy frameworks affecting renewable energy development in Mexico. This table 4 summarizes the qualitative data analysis approaches utilized in this study.

Table 4. Qualitative Analysis, Qualitative Data Analysis

Qualitative Analysis.	Qualitative data analysis employs:
• Thematic Analysis	Systematic and detailed identification and examination of recurrent themes in policy reports, industry reports, and case studies of green innovation and renewable energy development.
• Content Analysis	Statistical analysis of the content of policy reports to identify areas of focus, policy instruments, and strategic priorities over time.
• Process Tracing	Detailed analysis of single policy decisions and implementation processes to identify causal mechanisms linking green innovation to renewable energy outcomes.
• Stakeholder Analysis	Mapping and analysis of powerful actors in Mexico's renewable energy innovation system, their relations, functions, and policy and market influence.

Note. Table created by the authors.

To properly structure the analysis, it is essential to delineate the variables that direct the empirical examination of green innovation's influence on renewable energy development. This research guarantees methodological clarity and analytical rigor by delineating dependent, independent, and control variables. This table 5 delineates the research framework and variables utilized in this investigation. (See table5).

Table 5. Research Framework and Variables

Dependent Variables. Outcome variables of primary concern are:	**Independent Variables. Explanatory variables of primary significance are:**	**Control Variables: control for alternative explanations:**
• Renewable Energy Capacity Growth: Solar, wind, hydroelectric, geothermal, and biomass generation capacity additions annually.	• Green Innovation Expenditures: Government and private R&D expenditures on renewable energy technologies and support infrastructure.	• Economic Conditions: GDP growth, electricity demand growth, energy prices, and macroeconomic stability indicators.
• Renewable Energy Generation: Actual electricity output from renewables as percentage of total generation.	• Policy Framework Strength: Composite indicator of the stability, comprehensiveness, and effectiveness of renewable energy policies.	• Natural Resource Endowments: Solar irradiation, wind speed, hydroelectric potential, and geographical conditions affecting renewable energy development.
• Technology Cost Decreases: Levelized cost of electricity (LCOE) trends for different renewable energy technologies.	• Technology Transfer Activities: Clean energy foreign direct investment, global cooperation, and knowledge exchange activities.	• Institutional Quality: Indicators of governance, regulatory effectiveness, and measures of political stability.
• Innovation Output Metrics: Patent applications, research publications, and technology demonstration projects in renewable energy technologies.	• Human Capital Development: Research capabilities, education and training programs, and technical competence in clean energy technologies.	• International Factors: Costs of technology globally, climate policy pledges, and cooperation agreements internationally.

Note. Table created by the authors

To maintain methodological rigor, it is essential to recognize the constraints inherent in this research and to elucidate the methods implemented to enhance validity and reliability. This perspective promotes transparency and credibility in evaluating the effects of green innovation on renewable energy advancement. The subsequent table 6 delineates these limitations in conjunction with the techniques implemented to mitigate them. (See table 6)

Table 6. Research Limitations and Validity and Reliability Research validity

A few limitations hold for this research:	Validity and Reliability Research validity is enhanced by:
• Data Availability: There is limited comprehensive data available regarding green innovation activities in Mexico, especially on private sector R&D spending and initial-stage technology development. This is overcome through data triangulation of diverse sources and qualitative insights to supplement quantitative results.	• Construct Validity: Clearer operational definitions of key concepts and application of established measures where available.
• Attribution Issues: Identification of direct causal relationships between green innovation activity and renewable energy outcomes is made difficult by intervening factors and time gaps between innovation investments and deployment outcomes.	• Internal Validity: Application of different analytical methods and sensitivity checks to test robustness of findings.
• Dynamic Policy Context: Mexico's renewable energy policy context has experienced significant changes, making it problematic to consider long-term policy impacts. This issue is addressed in this work by using accurate periodization and probing policy stability effects.	• External Validity: Explicit specification of scope conditions and comparison with global experience to check generalizability.
• Limited Comparative Data: Comparative analysis is hindered by differences in data collection procedures and reporting conventions across countries. This is addressed using standardized international databases to the extent possible.	Reliability is guaranteed by: • Data Quality Checks: Verification of accuracy of data through the use of multiple sources and ensuring consistency.
• Ethical Considerations. This research adheres to prevailing social science standards of ethical research practice. Data sources are all publicly available or purchased appropriately. Analysis is objective and free from advocacy positions in reporting policy-relevant data. Research mentions funding institutions and potential conflicts of interest that can influence outcomes.	• Methodological Transparency: Detailed documentation of analytical steps to enable replication. • Peer Review: Invitation of academic and policy expertise to cross-check findings and interpretation.

Note. Table created by the authors.

ANALYSIS OF RESULTS

Results of Renewable Energy Development: Trends in Increasing Capacities.

The renewable energy industry in Mexico has experienced dramatic growth during the past decade, with renewable energy's total capacity at approximately

32,000 MW in 2024. The solar photovoltaic capacity has observed the most dramatic increase, rising from less than 50 MW in 2013 to more than 8,500 MW in 2024, with a more than 50% average annual growth rate (SENER, 2024). Wind capacity is currently around 7.7 GW, and 976 MW of geothermal power generation reflects Mexico's portfolio of diversified renewable energy. Policy development in recent times indicates continued ambition for renewables expansion.

Between 2025 and 2030, 22.7 gigawatts (GW) of new generation capacity are scheduled, of which 4.7GW would be solar power. That is a significant pickup from previous deployment rates and reflects growing confidence in renewable energy technology. However, growth patterns reflect regional and technological imbalances. Solar development has been focused in high irradiation north states, and wind development has been focused on southern states having good wind resources. Geographical concentration creates transmission issues and diminishes the broader economic benefits of renewable energy development.

Investment and Financial Flows

Investment in Mexico's renewable energy market shows green innovation opportunities and challenges. As of April 2024, there were about 15 new projects planned, with collective investment of USD20 billion. The Mexican government has also publicly embraced a USD10 billion investment by a Danish fund for a green hydrogen plant in southern Mexico as a sign of new opportunities in next-generation renewable technologies.

Private sector investment has been the primary driver of renewable energy development, with international investors attracted by Mexico's endowments and policy regimes. However, investment patterns reflect minimal emphasis on local technology development and innovation capacity building. Most projects utilize imported technologies with minimal local content requirements or knowledge transfer arrangements. The focus of investment in utility-scale projects has limited the potential for distributed innovation and local technology adaptation. Although large-scale projects benefit from economies of scale, they have fewer opportunities for domestic companies to innovate or gain technical know-how.

Technology Deployment Patterns

Analysis of technology deployments suggests the preference for proven, low-cost technologies over new or experimental ones. Solar PV deployments have predominantly been drawn from mainstream crystalline silicon technologies, with relatively low penetration of emerging technologies such as bifacial panels, tracker systems, or power system integration with energy storage. Wind farm deployments have concentrated on conventional horizontal-axis turbines by established global suppliers. Although turbine efficiency and unit size have grown since the early days, trends in deployment show few innovations or locally produced technologies.

Grid integration technologies have not received appropriate focus when they are crucial to the success of renewable energies. Limited investment in smart grids, energy storage equipment, and demand response technologies restrict the ability to maximize renewable energy applications and inhibits increased penetration of renewable energies.

Green Innovation Indicators. Research and Development Activities

Mexico's green technology in the field of renewable energy indicates strengths and weaknesses in research and development capabilities. Colleges like UNAM, CINVESTAV, and IPN have built robust programs of research in solar power, wind power, and bioenergy technologies. However, the research funding ratio is still low compared to developed countries and other emerging economies.

Patent statistics indicate modest but rising activity in renewable energy technology. Mexican patent applications in solar energy technology increased from 12 in 2018 to 34 in 2023, and wind energy patents increased from 8 to 19 over the same period. While small as absolute numbers, these statistics indicate significant percentage growth and represent nascent innovation capabilities.

Collaboration between schools of learning and enterprise remains poor, inhibiting the realization of research results to commercial activities. Most Mexican renewable energy companies employ foreign technologies rather than domestic innovations, denying opportunities for local R&D commercialization.

Technology Transfer and Adaptation

Global technology transfer has been the primary vehicle used to introduce green innovations in Mexico's renewable energy sector. Big renewable energy projects typically include cooperation among international technology providers and local developers, which creates space for knowledge transfer and capacity building. However, technology transfer patterns are reflective of inadequacies in indigenous innovation and adaptation capacity. Most transfers are in turnkey technology packages with very little local adaptation or customization for conditions or requirements. This action minimizes risk but also limits opportunities for local innovation capacity development and competitive advantage creation.

There are also examples of technology adaptation, particularly at small scale. Mexican companies have developed innovative models of financing for distributed solar systems and adapted installation practices to suit local building codes and climatic factors. These innovations offer potential for locally adapted technology development whenever there are appropriate incentives.

Human Capital Development

Human capital building is both a success story and ongoing challenge for green innovation within Mexico's renewable energy sector. Universities have introduced majors in renewable energy engineering, with over 50 institutions now offering specialized coursework in clean energy technologies. Technical training programs have adapted to include installation, maintenance, and operation of renewable energy systems. However, skills gaps still exist within more advanced technical areas like grid integration, energy storage, and system optimization. Lack of experienced professionals hinders project development and the quality of operations. Brain drains into international markets also undermines local capacity development since trained professionals tend to find opportunities in more developed renewable energy markets.

Mechanisms for professional accreditation and certification are underdeveloped, creating control of quality problems and limiting career advancement opportunities for technical professionals. Stronger professional development arrangements would enhance potential for innovation and sector competitiveness.

Analysis of the Policy Framework: Development of the Regulatory Environment

Mexico's renewable energy regulation environment has transformed significantly since 2013, with some periods of improvement and retreat. The 2013 Energy Reform created competitive electricity markets and renewable energy development models. The Energy Transition Law and the General Climate Change Law stipulated clear targets and policy instruments for clean energy deployment. However, policy implementation has been inconsistent, with changes in political leadership introducing changes to priorities and strategies. Recent policy change has focused on state-owned utility companies and raised questions about private sector investment in the growth of renewable energy. Such reforms introduce uncertainty that has the potential to serve as a limit to green innovation investment and construction.

Regulatory overcomplication is still a major issue, with conflicting jurisdictions and obscure procedures hindering project development and innovation rollout. Simplified permitting procedures and more transparent regulatory frameworks can facilitate greater adoption and deployment of green innovation.

Incentive Mechanisms

Mexico has utilized several incentive mechanisms to encourage renewable energy development, albeit partially successful in enhancing green innovation. Feed-in tariffs, renewable energy certificates, and tax credits have supported technology installation but provided minimal incentive to local innovation development. The Clean Energy Certificate (CEL) program creates market demand for clean energy but is technology-neutral, i.e., it does not differentiate between new and mature technologies. Technology-neutrality may curb incentives to use leading-edge innovations that can enhance longer-term competitiveness.

Financing support infrastructure remains underdeveloped, particularly for stage-one innovation and technology demonstration programs. Project financing through development banks exists for proven technologies but little assistance is provided to innovative approaches or locally developed technologies.

Framework of International Cooperation

International collaboration has played a key role in the evolution of green innovation in Mexico, with cooperation cutting across financing, capacity building, and technology transfer. Collaboration with the United States, European Union, and multilateral organizations has provided access to advanced technologies and best practices. However, cooperation frameworks are more inclined to emphasize technology adoption than indigenous innovation capacity building. Even though technology transfer has the benefit of offering short-run gains, less emphasis on local innovation capacity building might cap long-term competitiveness and technological self-reliance.

New possibilities for South-South cooperation may provide new models of technology development and deployment. Collaboration with other Latin American countries facing similar dilemmas can produce regional innovation networks and collaborative technology development.

Challenges and Barriers. Infrastructure Limitations

Grid bottlenecks are a fundamental restriction on the advancement of clean power and the implementation of green technology in Mexico. Bottlenecks in transmission limit renewable energy transportation capacity from abundant resource regions to consumption centers, hindering feasibility of projects and possible opportunities for innovation. Distribution grid modernization requirements are high, and modest smart grid infrastructure deployment constrains distributed energy resources and demand response innovations' prospects. Grid infrastructure investment has lagged behind renewable energy growth, forcing mounting restrictions on system optimization.

Storage infrastructure for energy is still lacking despite its indispensable value in renewable energy integration. Limited implementation of battery storage, pumped hydro, or other forms of storage technologies suppresses system flexibility as well as rates of renewable energy use.

Economic and Financial Hurdles

Access to funding is a significant barrier for deployment of green innovation, particularly for start-up technologies and small-scale projects. Traditional project finance approaches favor experienced developers and long-established technologies, limiting the scope for innovative solutions or start-up companies. Macroeconomic uncertainty and currency risk provide additional obstacles to long-term green innovation investment. Much of the renewable energy technology is foreign-currency-denominated and imported, rendering it vulnerable to exchange rate risks that affect project economics.

Patient capital for technology demonstration and development is limited and creates constraints on the commercialization of innovation. Venture capital and angel investing markets are yet underdeveloped for clean energy technologies, thereby limiting the chances of funding by innovative startups and early-stage companies.

Institutional and Governance Challenges

Institutional capacity constraints restrict effective policy enforcement as well as green innovation promotion. Technical skill sets in government offices are generally limited, affecting the policy planning and implementation monitoring quality. Federal, state, and local governments' difficulty in coordinating with each other adds complexity to regulation as well as lag in implementation. Different levels of government could have incompatible priorities or plans, promoting investor and innovation developer uncertainty.

Corruption and government issues impinge on certain aspects of renewable energy development, benefiting perhaps entrenched interests over innovative solutions. Open and competitive procurement is required in order to foster innovation and achieve best technology deployment.

Success Factors and Opportunities. Market-Driven Innovation

Competitive power markets have yielded cost-saving and efficiency-increasing incentives in renewable energy technologies. Marketplace competition encourages innovation as developers seek competitive advantage through improved technologies and creative project structures.

Corporate green power purchasing is a significant market driver, with companies making increasing commitments to purchasing green power. Corporate demand creates opportunity for new types of solutions including virtual power purchase agreements, renewable energy certificates, and integrated energy services. Potential for distributed generation opportunity for innovation in small-scale technologies, business models, and energy management systems exists. Residential solar and commercial solar markets offer potential for evolving regionally appropriate innovations.

Strengths in Resources

Mexico's higher renewable resources form natural advantages for technology deployment and innovation generation. Solar technologies are highly competitive due to intensive solar irradiation, and wind power can be developed using solid wind resources. Geographical variety enables portfolio approaches to developing renewable energy to mitigate risks and provide opportunities for complementary technology deployment. Differing regional resources create niches for specialized technologies and innovative approaches.

Being in proximity to huge markets, particularly the United States, provides technology transfer and regional integration possibilities. Cross-border electricity trading can open new markets for clean energy as well as drive innovation in system integration and transmission technology.

Emerging Technology Opportunities. Green hydrogen production is a massive prospect for future innovation within Mexico. Mexico's abundant renewable energy sources can be utilized to produce huge quantities of hydrogen for domestic use and for exportation. The government of Mexico has also publicly supported a USD10 billion investment by a Danish fund in a green hydrogen project as a growing realization of the potential. Energy storage technologies present great opportunities for innovation and market growth.

The cost of battery storage keeps coming down at a fast pace, opening up possibilities for grid-level and distributed storage uses that can accelerate the use of renewable energy.

Applications of digitalization and artificial intelligence in renewable energy present opportunities for innovative solutions in grid management, anticipatory maintenance, and system optimization. Mexican businesses and research centers may be able to specialize in these new fields.

DISCUSSION

Green Innovation Impact Assessment

The study shows a complex relationship between green innovation and the promotion of renewable energy in Mexico with high success but longstanding challenges. Green innovation has enhanced renewable energy growth primarily through cost decline in technology driven by global innovation and production scale rather than domestic innovation development. Mexico has benefited much from global technology spillovers, but the capacity for domestic innovation remains low.

The impacts of green innovation can be categorized into direct and indirect effects. Direct effects include technology improvement that reduces costs, enhances performance, and enables new applications. Solar PV cost declines of over 80% since 2010 have made solar power very competitive in Mexico, while improvements in wind turbines have opened larger areas to wind development. These cost and performance improvements directly enable more renewable power deployment.

Indirect effects include innovation spillovers, building of human capacity, and institutional learning that create capacity for future use and innovation. They may be difficult to quantify but form useful foundations for long-term renewable power development and technological competitiveness. Nevertheless, the study also shows areas where Mexico's green innovation strategy falls short. Internal development of innovation receives limited attention, which inhibits technological leadership and high-value economic growth opportunities. Dependence on technologies imported from abroad exposes vulnerabilities to supply chain shocks and reduces local value creation opportunities.

A critical barrier identified in this study is the lack of advanced technical skills within local communities, which limits local innovation capabilities and the effective deployment of renewable energy systems in Mexico. Without locally developed expertise in system installation, maintenance, and optimization, communities remain dependent on external providers and imported technologies, which restricts knowledge transfer and slows the development of domestic innovation ecosystems. In this context, CBE emerges as a key strategy to reduce dependence on imported technology by fostering local technical capacity, financial literacy, and adaptive skills that enable communities to engage directly in the renewable energy transition and localize technological adaptation (Mazzucato, 2018).

The development of these competencies not only supports the operational aspects of green energy systems but also creates local employment opportunities and strengthens SER by ensuring communities have the skills to manage and sustain renewable energy projects effectively. This aligns with the findings that robust local human capital development is essential for transitioning from technology adoption to innovation generation within emerging economies, positioning CBE as a lever for increasing local economic resilience while advancing the SDGs.

A clear linkage between SER and green energy adoption is critical for ensuring sustainable transitions in rural and semi-urban communities. SER provides the foundational stability for communities to absorb and adapt to technological changes in renewable energy, while GEFL supports households in making economically informed choices about clean energy investments. When these competencies are developed through CBE frameworks, they create pathways for communities to actively participate in the green economy while building resilience against climate and market volatility, thereby aligning local actions with global SDG commitments (Mazzucato, 2018).

While Mexico has advanced in renewable energy capacity through technology imports, the lack of locally developed competencies constrains the transition toward self-sustaining innovation ecosystems and bioeconomic development. CBE frameworks are essential to equip communities with the technical, financial, and ecological skills needed to maintain and adapt renewable energy systems, understand financial implications of clean energy, and engage in sustainable economic activities. Through the integration of GEFL and SER within CBE, individuals can actively contribute to SDG 7 and SDG 13 while fostering local green economies, reducing poverty, and enhancing climate resilience (Porter & Van der Linde, 1995; Barbieri et al., 2020).

As highlighted by Mazzucato (2018), mission-oriented innovation policies are crucial in driving technological development while also addressing society's broader goals.

Financial literacy in green energy is critical for empowering individuals, particularly in rural and semi-urban areas, to make informed energy decisions (Porter & Van der Linde, 1995). Understanding the financial implications of green energy technologies can help communities adopt sustainable practices and reduce energy costs. CBE provides the structure necessary to ensure that individuals not only understand green technologies but also possess the practical skills needed to apply this knowledge in real-world contexts (Rennings, 2000).

Bio economics, as a field, emphasizes the integration of biological, environmental, and economic systems to create a more sustainable and circular economy (Barbieri et al., 2020).

Policy Framework Effectiveness

Mexico's renewable policy framework is both positive and negative for green innovation. Competitive power markets and renewable targets were instituted to give substantial supports for technology deployment. However, policy implementation has been poor with the capriciousness of politics causing flip-flops in priorities and approaches that create uncertainty for investing in innovation. The policy framework has been more successful in facilitating technology deployment than innovation development. Market-based policies have worked to bring in private investment and drive capacity expansion, but have provided little incentive for indigenous innovation or local technology development. Technology-neutral approaches can miss chances to advance promising innovations or build domestic capability in areas of emerging technologies.

Institutional coordination and regulatory complexity limitations restrict policy effectiveness. Overlapping jurisdictions of multiple agencies create uncertainty and transaction costs that may uniquely affect innovative projects or methods that cannot be inserted into available frameworks. Smoother regulatory process and enhanced institutional coordination may facilitate innovation deployment. International comparison shows that Mexico's policy approach has been reasonably successful in applying renewable energy but has lagged behind the pioneers in innovation development. Germany, Denmark, and China have combined market-oriented policies with innovational support to build technological capabilities and international competitive strength.

Market Dynamics and Innovation Drivers

Market pressures in Mexico's renewable energy market provide opportunities and challenges for green innovation. Contestable electricity markets put pressure on cost reduction and efficiency improvement, inducing the adoption of more efficient technology. Market structures may, however, favor incumbent technologies over newer technologies involving higher risks or uncertainty. Renewable energy purchasing by businesses is a market driver that would finance innovative solutions. Large corporations will continue to seek renewable energy options that provide them with cost benefits, risk management, and sustainability benefits. Corporate demand would be able to finance innovative financing mechanisms, technology solutions, and service solutions.

Distributed generation markets are of particular interest for locally relevant innovation development. Small-scale solar systems need to be local building codes, climatic conditions, and consumer preferences adapted, and therefore, there is a chance for Mexican companies to develop niche solutions. Innovative financing approaches, installation technologies, and maintenance approaches can possibly provide competitive edge in local markets. However, market fragmentation and low standardization impede scaling innovation. Regional variation in regulation, utilization of utilities, and organization of the market increase complexity and cost in innovation deployment. Greater market integration and standardization can enhance opportunities for innovation development and diffusion.

Technology Transfer and Local Capacity Building

Technology transfer trends considered have implications and limitations for green innovation development in Mexico. International technology transfer has played a key role in enabling the rapid deployment of renewable energy, bringing in available technologies, and reducing deployment risk. Large-scale renewable energy developments typically involve partnerships with foreign technology providers, enabling the prospects for technology transfer and capacity building. But technology transfer patterns direct their attention more to turnkey solutions than to the development of joint innovation. Most of the transfers are comprehensive technology packages with little adaptation to local conditions or needs. Although this minimizes technical risks, it also prevents local learning and developing innovation capacity.

Successful cases of technology adaptation are identified, most prominently in distributed applications and specialized market niches. New local approaches to solar system funding, localized installation patterns to suit local building customs, and maintenance procedures adapted to local environments have been created by Mexican firms. These are examples of scope for locally applicable innovation given effective market incentives and institutional direction.

Technology transfer success in building local innovation capability relies much on the structure of partnerships and emphasis placed on knowledge sharing as opposed to technology deployment per se. Partnerships with clear elements of technology transfer, local content specifications, and capacity building are likely to be drivers of domestic innovation development.

Regional and International Context

Mexico's green innovation development is placed within a broader regional and global context presenting opportunities and constraints. Its integration with USMCA markets together with the United States and Canada presents access to state-of-the-art technology, investment, and technical abilities. The integration also creates competitive forces capable of constraining domestic innovation development incentives. Regional analysis indicates that Mexico has maintained moderate rates of success in renewable energy deployment relative to other Latin American countries but lags in innovation development. There are other nations, such as Brazil and

Chile, that have developed higher domestic capability in specific technology areas while Mexico is importing more technology.

The global context of rapidly declining costs of renewable energy technology has favored Mexico through improved project economics but can have reduced incentives for domestic innovation development. If foreign technologies are readily available at competitive costs, the business case for domestic innovation development is less compelling. Nevertheless, emerging technology sectors like energy storage, green hydrogen, and digitalization offer new challenges for innovation development that are less constrained by current international supply chains. Mexico can create competitive strengths in these emerging sectors through targeted innovation policies and strategic investments.

Barriers to Innovation and Deployment

The research identifies numerous challenges that constrain the development and application of green innovations in Mexico. The financial challenges include scarce availability of risk capital for early-stage innovations, poor patient capital for tech development, and currency risks that affect long-term investment strategies. Traditional practices in project finance favor established technologies and familiar developers, posing obstacles to innovative companies and approaches. Technical barriers include grid infrastructure restraints in limiting renewable integration and innovation deployment. Low transmission capacity, lack of smart grid infrastructure, and low energy storage installation constitute bottlenecks affecting system optimization and innovation opportunities.

Institutional barriers include regulatory complexity, technical capacity limits of the government, and poor coordination across different levels of government. Policy uncertainty and frequent changes of priorities are risks that could be particularly disruptive to technological innovation projects with longer developmental cycles. Human capital constraints affect innovation development as well as implementation. While there has been greater emphasis on education and training in renewable energy in Mexico, there are still gaps in advanced technical areas. Lack of adequate supply of experienced staff constrains project quality and innovation potential.

Success Factors and Best Practices

In spite of difficulties, the analysis points to various success factors that have facilitated green innovation and renewable energy growth in Mexico. Competitiveness in the market has generated incentives for efficiency enhancement and cost savings, stimulating uptake of more sophisticated technologies. Auction schemes and competitive electricity markets have effectively lured private investment and fueled capacity expansion.

International partnerships have provided access to latest technology and best practice, enabling rapid technology deployment and transfer of knowledge. Strategic collaboration with governments of developed countries, multilateral agencies, and private sector has enabled both project funding and building of capacities. Resource advantages like excellent solar irradiation and strong wind resources provide natural foundations for competitive renewable energy growth. Such resource endowments reduce technology risks and costs, making them more attractive to investors and developers.

Market opportunities in emerging markets like distributed generation, corporate renewable energy procurement, and export markets create incentives for innovation development and implementation. Such market opportunities may enable technology and business model development that is locally appropriate.

Future Challenges and Opportunities

The future for Mexico holds both promising opportunities and challenges related to green innovation and the development of renewable energy. Opportunities include new green technologies such as energy storage, green hydrogen, and digital management of the grid to facilitate development of innovation as well as industrial competitiveness. Government support for major green hydrogen investments is a demonstration of awareness of such opportunities.

Greater emphasis on nearshoring and supply chain resiliency could create opportunities for Mexico to increase renewable energy production capacity and become a clean energy technology hub for the region. Mexico's geographic features, trade relationships, and industrial base can be capitalized upon for further development of renewable energy supply chains. However, there are policy uncertainty, shortage of infrastructure, and shortage of institutional capacity that could constrain development and deployment of innovation. Policy changes and shifts in politics increase long-term investment and capacity-building risks.

Climate change impacts like changes in precipitation regimes, weather occurrences, and resources availability would influence renewable energy deployment and impose new adaptive technology and practice needs. Climate resilience and adaptation innovation could become increasingly important.

Implications for Theory and Practice

The Mexico case is extremely pertinent to understanding green innovation in developing country environments. The experience teaches that technology transfer and international cooperation can help with rapid deployment of renewable energy but will not automatically build local innovation capabilities. There are distinctive policies and institutional structures required to transform the deployment of technology into innovation development at home.

The case also illustrates the influence of incentive mechanisms and market structures on innovation performance. While competitive markets can induce efficient and cost-reducing innovation, they may not necessarily stimulate domestic innovation development, particularly in cases where foreign technologies are readily available. For this policy practice, the analysis is that successful green innovation development requires coordinated policies that combine market-based policies with complement policies aimed at innovation development, human capital building, and institution building. Technology-neutral policies may need to be complemented by targeted measures favorable to innovative ideas and domestic capability building.

The Mexican case also highlights the importance of policy consistency and long-term commitment in building innovation. Frequent regime switches and policy uncertainty risk undermining especially innovative projects that take longer times to develop and have more initial investments.

SOLUTIONS AND RECOMMENDATIONS

To strengthen green innovation and renewable energy deployment in Mexico, alliances between universities, technical centers, local governments, and private companies should be fostered to develop CBE programs that address practical skills for the energy transition. These partnerships can create technical training modules focused on the installation and maintenance of solar and wind technologies, the development of sustainable business models for renewable energy entrepreneurship, and the use of digital tools for energy management in rural and semi-urban communities. Additionally, implementing modular certification and training schemes can ensure the accessibility of CBE, enabling communities to acquire specialized competencies aligned with local labor markets while supporting SER and local economic development (Mazzucato, 2018; Porter & Van der Linde, 1995). These initiatives will reduce dependence on imported technologies by building local technical capacity, facilitating local employment opportunities, and empowering communities to actively participate in the green transition while advancing SDG targets (Rennings, 2000).

To translate these frameworks into actionable policy and community interventions, programs should be designed to teach specific competencies such as solar panel installation, maintenance, and microgrid management, integrated with modules on GEFL to enable informed financial decisions about adopting green energy.

For example, communities can participate in microfinancing programs for solar technologies while learning to manage cooperative renewable energy projects, contributing directly to SDG 7 and SDG 13. Bioeconomic principles can guide land use optimization and green entrepreneurship training to create measurable economic resilience in these communities, aligning with SDG 8 (Decent Work and Economic Growth) while reinforcing local circular economies (Barbieri et al., 2020; Porter & Van der Linde, 1995).

FUTURE RESEARCH DIRECTIONS

Future research should explore how Mexico can enhance local innovation capabilities while continuing to leverage global technology spillovers for renewable energy growth. While imported technologies have driven capacity expansion, there is a need to examine policies and incentives that foster domestic R&D, prototyping, and commercialization of green technologies within Mexican institutions and companies.

It would be valuable to study the role of competency-based education and technical training in building human capital for advanced areas like grid integration, energy storage, and smart grid management, particularly in rural and semi-urban areas where renewable energy can strengthen socio-ecosystem resilience.

Future research could also analyze innovative financing mechanisms such as green venture capital, blended finance, and risk-sharing models to support early-stage technology development and deployment, helping startups and small-scale innovators overcome financial barriers.

Another promising area is the exploration of policy consistency and long-term frameworks for green innovation, assessing how regulatory stability impacts private investment, technological learning, and capacity-building over time.

Studies should evaluate technology transfer structures that prioritize knowledge sharing and adaptation to local conditions, moving beyond turnkey solutions to models that stimulate domestic innovation and skill development.

Finally, emerging opportunities in green hydrogen production, energy storage, and digitalization of energy systems should be investigated, identifying how Mexico can build competitive advantages in these areas while aligning with climate goals and creating sustainable economic opportunities.

CONCLUSION

This comprehensive analysis on green innovation impacting renewable energy development in Mexico provides a multi-dimensional picture with outstanding achievement going hand in hand with ongoing challenges. Mexico has achieved the installation of high-capacity renewable energy, to the value of about 32,000 MW by 2024, with significantly strong growth in solar photovoltaic installation from less than 50 MW in 2013 to over 8,500 MW by 2024. This installation has, however, come in the form of imported rather than domestic origin innovation development.

The study reveals that green innovation has affected the development of renewable energy in Mexico largely through international technology spillovers and economies of scale, not domestic innovation generation. Although Mexico has accrued significant gains from international progress in renewable energy technology, indigenous innovation capability is limited, hindering technological leadership and high-value economic growth opportunities.

Policy instruments have played the key role in enabling the deployment of renewable energy through market-based reforms, competitive electricity markets, and renewable energy policies. Implementation of policy has not been stable, however, with political instability resulting in priority changes, creating uncertainty regarding innovation investment. Recent policy adjustments aimed at state-owned utilities have raised concerns regarding continued private sector involvement and innovation development.

Investment trends recognize opportunities and limitations alike, with approximately $30 billion invested in renewable energy projects but minimal emphasis on building innovation capacity and domestic technological development. Most projects employ turnkey foreign technology with low content and knowledge transfer agreement requirements.

Theoretical Contributions

This research contributes theoretically to the comprehension of green innovation within developing country environments in several important ways. In the first place, it depicts a situation in which the relationship between green innovation and renewable energy development is contingent upon institutional, market, as well as policy conditions that vary considerably by country. The triple helix model of innovation is best suited for exploring how the relationships between university, industry, and government influence the yield of innovation.

Second, the research illustrates the importance of differentiating between innovation development and innovation adoption in the analysis of green innovation impacts. The Mexican example illustrates that technology adoption may enable rapid deployment of renewable energy without necessarily building indigenous innovation capacity, highlighting the need for overt indigenous innovation development-support policies. Third, the analysis contributes to existing knowledge regarding the success of technology transfer in boosting local innovation capacity. The findings show that technology transfer contracts with an emphasis on turnkey solutions might limit local learning and capacity-building opportunities, but more collaborative approaches would enhance results in innovation development.

Policy Implications

The findings of the study have far-reaching policy development implications for Mexico and other emerging countries undertaking renewable energy transitions. First, efficient green innovation construction requires consistent policy frameworks that align market-based mechanisms with innovation development assistance, human capital creation, and institutional capacity empowerment. Secondly, long-term policy commitment and stability are essential in innovation development because continuous policy change leads to risks particularly borne by innovative projects that have longer development time and greater initial investments. Mexico's history of policy flip-flops underlines the necessity for building broad political consensus for policies on renewable energy and innovations.

Third, technology transfer policies should explicitly emphasize knowledge sharing, capacity building, and innovation development over technology deployment. Local content requirements, technology transfer commitments, and R&D initiatives undertaken in partnership could enhance the advantages of innovation development of international partnerships. Fourth, appropriate funding mechanisms that respond to the requirements for innovation development, e.g., early-stage venture capital, patient capital to prove technology, and risk-sharing tools to support innovative projects, are needed additions to traditional project finance techniques.

Practical Recommendations

Based on the research findings, the following practical recommendations are seen for enhancing green innovation and the development of renewable energy in Mexico:

Institutional Development: Establish an independent green innovation agency or strengthen existing institutions to organize the policy for innovation, provide technical expertise, and support technology development and deployment. The institution should have sufficient technical capacity as well as political mandate to organize across government ministries and drive public-private partnerships.

Innovation Finance: Develop specialized financial products for green innovation, including venture capital funds, innovation grants, and risk-sharing facilities to support early-stage technology development and demonstration projects. Public development banks can potentially play a key role in mobilizing patient capital for innovative solutions.

Human Capital Development and Investment: Enhance and upgrade education and training systems in renewable energy technologies, particularly for advanced-level technical skills in grid integration, energy storage, and system optimization. Professional certification and career development programs would enhance innovation capacity and competitiveness of the industry.

Technology Transfer Facilitation: Implement policies that optimize the return on international technology transfer for innovation development, including local content requirements, technology transfer requirements, and joint R&D programs. The policies are to balance the benefits of access to technology against the need for domestic capability.

Grid Infrastructure Modernization: Increase investment in grid infrastructure modernization, including smart grid technologies, expanding transmission capacity, and installing energy storage. These investments are facilitators for renewable energy integration and innovation deployment.

Regional Cooperation: Establish regional cooperation forums with other Latin American countries to share experiences, coordinate technology development, and create larger markets for innovative solutions. South-South cooperation can provide different models of technology development and deployment.

Limitations and Future Research

There are several limitations in this study that suggest directions for future research. First, constraints on data, particularly regarding private sector R&D expenditures and early-stage innovation activity, restrict extensive analysis of innovation inputs and outputs. Greater detail in available innovation data to future research might enable greater understanding of innovation processes and impacts. Second, the dynamic policy environment makes it challenging to quantify long-term policy impacts and results of innovation development. Longitudinal research that traces innovation development through periods of years might be able to provide useful insight into the effectiveness of different policy approaches.

Third, there is limited comparative analysis that constrains what is understood about how Mexico's experience can be exported to other developing countries pursuing similar energy transitions. Green innovation studies in a number of developing country settings might uncover generalizable best practices and lessons. Future research requirements include detailed examination of some innovation examples and success stories, exploration of the subnational government role in enabling innovation development, analysis of the impact of different technology transfer arrangements on innovation, and analysis of emerging opportunity technologies like green hydrogen and energy storage.

Final Reflections

Mexico's experience with green innovation and renewable energy development illustrates both opportunities and challenges for developing countries in the global energy transition. There are solid foundations for future development in Mexico's massive renewable energy endowments, broadening market opportunities, and policy commitment. However, realization of such opportunities in full is required through continuous commitment to the building of innovation, capacity building of institutions, as well as policy coordination.

International conditions of rising renewable energy technology improvement both challenge and present opportunities for countries like Mexico. In serving all markets, reduced cost and increased performance can also weaken incentives for domestic innovation development in those countries that are able to readily acquire foreign technology. This circumstance reflects the importance of strategic choices in where and how domestic capacity is to be built within the rapidly evolving clean energy technology regime.

The success of renewable energy development and green innovation, in the end, depends on the ability to match technological capability with local needs, resources, and capacities. Mexico's experience is that the matching requires sophisticated policy action, strong institutional capacity, and sustained political will. Future success depends on its ability to learn from the past to address shifting technological and market options. The lessons of Mexico's experience extend beyond the energy sector to broader issues of innovation building, technology transfer, and sustainable development in emerging economies. With an increasingly sustainability- and climate-focused world economy, the ability to build and apply green innovations becomes a defining element of economic competitiveness and development success.

Mexico's green technology and renewable energy experience is still ongoing, with brighter futures ahead in new technologies of innovation and expanding market applications. The experience offers valuable lessons for policymakers, industry executives, and researchers to facilitate further sustainable energy transitions in Mexico and beyond.

ACKNOWLEDGMENT

This research received no specific grant from any funding agency in the public, commercial, or not-for-profit sectors.

REFERENCES

Barbieri, N., Marzucchi, A., & Rizzo, U. (2020). Knowledge sources and impacts on subsequent inventions: Do green technologies differ from non-green ones? *Research Policy*, *49*(2), 103901. DOI: 10.1016/j.respol.2019.103901

Barney, J. (1991). Firm resources and sustained competitive advantage. *Journal of Management, 17*(1), 99–120. DOI: 10.1177/014920639101700108

Bauer, N., Hilaire, J., Brecha, R. J., Edmonds, J., Jiang, K., Kriegler, E., & van Vuuren, D. P. (2022). Assessing global fossil fuel availability in a scenario framework. *Energy*, 111147.

CENACE (Centro Nacional de Control de Energía). (2024). *Market Data and Statistics*. CENACE.

Chen, X., Yi, N., Zhang, L., & Li, D. (2022). Does institutional pressure foster corporate green innovation? Evidence from China's top 100 companies. *Journal of Cleaner Production*, *343*, 130924.

Creswell, J. W., & Plano Clark, V. L. (2017). *Designing and conducting mixed methods research*. Sage Publications.

Davis, F. D. (1989). Perceived usefulness, perceived ease of use, and user acceptance of information technology. *Management Information Systems Quarterly*, *13*(3), 319–340. DOI: 10.2307/249008

Elizondo, G. (2021). Mexico's energy policy reversal: Implications for renewable energy development. *Energy Policy*, *152*, 112209.

Elizondo Azuela, G., Barroso, L., Khanna, A., Wang, X., Wu, Y., & Cunha, G. (2014). *Performance of renewable energy auctions: Experience in Brazil, China and India*. World Bank Policy Research Working Paper 7062.

Etzkowitz, H., & Leydesdorff, L. (2000). The dynamics of innovation: From National Systems and "Mode 2" to a Triple Helix of university–industry–government relations. *Research Policy*, *29*(2), 109–123. DOI: 10.1016/S0048-7333(99)00055-4

Fraunhofer, I. S. E. (2020). *Photovoltaics Report 2020*. Fraunhofer Institute for Solar Energy Systems.

Gallagher, K. S. (2014). *The globalization of clean energy technology: Lessons from China*. MIT Press. DOI: 10.7551/mitpress/9805.001.0001

García-Olivares, A., Solé, J., & Osychenko, O. (2021). Transportation in a 100% renewable energy system. *Energy Conversion and Management, 158*, 266–285. DOI: 10.1016/j.enconman.2017.12.053

Griffith-Jones, S., Spratt, S., & Andrade, R. (2017). *Government development banks in Latin America: Innovation and investment for a sustainable future*. Inter-American Development Bank.

Güereca, L. P., Torres, N., & Juárez-López, C. R. (2020). The co-benefits approach and climate change policies in Mexico. *Current Opinion in Environmental Sustainability, 47*, 138–147.

Hart, S. L. (1995). A natural-resource-based view of the firm. *Academy of Management Review, 20*(4), 986–1014. DOI: 10.2307/258963

IMPI (Mexican Institute of Industrial Property). (2024). *Patent Statistics Database*. IMPI.

IRENA (International Renewable Energy Agency). (2024). *Global Energy Transformation: A Roadmap to 2050*. IRENA.

Jacobsson, S., & Johnson, A. (2000). The diffusion of renewable energy technology: An analytical framework and key issues for research. *Energy Policy, 28*(9), 625–640. DOI: 10.1016/S0301-4215(00)00041-0

Johnstone, N., Haščič, I., & Popp, D. (2010). Renewable energy policies and technological innovation: Evidence based on patent counts. *Environmental and Resource Economics, 45*(1), 133–155. DOI: 10.1007/s10640-009-9309-1

Lema, R., & Lema, A. (2012). Technology transfer? The rise of China and India in green technology sectors. *Innovation and Development, 2*(1), 23–44. DOI: 10.1080/2157930X.2012.667206

Lewis, J. I., & Wiser, R. H. (2007). Fostering a renewable energy technology industry: An international comparison of wind industry policy support mechanisms. *Energy Policy, 35*(3), 1844–1857. DOI: 10.1016/j.enpol.2006.06.005

Lundvall, B. A. (Ed.). (1992). *National systems of innovation: Towards a theory of innovation and interactive learning*. Pinter Publishers.

Martínez-Gordón, R., Morales-Ruvalcaba, C. F., Martínez-Díez, R., González-Albarrán, M. A., Tovar-Salinas, J. L., & Becerra-Pérez, L. A. (2022). A comprehensive review of photovoltaic solar energy in Mexico: Current status, trends, and future perspectives. *Energy Strategy Reviews*, *42*, 100882.

Mazzucato, M. (2018). Mission-oriented innovation policies: Challenges and opportunities. *Industrial and Corporate Change*, *27*(5), 803–815. DOI: 10.1093/icc/dty034

Nelson, R. R. (Ed.). (1993). *National innovation systems: a comparative analysis*. Oxford University Press. DOI: 10.1093/oso/9780195076165.001.0001

Nemet, G. F. (2006). Beyond the learning curve: Factors influencing cost reductions in photovoltaics. *Energy Policy*, *34*(17), 3218–3232. DOI: 10.1016/j.enpol.2005.06.020

OECD. (Organisation for Economic Co-operation and Development). (2011). *Invention and transfer of environmental technologies*. OECD Publishing.

Porter, M. E., & Van der Linde, C. (1995). Toward a new conception of the environment-competitiveness relationship. *The Journal of Economic Perspectives*, *9*(4), 97–118. DOI: 10.1257/jep.9.4.97

Rennings, K. (2000). Redefining innovation — Eco-innovation research and the contribution from ecological economics. *Ecological Economics*, *32*(2), 319–332. DOI: 10.1016/S0921-8009(99)00112-3

Rogers, E. M. (2003). *Diffusion of innovations* (5th ed.). Free Press.

Rogge, K. S., & Reichardt, K. (2016). Policy mixes for sustainability transitions: An extended concept and framework for analysis. *Research Policy*, *45*(8), 1620–1635. DOI: 10.1016/j.respol.2016.04.004

Schot, J., & Steinmueller, W. E. (2018). Three frames for innovation policy: R&D, systems of innovation and transformative change. *Research Policy*, *47*(9), 1554–1567. DOI: 10.1016/j.respol.2018.08.011

SENER (Secretaría de Energía). (2024). *Energy Statistics and Market Analysis*. Government of Mexico.

Watson, J., Byrnes, R., Ockwell, D., & Stua, M. (2015). Lessons from China: Building technological capabilities for low carbon technology transfer and development. *Climatic Change*, *131*(3), 387–399. DOI: 10.1007/s10584-014-1124-1

Wiser, R., & Bolinger, M. (2021). *2020 Wind Technologies Market Report*. Lawrence Berkeley National Laboratory.

Compilation of References

Abbass, M., Abbas, U., Jafri, R., Arif, S. M., & Akhai, S. (2025). AI and Machine Learning Applications in Sustainable Smart Cities. In *Sustainable Smart Cities and the Future of Urban Development* (pp. 1–32). IGI Global Scientific Publishing.

Abdallah, M., Talib, M. A., Feroz, S., Nasir, Q., Abdalla, H., & Mahfood, B. (2020). Artificial intelligence applications in solid waste management: A systematic research review. *Waste Management (New York, N.Y.)*, *109*, 231–246. DOI: 10.1016/j.wasman.2020.04.057 PMID: 32428727

Abeygunawardhana, A. G. D. T., Shalinda, R. M. M. M., Bandara, W. H. M. D., Anesta, W. D. S., Kasthurirathna, D., & Abeysiri, L. (2020, December). AI-driven smart bin for waste management. In *2020 2nd International Conference on Advancements in Computing (ICAC)* (Vol. 1, pp. 482-487). IEEE. DOI: 10.1109/ICAC51239.2020.9357151

Abir, A. Z. M. N. (2024, April). Industry 4.0 technologies for a data-driven, secure, green, and circular manufacturing economy. In *2024 IEEE Green Technologies Conference (GreenTech)* (pp. 26-31). IEEE. DOI: 10.1109/GreenTech58819.2024.10520440

Aboelata, A. (2021). Reducing Outdoor Air Temperature, Improving Thermal Comfort, and Saving Buildings' Cooling Energy Demand in Arid Cities – Cool Paving Utilization. *Sustainable Cities and Society*, *68*, 102762. Advance online publication. DOI: 10.1016/j.scs.2021.102762

Adadi, A., & Berrada, M. (2018). Peeking Inside the Black-Box: A Survey on Explainable Artificial Intelligence (XAI). *IEEE Access : Practical Innovations, Open Solutions*, *6*, 52138–52160. DOI: 10.1109/ACCESS.2018.2870052

Adapa, V. R. K. (2024a). AI for climate action: Leveraging Artificial Intelligence to address climate change challenges. *International Journal For Multidisciplinary Research*, *6*(5).

Adapa, V. R. K. (2024b). AI for climate action: Leveraging Artificial Intelligence to address climate change challenges. *International Journal For Multidisciplinary Research, 6*(5).

Addad, M., & Al-Taani, S. (2024). Leveraging AI for Energy-Efficient Smart Cities: Architectural and Urban Planning Solutions for Sustainable Growth—A Comparative Case Study of Amman City and International Examples. *Journal of Civil Engineering and Architecture, 18*, 620–632.

Aderibigbe, A. O., Ani, E. C., Ohenhen, P. E., Ohalete, N. C., & Daraojimba, D. O. (2023). Enhancing Energy Efficiency with Ai: A review of machine learning models in electricity Demand Forecasting. *Eng. Sci. Technol. j., 4*(6), 341–356.

Adewumi, A., Okoli, C. E., Usman, F. O., Olu-lawal, K. A., & Soyombo, O. T. (2024). Reviewing the impact of AI on renewable energy efficiency and management. *Int. J. Sci. Res. Arch., 11*(1), 1518–1527. DOI: 10.30574/ijsra.2024.11.1.0245

Aghili, S. A., Haji Mohammad Rezaei, A., Tafazzoli, M., Khanzadi, M., & Rahbar, M. (2025). Artificial Intelligence Approaches to Energy Management in HVAC Systems: A Systematic Review. *Buildings (Basel, Switzerland), 15*(7), 1008. Advance online publication. DOI: 10.3390/buildings15071008

Aguilar, J., & Garces-Jimenez, A. (2021). A Systematic Literature Review on the Use of Artificial Intelligence in Energy Self-Management in Smart Buildings. Renewable and Sustainable Energy Reviews, 151, Article 111530. DOI: 10.1016/j.rser.2021.111530

Ahmad, G., Aleem, F. M., Alyas, T., Abbas, Q., Nawaz, W., Ghazal, T. M., Aziz, A., Aleem, S., Tabassum, N., & Ibrahim, A. M. (2025). Intelligent waste sorting for urban sustainability using deep learning. *Scientific Reports, 15*(1), 27078. DOI: 10.1038/s41598-025-08461-w PMID: 40715219

Ahmad, M. W., Mourshed, M., & Rezgui, Y. (2017). Trees vs Neurons: Comparison between random forest and ANN for high-resolution prediction of building energy consumption. *Energy and Building, 147*, 77–89. DOI: 10.1016/j.enbuild.2017.04.038

Ahmed, A. (2025). *Predicting Electricity Consumption Using Time Series Algorithms and Utilizing Energy Storage: Mitigating Negative Price Impacts to Encourage New Investments in the Finnish Electricity Market*. Academic Press.

Ahmed, A. A. A., & Asadullah, A. (2020). Artificial intelligence and machine learning in waste management and recycling. *Engineering International, 8*(1), 43–52. DOI: 10.18034/ei.v8i1.498

Ahmed, S. F., Alam, M. S., Hassan, M., Rozbu, M. R., Ishtiak, T., Rafa, N., Mofijur, M., Shawkat Ali, A. B. M., & Gandomi, A. H. (2023). Deep learning modelling techniques: Current progress, applications, advantages, and challenges. *Artificial Intelligence Review, 56*(11), 13521–13617. DOI: 10.1007/s10462-023-10466-8

Aishwarya, A., Wadhwa, P., Owais, O., & Vashisht, V. (2021, January). A waste management technique to detect and separate non-biodegradable waste using machine learning and YOLO algorithm. In *2021 11th International Conference on Cloud Computing, Data Science & Engineering (Confluence)* (pp. 443-447). IEEE. DOI: 10.1109/Confluence51648.2021.9377163

Ajayi, F., Ademola, O. M., Amuda, K. F., & Alade, B. (2024). AI-driven decarbonization of buildings: Leveraging predictive analytics and automation for sustainable energy management. *World Journal of Advanced Research and Reviews, 24*(1), 61-79.

Akhtar, M. J., Mahum, R., Butt, F. S., Amin, R., El-Sherbeeny, A. M., Lee, S. M., & Shaikh, S. (2022). A robust framework for object detection in a traffic surveillance system. *Electronics (Basel), 11*(21), 3425. DOI: 10.3390/electronics11213425

Al Amin Gerary, A. (2024). Applications of Smart Technologies in Sustainable Buildings. *International Journal of Architectural Engineering and Urban Research, 7*(2), 41–59. DOI: 10.21608/ijaeur.2025.274865.1071

Alahi, M. E. E., Sukkuea, A., Tina, F. W., Nag, A., Kurdthongmee, W., Suwannarat, K., & Mukhopadhyay, S. C. (2023). Integration of IoT-Enabled Technologies and Artificial Intelligence (AI) for Smart City Scenario: Recent Advancements and Future Trends. *Sensors (Basel), 23*(11), 5206. Advance online publication. DOI: 10.3390/s23115206 PMID: 37299934

Alawi, A. E. B., Saeed, A. Y., Almashhor, F., Al-Shathely, R., & Hassan, A. N. (2021, July). Solid waste classification using deep learning techniques. In *2021 International Congress of Advanced Technology and Engineering (ICOTEN)* (pp. 1-5). IEEE.

Aldhanhani, T., Abraham, A., Hamidouche, W., & Shaaban, M. (2024). Future Trends in Smart Green IoV: Vehicle-to-Everything in the Era of Electric Vehicles. *IEEE Open Journal of Vehicular Technology, 5*, 278–297. DOI: 10.1109/OJVT.2024.3358893

Aldoseri, A., Al-Khalifa, K. N., & Hamouda, A. M. (2023). Re-Thinking Data Strategy and Integration for Artificial Intelligence: Concepts, Opportunities, and Challenges. *Applied Sciences (Basel, Switzerland), 13*(12), 7082. Advance online publication. DOI: 10.3390/app13127082

Aliero, M. S., Asif, M., Ghani, I., Pasha, M. F., & Jeong, S. R. (2022). Systematic Review Analysis on Smart Building: Challenges and Opportunities. *Sustainability (Basel), 14*(5), 3009. Advance online publication. DOI: 10.3390/su14053009

Alijoyo, F. A. (2024). AI-powered deep learning for sustainable industry 4.0 and internet of things: Enhancing energy management in smart buildings. *Alexandria Engineering Journal, 104*, 409–422. DOI: 10.1016/j.aej.2024.07.110

Ali, M. S., Sharma, A., Joy, T. A., & Halim, M. A. (2024). A comprehensive review of integrated energy management for future smart energy system. *Control Systems and Optimization Letters, 2*(1), 43–51. DOI: 10.59247/csol.v2i1.77

Ali, S., Bogarra, S., Riaz, M. N., Phyo, P. P., Flynn, D., & Taha, A. (2024). From time-series to hybrid models: Advancements in short-term load forecasting embracing smart grid paradigm. *Applied Sciences (Basel, Switzerland), 14*(11), 4442. DOI: 10.3390/app14114442

Aljohani, A. (2024). Deep learning-based optimization of energy utilization in IoT-enabled smart cities: A pathway to sustainable development. *Energy Reports, 12*, 2946–2957. DOI: 10.1016/j.egyr.2024.08.075

Allcott, H., & Greenstone, M. (2012). Is there an energy efficiency gap? *The Journal of Economic Perspectives, 26*(1), 3–28. DOI: 10.1257/jep.26.1.3

Alnaser, A. A., Maxi, M., & Elmousalami, H. (2024). AI-powered digital twins and internet of things for smart cities and sustainable building environment. *Applied Sciences (Basel, Switzerland), 14*(24), 12056. DOI: 10.3390/app142412056

Al-Obaidi, K. M., Hossain, M., Alduais, N A M., Al-Duais, H. S., Omrany, H., & Ghaffarianhoseini, A. (2022). A Review of Using IoT for Energy Efficient Buildings and Cities: A Built Environment Perspective. *Energies, 15*(16), 5991. Advance online publication. DOI: 10.3390/en15165991

Alonso, S. L. N., Forradellas, R. F. R., Morell, O. P., & Jorge-Vazquez, J. (2021). Digitalization, circular economy and environmental sustainability: The application of artificial intelligence in the efficient self-management of waste. *Sustainability, 13*(4), 1–19.

Alotaibi, I., Abido, M. A., Khalid, M., & Savkin, A. V. (2020). A comprehensive review of recent advances in smart grids: A sustainable future with renewable energy resources. *Energies, 13*(23), 6269. DOI: 10.3390/en13236269

Alqahtani, H., & Kumar, G. (2024). Cybersecurity in Electric and Flying Vehicles: Threats, Challenges, AI Solutions & Future Directions. *ACM Computing Surveys, 57*(4), 1–34. DOI: 10.1145/3697830

Al-Raeei, M. (2025). The smart future for sustainable development: Artificial intelligence solutions for sustainable urbanization. *Sustainable Development (Bradford), 33*(1), 508–517. DOI: 10.1002/sd.3131

Alsubaei, F. S., Al-Wesabi, F. N., & Hilal, A. M. (2022). Deep learning-based small object detection and classification model for garbage waste management in smart cities and iot environment. *Applied Sciences (Basel, Switzerland), 12*(5), 2281. DOI: 10.3390/app12052281

Amasuomo, E., & Baird, J. (2016). The concept of waste and waste management. *Journal of Management and Sustainability, 6*(4), 88. DOI: 10.5539/jms.v6n4p88

Amnuaylojaroen, T. (2025). Advancements and challenges of artificial intelligence in climate modeling for sustainable urban planning. *Frontiers in Artificial Intelligence, 8*, 1517986. DOI: 10.3389/frai.2025.1517986 PMID: 40463823

Anandakumar, H., Arulmurugan, R., & Onn, C. C. (2019). *Computational intelligence and sustainable systems*. EAI/Springer Innovations in Communication and Computing. DOI: 10.1007/978-3-030-02674-5

Andoni, M., Robu, V., Flynn, D., Abram, S., Geach, D., Jenkins, D., McCallum, P., & Peacock, A. (2019). Blockchain technology in the energy sector: A systematic review of challenges and opportunities. *Renewable & Sustainable Energy Reviews, 100*, 143–174. DOI: 10.1016/j.rser.2018.10.014

Anthony, L. F. W., Kanding, B., & Selvan, R. (2020). Carbontracker: Tracking and predicting the carbon footprint of training deep learning models. arXiv preprint arXiv:2007.03051.

Antonialli, F., Martinesco, A., & Mira-Bonnardel, S. (2022). Artificial Intelligence as a Determinant for Reshaping the Automotive Industry and Urban Mobility Services. *International Journal of Automotive Technology and Management, 22*(3), 324–351. DOI: 10.1504/IJATM.2022.124828

Apata, O., Bokoro, P. N., & Sharma, G. (2023). The Risks and Challenges of Electric Vehicle Integration into Smart Cities. *Energies, 16*(14), 5274. Advance online publication. DOI: 10.3390/en16145274

Aprizal, A., Wiranatakusuma, D. B., Rizki, M., & Anugrah, R. A. (2025). The role of technology in climate resilience: A systematic literature review and mapping study approach. *Journal of Physics: Conference Series, 2989*(1), 12037. DOI: 10.1088/1742-6596/2989/1/012037

Arévalo, P., & Jurado, F. (2024). Impact of Artificial Intelligence on the Planning and Operation of Distributed Energy Systems in Smart Grids. *Energies, 17*(17), 4501. Advance online publication. DOI: 10.3390/en17174501

Arévalo, P., Ochoa-Correa, D., & Villa-Ávila, E. (2024). A Systematic Review on the Integration of Artificial Intelligence into Energy Management Systems for Electric Vehicles: Recent Advances and Future Perspectives. *World Electric Vehicle Journal, 15*(8), 364. Advance online publication. DOI: 10.3390/wevj15080364

Arfeen, Z. A., Khairuddin, A. B., Larik, R. M., & Saeed, M. S. (2019). Control of distributed generation systems for microgrid applications: A technological review. *International Transactions on Electrical Energy Systems, 29*(9), e12072. DOI: 10.1002/2050-7038.12072

Aslan, Ö., Aktuğ, S. S., Ozkan-Okay, M., Yilmaz, A. A., & Akin, E. (2023). A comprehensive review of cyber security vulnerabilities, threats, attacks, and solutions. *Electronics (Basel), 12*(6), 1333. DOI: 10.3390/electronics12061333

Badihi, H., Zhang, Y., Jiang, B., Pillay, P., & Rakheja, S. (2022). A comprehensive review on signal-based and model-based condition monitoring of wind turbines: Fault diagnosis and lifetime prognosis. *Proc. IEEE Inst. Electr. Electron. Eng., 110*(6), 754–806. DOI: 10.1109/JPROC.2022.3171691

Baghaee, H. R., Mlakić, D., Nikolovski, S., & Dragicević, T. (2019). Support vector machine-based islanding and grid fault detection in active distribution networks. *IEEE Journal of Emerging and Selected Topics in Power Electronics, 8*(3), 2385–2403. DOI: 10.1109/JESTPE.2019.2916621

Baharetha, S., Soliman, A. M., Hassanain, M. A., Alshibani, A., & Ezz, M. S. (2024). Assessment of the Challenges Influencing the Adoption of Smart Building Technologies. *Frontiers in Built Environment, 9*, 1334005. Advance online publication. DOI: 10.3389/fbuil.2023.1334005

Bansal, M. (2024). Climate-resilience readiness of emerging clean energy technologies for India's energy transition: A deep dive into select technologies. *IOP Conference Series. Earth and Environmental Science, 1395*(1), 12015. DOI: 10.1088/1755-1315/1395/1/012015

Barbieri, N., Marzucchi, A., & Rizzo, U. (2020). Knowledge sources and impacts on subsequent inventions: Do green technologies differ from non-green ones? *Research Policy, 49*(2), 103901. DOI: 10.1016/j.respol.2019.103901

Barney, J. (1991). Firm resources and sustained competitive advantage. *Journal of Management, 17*(1), 99–120. DOI: 10.1177/014920639101700108

Barredo Arrieta, A., Díaz-Rodríguez, N., Del Ser, J., Bennetot, A., Tabik, S., Barbado, A., Garcia, S., Gil-Lopez, S., Molina, D., Benjamins, R., Chatila, R., & Herrera, F. (2020). Explainable Artificial Intelligence (XAI): Concepts, taxonomies, opportunities and challenges toward responsible AI. *Information Fusion*, *58*, 82–115. DOI: 10.1016/j.inffus.2019.12.012

Barua, A., Karim, F., Islam, M. M., Das, N., Sumon, M F I., Rahman, A., Debnath, P., Karmakar, M., & Khan, M. A. (2025). Optimizing Energy Consumption Patterns in Southern California: An AI-Driven Approach to Sustainable Resource Management. *Journal of Ecohumanism*, *4*(1), 88–102. DOI: 10.62754/joe.v4i1.6087

Bathla, G., Bhadane, K., Singh, R. K., Kumar, R., Aluvalu, R., Krishnamurthi, R., Kumar, A., Thakur, R. N., & Basheer, S. (2022). Autonomous Vehicles and Intelligent Automation: Applications, Challenges, and Opportunities. *Mobile Information Systems*, *2022*, 7632892. Advance online publication. DOI: 10.1155/2022/7632892

Bauer, N., Hilaire, J., Brecha, R. J., Edmonds, J., Jiang, K., Kriegler, E., & van Vuuren, D. P. (2022). Assessing global fossil fuel availability in a scenario framework. *Energy*, 111147.

Bazilian, M., Nakhooda, S., & Van de Graaf, T. (2014). Energy governance and poverty. *Energy Research & Social Science*, *1*, 217–225. DOI: 10.1016/j.erss.2014.03.006

Beloglazov, A., Buyya, R., Lee, Y. C., & Zomaya, A. (2011). A taxonomy and survey of energy-efficient data centers and cloud computing systems. *Advances in Computers*, *82*, 47–111. DOI: 10.1016/B978-0-12-385512-1.00003-7

Bender, E. M., Gebru, T., McMillan-Major, A., & Shmitchell, S. (2021, March). On the dangers of stochastic parrots: Can language models be too big? In *Proceedings of the 2021 ACM conference on fairness, accountability, and transparency* (pp. 610-623). DOI: 10.1145/3442188.3445922

Beretta, M., Julian, A., Sepulveda, J., Cusidó, J., & Porro, O. (2021). An Ensemble Learning solution for predictive maintenance of wind turbines main bearing. *Sensors (Basel)*, *21*(4), 1512. DOI: 10.3390/s21041512 PMID: 33671601

Bhattacharjee, P., More, A., Kathalkar, S., & Student, M. E. (2022). Effective Management of Solid waste by using GIS and Artificial Intelligence Based Support System. *International Journal of Mechanical Engineering*, *7*, 974–5823.

Bhattacharya, S., & Lane, N. D. (2016, November). Sparsification and separation of deep learning layers for constrained resource inference on wearables. In *Proceedings of the 14th ACM Conference on Embedded Network Sensor Systems CD-ROM* (pp. 176-189). DOI: 10.1145/2994551.2994564

Bhatt, C., Kumar, I., Vijayakumar, V., Singh, K. U., & Kumar, A. (2021). The state of the art of deep learning models in medical science and their challenges. *Multimedia Systems, 27*(4), 599–613. DOI: 10.1007/s00530-020-00694-1

Bilal, M., Ali, G., Iqbal, M. W., Anwar, M., Malik, M. S. A., & Kadir, R. A. (2022). Auto-prep: Efficient and automated data preprocessing pipeline. *IEEE Access : Practical Innovations, Open Solutions, 10*, 107764–107784. DOI: 10.1109/ACCESS.2022.3198662

Bodemer, O. (2024). *Empowering prosumers in the energy sector: Integrating AI and Blockchain for enhanced photovoltaic and battery systems. TechRxiv*. DOI: 10.36227/techrxiv.170630603.32096380/v1

Bollinger, B., & Gillingham, K. (2012). Peer effects in the diffusion of solar photovoltaic panels. *Marketing Science, 31*(6), 900–912. DOI: 10.1287/mksc.1120.0727

Bontorin, S. (2025, May 26). *Humanoid robots learn waste sorting from real workers using VR headsets.* Recycling Product News. https://www.recyclingproductnews.com/article/43317/humanoid-robots-learn-waste-sorting-from-real-workers-using-vr-headsets

Borah, G. (2025). Emerging Trends in Smart Green Building Technologies. *Heat Transfer Enhancement Techniques: Thermal Performance, Optimization and Applications*, 391-415.

Boumaiza, A., & Sanfilippo, A. (2022). *Data forecasting with application to blockchain-based local energy markets*. 2022 13th International Renewable Energy Congress (IREC), Brussels, Belgium. IEEE. DOI: 10.1109/IREC56325.2022.10001989

Bouzarovski, S., & Simcock, N. (2017). Spatializing energy justice. *Energy Policy, 107*, 640–648. DOI: 10.1016/j.enpol.2017.03.064

Bracke, P., Datta, A., Jung, C., & Sen, S. (2019). *Machine learning explainability in finance: An application to default risk analysis* (Staff Working Paper No. 816). Bank of England. https://www.bankofengland.co.uk/working-paper/2019/machine-learning-explainability-in-finance-an-application-to-default-risk-analysis

Braun, V., & Clarke, V. (2006). Using thematic analysis in psychology. *Qualitative Research in Psychology, 3*(2), 77–101. DOI: 10.1191/1478088706qp063oa

Buyya, R., Beloglazov, A., & Abawajy, J. (2010). Energy-efficient management of data center resources for cloud computing: a vision, architectural elements, and open challenges. arXiv preprint arXiv:1006.0308.

Cai, H., Zhu, L., & Han, S. (2018). ProxylessNAS: Direct neural architecture search on target task and hardware. arXiv preprint arXiv:1812.00332.

Camacho, J. D. J., Aguirre, B., Ponce, P., Anthony, B., & Molina, A. (2024). Leveraging artificial intelligence to bolster the energy sector in smart cities: A literature review. *Energies, 17*(2), 353. DOI: 10.3390/en17020353

Carley, S. (2009). State renewable energy electricity policies: An empirical evaluation of effectiveness. *Energy Policy, 37*(8), 3071–3081. DOI: 10.1016/j.enpol.2009.03.062

Carreón Sosa, R., & Bonilla Vargas, D. (2022). Energías renovables, PIB, mercados financieros e investigación: La experiencia de América Latina (2000-2019*). Revista Mexicana de Economía y Finanzas, 17*(4), e791. Advance online publication. DOI: 10.21919/remef.v17i4.791

Castaño, J., Martínez-Fernández, S., Franch, X., & Bogner, J. (2023, October). Exploring the carbon footprint of hugging face's ml models: A repository mining study. In *2023 ACM/IEEE International Symposium on Empirical Software Engineering and Measurement (ESEM)* (pp. 1-12). IEEE. DOI: 10.1109/ESEM56168.2023.10304801

Castro, R. C. C., Al Justin, S. G., Conato, C., Cruz, J. D., Alvaran, J. R., & Oblanca, V. J. (2020, December). Development of waste management system using the concept of 'Basura advantage points' through artificial neural network. In *2020 IEEE 12th International Conference on Humanoid, Nanotechnology, Information Technology, Communication and Control, Environment, and Management (HNICEM)* (pp. 1-5). IEEE.

Cavus, M. (2024). Integration smart grids, distributed generation, and cybersecurity: Strategies for securing and optimizing future energy systems. In *Preprints*.

Cavus, M. (2025). Advancing Power Systems with Renewable Energy and Intelligent Technologies: A Comprehensive Review on Grid Transformation and Integration. *Electronics (Basel), 14*(6), 1159. DOI: 10.3390/electronics14061159

Cavus, M., Dissanayake, D., & Bell, M. (2025). Next Generation of Electric Vehicles: AI-Driven Approaches for Predictive Maintenance and Battery Management. *Energies, 18*(5), 1041. Advance online publication. DOI: 10.3390/en18051041

Cebesoy, Ü. B., & Önger, S. (2025). How Artificial Intelligence (AI) Tools Can Be Used to Support Climate Change Education. Intelligent Solutions to Evaluate Climate Change Impacts, 281.

CENACE (Centro Nacional de Control de Energía). (2024). *Market Data and Statistics*. CENACE.

Chamola, V., Hassija, V., Sulthana, A. R., Ghosh, D., Dhingra, D., & Sikdar, B. (2023). A review of trustworthy and explainable artificial intelligence (XAI). *IEEE Access : Practical Innovations, Open Solutions*, *11*, 78994–79015. DOI: 10.1109/ACCESS.2023.3294569

Chandio, D., & Yu, J. (2023). *Decentralized energy trading using blockchain technology: A novel approach to democratizing the energy market.* 2023 IEEE 8th International Conference on Engineering Technologies and Applied Sciences (ICETAS), Bahrain, Bahrain. IEEE. DOI: 10.1109/ICETAS59148.2023.10346334

Chanussot, L., Das, A., Goyal, S., Lavril, T., Shuaibi, M., Riviere, M., Tran, K., Heras-Domingo, J., Ho, C., Hu, W., Palizhati, A., Sriram, A., Wood, B., Yoon, J., Parikh, D., Zitnick, C. L., & Ulissi, Z. (2021). Open catalyst 2020 (OC20) dataset and community challenges. *ACS Catalysis*, *11*(10), 6059–6072. DOI: 10.1021/acscatal.0c04525

Chatti, W., Majeed, M. T., Khoj, H., Miraz, M. H., & Ali, A. (2024). Towards Smart and Sustainable Transportation: The Role of Artificial Intelligence and New Technologies in Mitigating Passenger Car CO_2 Emissions in European Countries. *Environment, Development and Sustainability*, *26*, 1–25. DOI: 10.1007/s10668-024-05685-0

Chen, W., Men, Y., Fuster, N., Osorio, C., & Juan, A. A. (2024). Artificial Intelligence in Logistics Optimization with Sustainable Criteria: A Review. *Sustainability (Basel)*, *16*(21), 9145. Advance online publication. DOI: 10.3390/su16219145

Chen, W., Milosevic, Z., Rabhi, F. A., & Berry, A. (2023). Real-time analytics: Concepts, architectures, and ML/AI considerations. *IEEE Access : Practical Innovations, Open Solutions*, *11*, 71634–71657. DOI: 10.1109/ACCESS.2023.3295694

Chen, X., Yi, N., Zhang, L., & Li, D. (2022). Does institutional pressure foster corporate green innovation? Evidence from China's top 100 companies. *Journal of Cleaner Production*, *343*, 130924.

Chiang, S., Zito, J., Rao, V. R., & Vannucci, M. (2024). Time-series analysis. In *Statistical methods in epilepsy* (pp. 166–200). Chapman and Hall/CRC. DOI: 10.1201/9781003254515-8

Chlingaryan, A., Sukkarieh, S., & Whelan, B. (2018). Machine learning approaches for crop yield prediction and nitrogen status estimation in precision agriculture: A review. *Computers and Electronics in Agriculture*, *151*, 61–69. DOI: 10.1016/j.compag.2018.05.012

Chou, D. C., & Chou, A. Y. (2012). Awareness of Green IT and its value model. *Computer Standards & Interfaces*, *34*(5), 447–451. DOI: 10.1016/j.csi.2012.03.001

Chowdhary, S., Selvakumar, P., Sangeetha, B., Sarvaani, V. V., Anute, N., & Sharma, M. (2025). The Role of Social Capital in Entrepreneurial Development. In Martínez-Falcó, J., Sánchez-García, E., & Marco-Lajara, B. (Eds.), *Support Networks for Entrepreneurs: Social Capital and Strategic Collaborations* (pp. 311–338). IGI Global Scientific Publishing. DOI: 10.4018/979-8-3693-9836-4.ch013

Chowdhury, B. R. (2025). AI-Powered Forecasting and Optimization of Energy Consumption in the USA: Machine Learning Approaches for Sustainable Urban and Institutional Development. *Euro Vantage journals of Artificial intelligence, 2*(2), 21-38.

Chu, K. F., Lam, A. Y. S., & Li, V. O. K. (2022). Joint Rebalancing and Vehicle-to-Grid Coordination for Autonomous Vehicle Public Transportation System. *IEEE Transactions on Intelligent Transportation Systems*, *23*(7), 7156–7169. DOI: 10.1109/TITS.2021.3067044

CMF. (2021). Protección al Consumidor Financiero Marco Político Internacional y Desarrollos. Comisión para el Mercado Financiero. https://www.cmfchile.cl/portal/prensa/615/articles-49087_doc_pdf.pdf

Coates, J. M., Santhanalakshmi, K., Dinesh, N., Gandhimathi, S., & Chandel, P. S. Mohit & Selvakumar P, (2025). Assessment and Evaluation in Transnational Staff Development. In Training and Development in Transnational Higher Education (pp. 387–406). IGI Global Scientific Publishing. DOI: 10.4018/979-8-3373-1897-4.ch019

Consumers International. (2024). *Finanzas digitales transparentes para los consumidores*. https://www.consumersinternational.org/media/541545/transparent-digital-finance-for-consumers-sp.pdf

Coskuner, G., Jassim, M. S., Zontul, M., & Karateke, S. (2021). Application of artificial intelligence neural network modeling to predict the generation of domestic, commercial and construction wastes. *Waste Management & Research*, *39*(3), 499–507. DOI: 10.1177/0734242X20935181 PMID: 32586206

Cowls, J., Tsamados, A., Taddeo, M., & Floridi, L. (2023). The AI gambit: Leveraging artificial intelligence to combat climate change-opportunities, challenges, and recommendations. *AI & Society*, *38*(1), 283–307. DOI: 10.1007/s00146-021-01294-x PMID: 34690449

Creswell, J. W., & Plano Clark, V. L. (2017). *Designing and conducting mixed methods research*. Sage Publications.

Crisp, D., Pollock, H. R., & Rosenberg, R. (2017). The on-orbit performance of the Orbiting Carbon Observatory-2 (OCO-2) instrument and its initial results. *Atmospheric Measurement Techniques, 10*, 59–81. DOI: 10.5194/amt-10-59-2017

Cui, Y., Huang, C. J., Zhang, Y., Wang, L., Fan, L., Zhou, X., & Yang, Q. (2024). A survey on contribution evaluation in vertical federated learning. arXiv preprint arXiv:2405.02364.

Das, H. P., Lin, Y. W., Agwan, U., Spangher, L., Devonport, A., Yang, Y., Drgoňa, J., Chong, A., Schiavon, S., & Spanos, C. J. (2024). Machine learning for smart and energy-efficient buildings. *Environmental Data Science, 3*, e1. DOI: 10.1017/eds.2023.43

Davis, F. D. (1989). Perceived usefulness, perceived ease of use, and user acceptance of information technology. *Management Information Systems Quarterly, 13*(3), 319–340. DOI: 10.2307/249008

De la Torre, R., Corlu, C. G., Faulin, J., Onggo, B. S., & Juan, A. A. (2021). Simulation, Optimization, and Machine Learning in Sustainable Transportation Systems: Models and Applications. *Sustainability (Basel), 13*(3), 1551. Advance online publication. DOI: 10.3390/su13031551

Deif, A. M. (2011). A system model for green manufacturing. *Journal of Cleaner Production, 19*(14), 1553–1559. DOI: 10.1016/j.jclepro.2011.05.022

Delgado, R., & López, M. (2020). Financial literacy and distributed generation adoption in Mexico: Evidence from household solar programs. *Energy Policy, 138*, 111250. DOI: 10.1016/j.enpol.2019.111250

Derkenbaeva, E., Vega, S. H., Hofstede, G. J., & van Leeuwen, E. (2022). Positive Energy Districts: Mainstreaming Energy Transition in Urban Areas. *Renewable & Sustainable Energy Reviews, 153*, 111782. Advance online publication. DOI: 10.1016/j.rser.2021.111782

Deshpande, V., & Anasica, S. (2024). Optimizing Grid Reliability with AI-Enhanced Maintenance Strategies. *Acta Energetica, 02*, 23–34.

Devarakota, P., Tsesmetzis, N., Alpak, F. O., Gala, A., & Hohl, D. (2025). AI and the Net-Zero Journey: Energy Demand, Emissions, and the Potential for Transition. arXiv preprint arXiv:2507.10750.

Devi, J. V., Argiddi, R. V., Renuka, P., Janagi, K., Hari, B. S., & Boopathi, S. (2024). "Study on Integrated Neural Networks and Fuzzy Logic Control for Autonomous Electric Vehicles." In Semantic Web Technologies and Applications in Artificial Intelligence of Things, Chapter 6, pp. 88–112. DOI: DOI: 10.4018/979-8-3693-1487-6.ch006

Dhaliwal, A. (2022). "Towards AI-Driven Transport and Logistics." In Digital Transformation in the Viral Age. *Lecture Notes in Business Information Processing*, *508*, 119–131. DOI: 10.1007/978-3-031-60003-6_8

Dietrich, K., Dumont, M. J., Del Rio, L. F., & Orsat, V. (2017). Producing PHAs in the bioeconomy—Towards a sustainable bioplastic. *Sustainable Production and Consumption*, *9*, 58–70. DOI: 10.1016/j.spc.2016.09.001

Dikshit, S., Atiq, A., Shahid, M., Dwivedi, V., & Thusu, A. (2023). The Use of Artificial Intelligence to Optimize the Routing of Vehicles and Reduce Traffic Congestion in Urban Areas. *Energy Web*, *23*(1), e4613. Advance online publication. DOI: 10.4108/ew.4613

Dimnet, E. (2022). 3 Key notions for road transport sustainability: Resilience, Climate action and Energy transition. *Rom. J. Transp. Infrastruct.*, *11*(1), 1–20. DOI: 10.2478/rjti-2022-0002

Doshi-Velez, F., & Kim, B. (2017). *Towards a rigorous science of interpretable machine learning*. arXiv.

Du, K.-L., Jiang, B., Lu, J., Hua, J., & Swamy, M. N. S. (2024). Exploring Kernel Machines and Support Vector Machines: Principles, Techniques, and Future Directions. *Mathematics, 12*(24).

Egbuna, I. K., Salihu, F. B., Okara, C. C., Olayiwola, D. E., Smart, E. E., Anifowose, O., & Mbamalu, P. O. (2025). Advances in AI-powered energy management systems for renewable-integrated smart grids. *World J. Adv. Eng. Eng. Technol. Sci.*, *15*(2), 2300–2325. DOI: 10.30574/wjaets.2025.15.2.0685

Ejidike, C. C., & Mewomo, M. C. (2023). Benefits of Adopting Smart Building Technologies in Building Construction of Developing Countries: Review of Literature. *SN Applied Sciences*, *5*(2), 52. Advance online publication. DOI: 10.1007/s42452-022-05262-y

Ejjami, R. (2024). Integrating artificial intelligence for enhanced grid stability and renewable energy management in France. *International Journal For Multidisciplinary Research, 6*(3).

Ekechukwu, D. E., & Simpa, P. (2024). A comprehensive review of renewable energy integration for climate resilience. *Eng. Sci. Technol. j., 5*(6), 1884–1908.

Elizondo Azuela, G., Barroso, L., Khanna, A., Wang, X., Wu, Y., & Cunha, G. (2014). *Performance of renewable energy auctions: Experience in Brazil, China and India*. World Bank Policy Research Working Paper 7062.

Elizondo, G. (2021). Mexico's energy policy reversal: Implications for renewable energy development. *Energy Policy, 152*, 112209.

Elshebli, M. (2024). Extreme Weather and hydropower System: Production, Resilience, and Sustainability. In *Research Square*. DOI: 10.21203/rs.3.rs-2756965/v2

Elsken, T., Metzen, J. H., & Hutter, F. (2019). Neural architecture search: A survey. *Journal of Machine Learning Research, 20*(55), 1–21.

Enemosah, A., & Chukwunweike, J. (2022). Next-Generation SCADA Architectures for Enhanced Field Automation and Real-Time Remote Control in Oil and Gas Fields. *Int J Comput Appl Technol Res, 11*(12), 514–529.

Ericson, S., Grue, N., Hotchkiss, E., & Brown, M. (2023a). *Applications of measuring and valuing resilience in energy systems*. Office of Scientific and Technical Information (OSTI).

Ericson, S., Grue, N., Hotchkiss, E., & Brown, M. (2023b). *Applications of measuring and valuing resilience in energy systems*. Office of Scientific and Technical Information (OSTI).

Etzkowitz, H., & Leydesdorff, L. (2000). The dynamics of innovation: From National Systems and "Mode 2" to a Triple Helix of university–industry–government relations. *Research Policy, 29*(2), 109–123. DOI: 10.1016/S0048-7333(99)00055-4

European Commission. (2021). Proposal for a regulation of the European Parliament and of the Council laying down harmonised rules on artificial intelligence (artificial intelligence act). *COM(2021) 206 final*.

Fakhabi, M. M., Hamidian, S. M., & Aliehyaei, M. (2024). Exploring the role of the Internet of Things in green buildings. *Energy Science & Engineering, 12*(9), 3779–3822. DOI: 10.1002/ese3.1840

Fang, B., Yu, J., Chen, Z., Osman, A. I., Farghali, M., Ihara, I., Hamza, E. H., Rooney, D. W., & Yap, P. S. (2023). Artificial intelligence for waste management in smart cities: A review. *Environmental Chemistry Letters, 21*(4), 1959–1989. DOI: 10.1007/s10311-023-01604-3 PMID: 37362015

Farahani, R., Azimi, Z., Timmerer, C., & Prodan, R. (2024). Towards ai-assisted sustainable adaptive video streaming systems: Tutorial and survey. arXiv preprint arXiv:2406.02302.

Faucheux, S., & Nicolaï, I. (2011). IT for green and green IT: A proposed typology of eco-innovation. *Ecological Economics*, *70*(11), 2020–2027. DOI: 10.1016/j.ecolecon.2011.05.019

Faure, S. (2024). Innovative financing strategies for involving the private sector in the mainstreaming of Nature-Based Solutions-A case study in Barcelona (Master's thesis).

Ferdaus, M. M., Abdelguerfi, M., Ioup, E., Niles, K. N., Pathak, K., & Sloan, S. (2024). Towards trustworthy ai: A review of ethical and robust large language models. arXiv preprint arXiv:2407.13934.

Filippini, M., Leippold, M., & Wekhof, T. (2024). Sustainable finance literacy and the determinants of sustainable investing. *Journal of Banking & Finance*, *163*, 107167. DOI: 10.1016/j.jbankfin.2024.107167

Flammer, C. (2021). Corporate green bonds. *Journal of Financial Economics*, *142*(2), 499–516. DOI: 10.1016/j.jfineco.2021.01.010

Floridi, L., Cowls, J., Beltrametti, M., Nativi, S., Schafer, B., Taddeo, M., & Vayena, E. (2018). AI4People—An Ethical Framework for a Good AI Society: Opportunities, Risks, Principles, and Recommendations. *Minds and Machines*, *28*(4), 689–707. DOI: 10.1007/s11023-018-9482-5 PMID: 30930541

Fotovvatikhah, F., Ahmedy, I., Noor, R. M., & Munir, M. U. (2025). A Systematic Review of AI-Based Techniques for Automated Waste Classification. *Sensors (Basel)*, *25*(10), 3181. DOI: 10.3390/s25103181 PMID: 40431972

Fowlie, M., Greenstone, M., & Wolfram, C. (2018). Do energy efficiency investments deliver? Evidence from the weatherization assistance program. *The Quarterly Journal of Economics*, *133*(3), 1597–1644. DOI: 10.1093/qje/qjy005

Foxon, T. J. (2013). Transition pathways for a UK low carbon electricity future. *Energy Policy*, *52*, 10–24. DOI: 10.1016/j.enpol.2012.04.001

Fraunhofer, I. S. E. (2020). *Photovoltaics Report 2020*. Fraunhofer Institute for Solar Energy Systems.

Friede, G., Busch, T., & Bassen, A. (2015). ESG and financial performance: Aggregated evidence from more than 2000 empirical studies. *Journal of Sustainable Finance & Investment*, *5*(4), 210–233. DOI: 10.1080/20430795.2015.1118917

Gallagher, K. S. (2014). *The globalization of clean energy technology: Lessons from China*. MIT Press. DOI: 10.7551/mitpress/9805.001.0001

Gan, K. E., Taikan, O., Gan, T. Y., Weis, T., Yamazaki, D., & Schüttrumpf, H. (2023). Enhancing renewable energy systems, contributing to Sustainable Development Goals of United Nation and building resilience against climate change impacts. *Energy Technology (Weinheim)*, *11*(11), 2300275. DOI: 10.1002/ente.202300275

García-Carrasco, J., Maté, A., & Trujillo, J. (2023). A data-driven methodology for guiding the selection of preprocessing techniques in a machine learning pipeline. *International Conference on Advanced Information Systems Engineering*, 34–42. DOI: 10.1007/978-3-031-34674-3_5

García-Olivares, A., Solé, J., & Osychenko, O. (2021). Transportation in a 100% renewable energy system. *Energy Conversion and Management*, *158*, 266–285. DOI: 10.1016/j.enconman.2017.12.053

Gautam, M. (2023). Deep Reinforcement learning for resilient power and energy systems: Progress, prospects, and future avenues. *Electricity*, *4*(4), 336–380. DOI: 10.3390/electricity4040020

Geetha, S., Saha, J., Dasgupta, I., Bera, R., Lawal, I. A., & Kadry, S. (2022). Design of Waste Management System Using Ensemble Neural Networks. *Designs*, *2022*(6), 27. DOI: 10.3390/designs6020027

Ge, M., Feng, Z., & Meng, Q. (2024). Urban planning and green building technologies based on artificial intelligence: Principles, applications, and global case study analysis. *Applied Science and Engineering Journal for Advanced Research*, *3*(5), 18–27.

Ghassemi, M., Oakden-Rayner, L., & Beam, A. L. (2021). The false hope of current approaches to explainable artificial intelligence in health care. *The Lancet. Digital Health*, *3*(11), e745–e750. DOI: 10.1016/S2589-7500(21)00208-9 PMID: 34711379

Ghorfi, T., Laraqui, S., & Nachit, H. (2025). AI for energy management: Driving efficiency and sustainability in the MENA region. In *AI in the Middle East for Growth and Business: A Transformative Force* (pp. 145–175). Springer Nature Switzerland. DOI: 10.1007/978-3-031-75589-7_10

Gilchrist, D., Yu, J., & Zhong, R. (2021). The limits of green finance: A survey of literature in the context of green bonds and green loans. *Sustainability (Basel)*, *13*(2), 478. DOI: 10.3390/su13020478

Goldthau, A. (2014). Rethinking the governance of energy infrastructure: Scale, decentralization and polycentrism. *Energy Research & Social Science*, *1*, 134–140. DOI: 10.1016/j.erss.2014.02.009

Golkar, A., & Sweeting, M. N. (2016). CubeSat mission design for Earth observation. *Acta Astronautica*, *114*, 112–124.

Gomes, C. P. (2010, February). Computational sustainability: Computational methods for a sustainable environment, economy, and society. In Frontiers of Engineering: Reports on Leading-Edge Engineering from the 2009 Symposium (pp. 27-40). Washington, DC: National Academies Press.

Gómez, D. O., Bermúdez, M. T., & Morales, W. H. (2021). *Solid Domestic Waste classification using Image Processing and Machine Learning* (Doctoral dissertation, Universidad EAFIT).

Gonczarowski, Y. A., & Weinberg, S. M. (2021). The sample complexity of up-to-ε multi-dimensional revenue maximization. *Journal of the Association for Computing Machinery*, *68*(3), 1–28. DOI: 10.1145/3439722

González-Briones, A., Palomino-Sánchez, J., Vale, Z., Ramos, C., & Corchado, J. M. (2025). Evolution of Building Energy Management Systems for greater sustainability through explainable artificial intelligence models. *Engineering Applications of Artificial Intelligence*, *147*, 110324. DOI: 10.1016/j.engappai.2025.110324

Google, A. I. (2021). *Blog*. AI Helps Track Wildfires in Real Time.

Görücü, S., Morais, L. A., & Panagiotidou, G. (2025, April). A Critical Analysis of Machine Learning Eco-feedback Tools through the Lens of Sustainable HCI. In *Proceedings of the 2025 CHI Conference on Human Factors in Computing Systems* (pp. 1-18).

Greenberg, S., Mills, E., Tschudi, B., Rumsey, P., & Myatt, B. (2006). Best practices for data centers: Lessons learned from benchmarking 22 data centers. Proceedings of the ACEEE summer study on energy efficiency in buildings in Asilomar, CA. ACEEE, August, 3, 76-87.

Griffith-Jones, S., Spratt, S., & Andrade, R. (2017). *Government development banks in Latin America: Innovation and investment for a sustainable future*. Inter-American Development Bank.

Güereca, L. P., Torres, N., & Juárez-López, C. R. (2020). The co-benefits approach and climate change policies in Mexico. *Current Opinion in Environmental Sustainability*, *47*, 138–147.

Gunning, D., & Aha, D. (2019). DARPA's Explainable Artificial Intelligence (XAI) Program. *AI Magazine*, *40*(2), 44–58. DOI: 10.1609/aimag.v40i2.2850

Gupta, S., Modgil, S., Bhattacharyya, S., & Bose, I. (2022). Artificial intelligence for decision support systems in the field of operations research: Review and future scope of research. *Annals of Operations Research*, *308*(1), 215–274. DOI: 10.1007/s10479-020-03856-6

Haben, S., Arora, S., Giasemidis, G., Voss, M., & Greetham, D. V. (2021). Review of low voltage load forecasting: Methods, applications, and recommendations. *Applied Energy*, *304*, 117798. DOI: 10.1016/j.apenergy.2021.117798

Hamdan, A., Ibekwe, K. I., Ilojianya, V. I., Sonko, S., & Etukudoh, E. A. (2024). AI in renewable energy: A review of predictive maintenance and energy optimization. *Int. J. Sci. Res. Arch.*, *11*(1), 718–729. DOI: 10.30574/ijsra.2024.11.1.0112

Hammad, A., & Abu-Zaid, R. (2024). Applications of AI in decentralized computing systems: Harnessing artificial intelligence for enhanced scalability, efficiency, and autonomous decision-making in distributed architectures. *Applied Research in Artificial Intelligence and Cloud Computing*, *7*(6), 161–187.

Hanafi, A. M., Moawed, M. A., & Abdellatif, O. E. (2024). Advancing sustainable energy management: A comprehensive review of artificial intelligence techniques in building. *Engineering Research Journal (Shoubra)*, *53*(2), 26–46. DOI: 10.21608/erjsh.2023.226854.1196

Han, S., Pool, J., Tran, J., & Dally, W. (2015). Learning both weights and connections for efficient neural network. *Advances in Neural Information Processing Systems*, 28.

Harputlugil, T., & de Wilde, P. (2021). The Interaction Between Humans and Buildings for Energy Efficiency: A Critical Review. *Energy Research & Social Science*, *71*, 101828. Advance online publication. DOI: 10.1016/j.erss.2020.101828

Hart, S. L. (1995). A natural-resource-based view of the firm. *Academy of Management Review*, *20*(4), 986–1014. DOI: 10.2307/258963

Hassan, A. M., & Megahed, N. A. (2022). Improvingurban energyresiliencewith an integrative framework based on machine learningmethods. *Arch. Eng.*, *7*(4), 17–35. DOI: 10.23968/2500-0055-2022-7-4-17-35

He, D., Li, K., Chen, Y., Miao, J., Li, X., Shan, S., & Ren, R. (2021). Obstacle detection in dangerous railway track areas by a convolutional neural network. *Measurement Science & Technology*, *32*(10), 105401. DOI: 10.1088/1361-6501/abfdde

Heidari, M., Thangavel, S., Al Naamani, E., & Khashehchi, M. (2024). Emerging trends in smart green building technologies. *Sustainable Technologies for Energy Efficient Buildings*, 313-336.

He, K., Gkioxari, G., Dollár, P., & Girshick, R. (2017). Mask r-cnn. In *Proceedings of the IEEE international conference on computer vision* (pp. 2961-2969).

He, K., Zhang, X., Ren, S., & Sun, J. (2016). Deep residual learning for image recognition. In *Proceedings of the IEEE conference on computer vision and pattern recognition* (pp. 770-778).

Henderson, P., Hu, J., Romoff, J., Brunskill, E., Jurafsky, D., & Pineau, J. (2020). Towards the systematic reporting of the energy and carbon footprints of machine learning. *Journal of Machine Learning Research, 21*(248), 1–43.

Hennessy, J. L., & Patterson, D. A. (2011). *Computer architecture: a quantitative approach*. Elsevier.

Hilty, L. M., & Aebischer, B. (2014). ICT for sustainability: An emerging research field. ICT innovations for Sustainability, 3-36.

Himeur, Y., Elnour, M., Fadli, F., Meskin, N., Petri, I., Rezgui, Y., Bensaali, F., & Amira, A. (2023). AI-Big Data Analytics for Building Automation and Management Systems: A Survey, Actual Challenges and Future Perspectives. *Artificial Intelligence Review, 56*(6), 4929–5021. DOI: 10.1007/s10462-022-10286-2 PMID: 36268476

Hinton, G., Vinyals, O., & Dean, J. (2015). Distilling the knowledge in a neural network. arXiv preprint arXiv:1503.02531.

Hirsch, A., Parag, Y., & Guerrero, J. (2018). Microgrids: A review of technologies, key drivers, and outstanding issues. *Renewable & Sustainable Energy Reviews, 90*, 402–411. DOI: 10.1016/j.rser.2018.03.040

Hoffmann, J., Borgeaud, S., Mensch, A., Buchatskaya, E., Cai, T., Rutherford, E., . . . Sifre, L. (2022). Training compute-optimal large language models. arXiv preprint arXiv:2203.15556.

Holzinger, A., Biemann, C., Pattichis, C. S., & Kell, D. B. (2025). What do we need to build explainable AI systems for the medical domain? arXiv.

Howard, J., & Gugger, S. (2020). Fastai: A layered API for deep learning. *Information (Basel), 11*(2), 108. DOI: 10.3390/info11020108

Huang, G., Liu, Z., Van Der Maaten, L., & Weinberger, K. Q. (2017). Densely connected convolutional networks. In *Proceedings of the IEEE conference on computer vision and pattern recognition* (pp. 4700-4708).

Hubballi, R. B., Selvakumar, P., Seenivasan, R., Basava Aradhya, S. G., Dinesh, N., & Seelam, P. K. (2025). Overview of Current AI Technologies in Education. In Kyei-Blankson, L., & Ntuli, E. (Eds.), *Transformative AI Practices for Personalized Learning Strategies* (pp. 1–26). IGI Global Scientific Publishing., DOI: 10.4018/979-8-3693-8744-3.ch001

Hu, H., Yu, S. S., & Trinh, H. (2024). A review of uncertainties in power systems—Modeling, impact, and mitigation. *Designs*, *8*(1), 10. DOI: 10.3390/designs8010010

Hupkes, D., Giulianelli, M., Dankers, V., Artetxe, M., Elazar, Y., Pimentel, T., Christodoulopoulos, C., Lasri, K., Saphra, N., Sinclair, A., Ulmer, D., Schottmann, F., Batsuren, K., Sun, K., Sinha, K., Khalatbari, L., Ryskina, M., Frieske, R., Cotterell, R., & Jin, Z. (2023). A taxonomy and review of generalization research in NLP. *Nature Machine Intelligence*, *5*(10), 1161–1174. DOI: 10.1038/s42256-023-00729-y

Husainy, A. S., Mangave, S. S., Ingale, A. S., Patil, Y. R., & Koganole, M. (2024). Innovation ecosystems and green building techniques for a sustainable future: Leveraging advanced technologies. *The Asian Review of Civil Engineering*, *13*(2), 1–10. DOI: 10.70112/tarce-2024.13.2.4233

Hussain, M., Zhang, T., Jamil, I., Soomro, A. A., & Hussain, I. (2024). Application of machine learning approaches to prediction of corrosion defects in energy pipelines. *Advances in Corrosion Modelling*, 127–166.

Hussain, A., Draz, U., Ali, T., Tariq, S., Irfan, M., Glowacz, A., Antonino Daviu, J. A., Yasin, S., & Rahman, S. (2020). Waste management and prediction of air pollutants using IoT and machine learning approach. *Energies*, *13*(15), 3930. DOI: 10.3390/en13153930

Huston, S. J. (2010). Measuring financial literacy. *The Journal of Consumer Affairs*, *44*(2), 296–316. DOI: 10.1111/j.1745-6606.2010.01170.x

IBM Research. (2020). Flood Forecasting with AI in India.

IBM. (2023, March 29). *What is explainable AI?* IBM Think. https://www.ibm.com/think/topics/explainable-ai

Ihsanullah, I., Alam, G., Jamal, A., & Shaik, F. (2022). Recent advances in applications of artificial intelligence in solid waste management: A review. *Chemosphere*, *309*, 136631. DOI: 10.1016/j.chemosphere.2022.136631 PMID: 36183887

IMPI (Mexican Institute of Industrial Property). (2024). *Patent Statistics Database*. IMPI.

International Energy Agency. (2024). *Mexico energy outlook 2024*. IEA Publications.

International Renewable Energy Agency. (2024). *Renewable energy statistics 2024*. IRENA Publications.

IRENA (International Renewable Energy Agency). (2024). *Global Energy Transformation: A Roadmap to 2050*. IRENA.

Isenkul, M. E. (2025). Energy-aware deep learning for real-time video analysis through pruning, quantization, and hardware optimization. *Journal of Real-Time Image Processing*, *22*(3), 125. DOI: 10.1007/s11554-025-01703-0

Islam, R. (2024). Sustainable Entrepreneurship in the Digital Age: The Role of AI in Green Business Practices. Academic Press.

Islam, A., & Othman, F. (2024). Renewable energy MicroGrid power forecasting: AI techniques with environmental perspective. In *Research Square*. DOI: 10.21203/rs.3.rs-4260337/v1

Islam, M. M., Yu, T., Giannoccaro, G., Mi, Y., La Scala, M., Rajabi, M. N., & Wang, J. (2024). Improving reliability and stability of the power systems: A comprehensive review on the role of energy storage systems to enhance flexibility. *IEEE Access : Practical Innovations, Open Solutions*.

Iyer, L. S. (2021). AI Enabled Applications Towards Intelligent Transportation. *Transportation Engineering*, *5*, 100083. Advance online publication. DOI: 10.1016/j.treng.2021.100083

Jacob, D. J., Scarpelli, T. R., & Sulprizio, M. P.. (2022). Quantifying methane emissions from the largest oil-producing basin in the United States from space. *Science Advances*, *8*(5), eabj4351.

Jacobsson, S., & Johnson, A. (2000). The diffusion of renewable energy technology: An analytical framework and key issues for research. *Energy Policy*, *28*(9), 625–640. DOI: 10.1016/S0301-4215(00)00041-0

Jagatheesaperumal, S. K., Bibri, S. E., Huang, J., Rajapandian, J., & Parthiban, B. (2024). Artificial Intelligence of Things for Smart Cities: Advanced Solutions for Enhancing Transportation Safety. *Computational Urban Science*, *4*(1), 10. Advance online publication. DOI: 10.1007/s43762-024-00120-6

James, C. (2021). *Comparative Analysis of Traditional vs. AI-Based Financial Forecasting Techniques*. February.

Jenner, S., Groba, F., & Indvik, J. (2013). Assessing the strength and effectiveness of renewable electricity feed-in tariffs in European Union countries. *Energy Policy*, *52*, 385–401. DOI: 10.1016/j.enpol.2012.09.046

Jiao, J. (2020). Application and prospect of artificial intelligence in smart grid. *IOP Conference Series. Earth and Environmental Science, 510*(2), 22012. DOI: 10.1088/1755-1315/510/2/022012

Jiménez-González, A., & Vega-Azamar, R. E. (2019). Community renewable energy projects in rural Mexico: Success factors and financial literacy impacts. *Renewable Energy, 135*, 108–117. DOI: 10.1016/j.renene.2018.11.098

Johnstone, N., Haščič, I., & Popp, D. (2010). Renewable energy policies and technological innovation: Evidence based on patent counts. *Environmental and Resource Economics, 45*(1), 133–155. DOI: 10.1007/s10640-009-9309-1

Joshi, A., Capezza, S., Alhaji, A., & Chow, M.-Y. (2023). Survey on AI and machine learning techniques for microgrid energy management systems. *IEEE/CAA Journal of Automatica Sinica, 10*(7), 1513–1529.

Jouppi, N. P., Young, C., Patil, N., Patterson, D., Agrawal, G., Bajwa, R., & Yoon, D. H. (2017, June). In-datacenter performance analysis of a tensor processing unit. In *Proceedings of the 44th annual international symposium on computer architecture* (pp. 1-12). DOI: 10.1145/3079856.3080246

Jude, A. B., Singh, D., Islam, S., Jameel, M., Srivastava, S., Prabha, B., & Kshirsagar, P. R. (2021). An artificial intelligence based predictive approach for smart waste management. *Wireless Personal Communications*, 123456789.

Jumper, J., Evans, R., Pritzel, A., Green, T., Figurnov, M., Ronneberger, O., ... Hassabis, D. (2021). Highly accurate protein structure prediction with AlphaFold. *Nature, 596*(7873), 583-589.

Kaack, L. H., Donti, P. L., Strubell, E., Kamiya, G., Creutzig, F., & Rolnick, D. (2022). Aligning artificial intelligence with climate change mitigation. *Nature Climate Change, 12*(6), 518–527. DOI: 10.1038/s41558-022-01377-7

Kabalci, Y. (2016). A survey on smart metering and smart grid communication. *Renewable & Sustainable Energy Reviews, 57*, 302–318. DOI: 10.1016/j.rser.2015.12.114

Kabeyi, M. J. B., & Olanrewaju, O. A. (2022). Sustainable energy transition for renewable and low carbon grid electricity generation and supply. *Frontiers in Energy Research, 9*, 743114. DOI: 10.3389/fenrg.2021.743114

Kahneman, D., & Tversky, A. (1979). Prospect theory: An analysis of decision under risk. *Econometrica, 47*(2), 263–291. DOI: 10.2307/1914185

Kanna, V., & Shibi. (2025). AI-driven Smart Energy Management Systems for optimized renewable energy utilization in urban smart grids. In *Research Square*.

Kareem, B. A., Zubaidi, S. L., Al-Ansari, N., & Muhsen, Y. R. (2024). Review of recent trends in the hybridisation of preprocessing-based and parameter optimisation-based hybrid models to forecast univariate streamflow. *CMES-Computer Modeling in Engineering & Sciences*, *138*(1), 1–41. DOI: 10.32604/cmes.2023.027954

Khan, A. A., Laghari, A. A., Rashid, M., Li, H., & Javed, A. R. (2023). Artificial Intelligence and Blockchain Technology for Secure Smart Grid and Power Distribution Automation: A State-of-the-Art Review. *Sustainable Energy Technologies and Assessments*, *57*, 103282. Advance online publication. DOI: 10.1016/j.seta.2023.103282

Khan, M. A., Saleh, A. M., Waseem, M., & Sajjad, I. A. (2023). Artificial Intelligence Enabled Demand Response: Prospects and Challenges in Smart Grid Environment. *IEEE Access : Practical Innovations, Open Solutions*, *11*, 1477–1505. DOI: 10.1109/ACCESS.2022.3231444

Khan, S., Mazhar, T., Khan, M. A., Shahzad, T., Ahmad, W., Bibi, A., Saeed, M. M., & Hamam, H. (2024a). Comparative analysis of deep neural network architectures for renewable energy forecasting: Enhancing accuracy with meteorological and time-based features. *Discover Sustainability*, *5*(1), 533. DOI: 10.1007/s43621-024-00783-5

Khorasany, M., Mishra, Y., & Ledwich, G. (2018). Market framework for local energy trading: A review of potential designs and market clearing approaches. *IET Generation, Transmission & Distribution*, *12*(22), 5899–5908. DOI: 10.1049/iet-gtd.2018.5309

Kiasari, M., Ghaffari, M., & Aly, H. H. (2024). A comprehensive review of the current status of smart grid technologies for renewable energies integration and future trends: The role of machine learning and energy storage systems. *Energies*, *17*(16), 4128. DOI: 10.3390/en17164128

Kidney, S. (2015). *Green bonds: The state of the market 2015*. Climate Bonds Initiative. https://www.climatebonds.net/files/files/CBI-HSBC%20report%207July%20JG01.pdf

Kiliçarslan, S. (2023). PSO+ GWO: A hybrid particle swarm optimization and Grey Wolf optimization based Algorithm for fine-tuning hyper-parameters of convolutional neural networks for Cardiovascular Disease Detection. *Journal of Ambient Intelligence and Humanized Computing*, *14*(1), 87–97. DOI: 10.1007/s12652-022-04433-4

Kitchenham, B., & Charters, S. (2007). Guidelines for performing systematic literature reviews in software engineering. *Technical report*, Ver. 2.3 EBSE Technical Report.

Klapper, L., Lusardi, A., & Van Oudheusden, P. (2019). *Financial literacy around the world: Insights from the Standard & Poor's Ratings Services Global Financial Literacy Survey.* World Bank Group. https://doi.org/DOI: 10.1596/1813-9450-7255

Koomey, J. G. (2008). Worldwide electricity used in data centers. *Environmental Research Letters, 3*(3), 034008. DOI: 10.1088/1748-9326/3/3/034008

Kotte, K. R., Attaluri, V., & Selvakumar, P. (2025). Revolutionising Retail Banking with AI and Virtual Technologies AI-Driven Financial Training. In Nair, A. J., Manohar, S., Limbu, Y. B., & Huhmann, B. A. (Eds.), *Intersecting Natural Language Processing and FinTech Innovations in Service Marketing* (pp. 421–440). IGI Global Scientific Publishing., DOI: 10.4018/979-8-3693-9944-6.ch020

Kumar, S. (2023). *Role of Interconnectors in Maintaining Security of Electricity Supply with Increasing Share of Renewables.* Academic Press.

Kumar, K., Pande, S. V., Kumar, T. C. A., Saini, P., Chaturvedi, A., Reddy, P. C. S., & Shah, K. B. (2023). Intelligent controller design and fault prediction using machine learning model. *International Transactions on Electrical Energy Systems, 2023*(1), 1056387. DOI: 10.1155/2023/1056387

Kurani, A., Doshi, P., Vakharia, A., & Shah, M. (2023). A comprehensive comparative study of artificial neural network (ANN) and support vector machines (SVM) on stock forecasting. *Annals of Data Science, 10*(1), 183–208. DOI: 10.1007/s40745-021-00344-x

Lacoste, A., Luccioni, A., Schmidt, V., & Dandres, T. (2019). Quantifying the carbon emissions of machine learning. arXiv preprint arXiv:1910.09700.

Ladi, T., Jabalameli, S., & Sharifi, A. (2022). Applications of machine learning and deep learning methods for climate change mitigation and adaptation. *Environment and Planning. B, Urban Analytics and City Science, 49*(4), 1314–1330. DOI: 10.1177/23998083221085281

Languric, M., & Zaki, L. (2022). Migrating monolithic system to domain-driven microservices: Developing a generalized migration strategy for an architecture built on microservices. Academic Press.

Lannelongue, L., Grealey, J., & Inouye, M. (2021). Green algorithms: Quantifying the carbon footprint of computation. *Advancement of Science, 8*(12), 2100707. DOI: 10.1002/advs.202100707 PMID: 34194954

Leal Filho, W., Yang, P., Eustachio, J. H. P. P., Azul, A. M., Gellers, J. C., Gielczyk, A., Dinis, M. A. P., & Kozlova, V. (2023). Deploying digitalisation and artificial intelligence in sustainable development research. *Environment, Development and Sustainability*, *25*(6), 4957–4988. DOI: 10.1007/s10668-022-02252-3 PMID: 35313685

Lee, B. C., Brooks, D., van Benthem, A., Elgamal, M., Gupta, U., Hills, G., Liu, V., Phan, L. T. X., Pierce, B., Stewart, C., Strubell, E., Wei, G.-Y., Wierman, A., Yao, Y., & Yu, M. (2025). A view of the sustainable computing landscape. *Patterns (New York, N.Y.)*, *6*(7), 101296. DOI: 10.1016/j.patter.2025.101296 PMID: 40926963

Lee, J., Ni, J., Singh, J., Jiang, B., Azamfar, M., & Feng, J. (2020). Intelligent maintenance systems and predictive manufacturing. *Journal of Manufacturing Science and Engineering*, *142*(11), 110805. DOI: 10.1115/1.4047856

Lema, R., & Lema, A. (2012). Technology transfer? The rise of China and India in green technology sectors. *Innovation and Development*, *2*(1), 23–44. DOI: 10.1080/2157930X.2012.667206

Lewis, J. I., & Wiser, R. H. (2007). Fostering a renewable energy technology industry: An international comparison of wind industry policy support mechanisms. *Energy Policy*, *35*(3), 1844–1857. DOI: 10.1016/j.enpol.2006.06.005

Li, Y. (2024). Advanced intelligent optimization algorithms for multi-objective optimal power flow in future power systems: A review. arXiv preprint arXiv:2404.09203.

Lim, K., Chang, J., Mudge, T., Ranganathan, P., Reinhardt, S. K., & Wenisch, T. F. (2009). Disaggregated memory for expansion and sharing in blade servers. *ACM SIGARCH Computer Architecture News, 37*(3), 267-278.

Lin, C., & Shahrad, M. (2024). Bridging the sustainability gap in serverless through observability and carbon-aware pricing. *ACM SIGENERGY Energy Informatics Review*, *4*(5), 120–126. DOI: 10.1145/3727200.3727218

Liu, N., Tang, Q., Zhang, J., Fan, W., & Liu, J. (2014). A hybrid forecasting model with parameter optimization for short-term load forecasting of micro-grids. *Applied Energy*, *129*, 336–345. DOI: 10.1016/j.apenergy.2014.05.023

Long, L. D. (2023). An AI-Driven Model for Predicting and Optimizing Energy-Efficient Building Envelopes. *Alexandria Engineering Journal*, *72*, 480–501. DOI: 10.1016/j.aej.2023.08.041

Longo, L., Brcic, M., Cabitza, F., Choi, J., Confalonieri, R., Del Ser, J., Guidotti, R., Hayashi, Y., Herrera, F., Holzinger, A., Jiang, R., Khosravi, H., Lecue, F., Malgieri, G., Páez, A., Samek, W., Schneider, J., Speith, T., & Stumpf, S. (2024). Explainable Artificial Intelligence (XAI) 2.0: A manifesto of open challenges and interdisciplinary research directions. *Information Fusion*, *106*, 102301. DOI: 10.1016/j.inffus.2024.102301

Luccioni, A. S., Viguier, S., & Ligozat, A. L. (2023). Estimating the carbon footprint of bloom, a 176b parameter language model. *Journal of Machine Learning Research*, *24*(253), 1–15.

Lukic Vujadinovic, V., Damnjanovic, A., Cakic, A., Petkovic, D. R., Prelevic, M., Pantovic, V., Stojanovic, M., Vidojevic, D., Vranjes, D., & Bodolo, I. (2024). AI-Driven Approach for Enhancing Sustainability in Urban Public Transportation. *Sustainability (Basel)*, *16*(17), 7763. Advance online publication. DOI: 10.3390/su16177763

Lundberg, S. M., & Lee, S.-I. (2017). A Unified Approach to Interpreting Model Predictions. In Guyon, I., Luxburg, U. V., Bengio, S., Wallach, H., Fergus, R., Vishwanathan, S., & Ford, R. (Eds.), Advances in Neural Information Processing Systems: Vol. 30. *NIPS 2017)* (pp. 4765–4774). Curran Associates, Inc. https://proceedings.neurips.cc/paper/2017/file/8a20a8621978632d76c43dfd28b67767-Paper.pdf

Lund, H., Østergaard, P. A., Connolly, D., & Mathiesen, B. V. (2017). Smart energy and smart energy systems. *Energy*, *137*, 556–565. DOI: 10.1016/j.energy.2017.05.123

Lundvall, B. A. (Ed.). (1992). *National systems of innovation: Towards a theory of innovation and interactive learning*. Pinter Publishers.

Lusardi, A., & Mitchell, O. S. (2014). The economic importance of financial literacy: Theory and evidence. *Journal of Economic Literature*, *52*(1), 5–44. DOI: 10.1257/jel.52.1.5

Machlev, R., Heistrene, L., Perl, M., Levy, K. Y., Belikov, J., Mannor, S., & Levron, Y. (2022). Explainable Artificial Intelligence (XAI) techniques for energy and power systems: Review, challenges and opportunities. *Energy and AI*, *9*, 100169. DOI: 10.1016/j.egyai.2022.100169

Maheswaran, K., Selvakumar, P., Shaikh, M., Shashank, G., Gupta, S., & Manjunath, T. C. (2025). The Evolution of Smart Urban Mobility With Emerging Technologies. In Ponnusamy, S., Assaf, M., Youssef, B., Jeon, G., & Kalyanaraman, S. (Eds.), *Leveraging VANETs and Blockchain Technology for Urban Mobility* (pp. 1–28). IGI Global Scientific Publishing. DOI: 10.4018/979-8-3373-0265-2.ch001

Mahmood, S., Sun, H., Ali Alhussan, A., Iqbal, A., & El-Kenawy, E. S. M. (2024). Active learning-based machine learning approach for enhancing environmental sustainability in green building energy consumption. *Scientific Reports*, *14*(1), 19894. DOI: 10.1038/s41598-024-70729-4 PMID: 39191844

Majchrowska, S., Mikołajczyk, A., Ferlin, M., Klawikowska, Z., Plantykow, M. A., Kwasigroch, A., & Majek, K. (2022). Deep learning-based waste detection in natural and urban environments. *Waste Management (New York, N.Y.)*, *138*, 274–284. DOI: 10.1016/j.wasman.2021.12.001 PMID: 34920243

Malka, L., Bidaj, F., Kuriqi, A., Jaku, A., Roçi, R., & Gebremedhin, A. (2023). Energy system analysis with a focus on future energy demand projections: The case of Norway. *Energy*, *272*, 127107. DOI: 10.1016/j.energy.2023.127107

Malmodin, J., & Lundén, D. (2018). The energy and carbon footprint of the global ICT and E&M sectors 2010–2015. *Sustainability (Basel)*, *10*(9), 3027. DOI: 10.3390/su10093027

Manduva, V. C. (2023). AI-Driven Edge Computing in the Cloud Era: Challenges and Opportunities. *International Journal of Modern Computing*, *11*(2), 88–102. DOI: 10.18535/ijsrm/v11i02.ec5

Mann, B., Ryder, N., Subbiah, M., Kaplan, J., Dhariwal, P., Neelakantan, A., . . . Agarwal, S. (2020). Language models are few-shot learners. arXiv preprint arXiv:2005.14165, 1(3), 3.

Martineau, K. (2020, August 7). *Shrinking deep learning's carbon footprint*. MIT News. https://news.mit.edu/2020/shrinking-deep-learning-carbon-footprint-0807

Martínez-Gordón, R., Morales-Ruvalcaba, C. F., Martínez-Díez, R., González-Albarrán, M. A., Tovar-Salinas, J. L., & Becerra-Pérez, L. A. (2022). A comprehensive review of photovoltaic solar energy in Mexico: Current status, trends, and future perspectives. *Energy Strategy Reviews*, *42*, 100882.

Martiny, A. (2023). Towards sustainable ai: Monitoring and analysis of carbon emissions in machine learning algorithms (Doctoral dissertation, Politecnico di Torino).

Mason, K., & Grijalva, S. (2019). A review of reinforcement learning for autonomous building energy management. *Computers & Electrical Engineering*, *78*, 300–312. DOI: 10.1016/j.compeleceng.2019.07.019

Mataifa, H., Krishnamurthy, S., & Kriger, C. (2022). Volt/var optimization: A survey of classical and heuristic optimization methods. *IEEE Access : Practical Innovations, Open Solutions*, *10*, 13379–13399. DOI: 10.1109/ACCESS.2022.3146366

Mazhar, T., Irfan, H. M., Haq, I., Ullah, I., Ashraf, M., Shloul, T. A., Ghadi, Y. Y., Imran, , & Elkamchouchi, D. H. (2023). Analysis of Challenges and Solutions of IoT in Smart Grids Using AI and Machine Learning Techniques: A Review. *Electronics (Basel), 12*(1), 242. Advance online publication. DOI: 10.3390/electronics12010242

Mazzucato, M. (2018). Mission-oriented innovation policies: Challenges and opportunities. *Industrial and Corporate Change, 27*(5), 803–815. DOI: 10.1093/icc/dty034

McGuinness, M. (2024). Financial literacy: Commission and OECD-INFE publish joint framework to improve individuals' financial skills. European External Action Service (EEAS). https://www.eeas.europa.eu/eeas/financial-literacy-commission-and-oecd-infe-publish-joint-framework-improve-individuals-financial_en

Mei, L., & Stamp, M. (2025). Energy Considerations for Large Pretrained Neural Networks. *arXiv preprint arXiv:2506.01311.*

Mengelkamp, E., Gärttner, J., Rock, K., Kessler, S., Orsini, L., & Weinhardt, C. (2018). Designing microgrid energy markets: A case study: The Brooklyn Microgrid. *Applied Energy, 210*, 870–880. DOI: 10.1016/j.apenergy.2017.06.054

Mills, B., & Schleich, J. (2012). Residential energy-efficient technology adoption, energy conservation, knowledge, and attitudes: An analysis of European countries. *Energy Policy, 49*, 616–628. DOI: 10.1016/j.enpol.2012.07.008

Mirzania, E., Achite, M., Elshaboury, N., Katipoğlu, O. M., & Saroughi, M. (2024). Prediction of monthly groundwater level using a new hybrid intelligent approach in the Tabriz plain, Iran. *Neural Computing & Applications, 36*(20), 12609–12624. DOI: 10.1007/s00521-024-09681-3

Mittelstadt, B. (2019). Principles alone cannot guarantee ethical AI. *Nature Machine Intelligence, 1*(11), 501–507. DOI: 10.1038/s42256-019-0114-4

Moher, D., Liberati, A., Tetzlaff, J., & Altman, D. G.Prisma Group. (2009). Preferred reporting items for systematic reviews and meta-analyses: The PRISMA statement. *PLoS Medicine, 6*(7), e1000097. DOI: 10.1371/journal.pmed.1000097 PMID: 19621072

Mohsen, B. M. (2024). AI-Driven Optimization of Urban Logistics in Smart Cities: Integrating Autonomous Vehicles and IoT for Efficient Delivery Systems. *Sustainability (Basel), 16*(24), 11265. Advance online publication. DOI: 10.3390/su162411265

Molchanov, P., Tyree, S., Karras, T., Aila, T., & Kautz, J. (2016). Pruning convolutional neural networks for resource efficient inference. arXiv preprint arXiv:1611.06440.

Molnar, C. (2019). *Interpretable machine learning: A guide for making black box models explainable*. Leanpub.

Mondal, S., Das, S., & Vrana, V. G. (2024). Exploring the role of artificial intelligence in achieving a net zero carbon economy in emerging economies: A combination of PLS-SEM and fsQCA approaches to digital inclusion and climate resilience. *Sustainability (Basel), 16*(23), 10299. DOI: 10.3390/su162310299

Mundaca, L., & Richter, J. L. (2015). Assessing 'green energy economy' stimulus packages: Evidence from the U.S. programs targeting renewable energy. *Renewable & Sustainable Energy Reviews, 42*, 1174–1186. DOI: 10.1016/j.rser.2014.10.060

Murugesan, S. (2008). Harnessing green IT: Principles and practices. *IT Professional, 10*(1), 24–33. DOI: 10.1109/MITP.2008.10

Mystakidis, A., Koukaras, P., Tsalikidis, N., Ioannidis, D., & Tjortjis, C. (2024). Energy forecasting: A comprehensive review of techniques and technologies. *Energies, 17*(7), 1662. DOI: 10.3390/en17071662

Na, S., Heo, S., Han, S., Shin, Y., & Lee, M. (2022). Development of an artificial intelligence model to recognise construction waste by applying image data augmentation and transfer learning. *Buildings (Basel, Switzerland), 12*(2), 175. DOI: 10.3390/buildings12020175

Nelson, R. R. (Ed.). (1993). *National innovation systems: a comparative analysis*. Oxford University Press. DOI: 10.1093/oso/9780195076165.001.0001

Nemet, G. F. (2006). Beyond the learning curve: Factors influencing cost reductions in photovoltaics. *Energy Policy, 34*(17), 3218–3232. DOI: 10.1016/j.enpol.2005.06.020

Nguyen, X. C., Nguyen, T. T. H., La, D. D., Kumar, G., Rene, E. R., Nguyen, D. D., Chang, S. W., Chung, W. J., Nguyen, X. H., & Nguyen, V. K. (2021). Development of machine learning-based models to forecast solid waste generation in residential areas: A case study from Vietnam. *Resources, Conservation and Recycling, 167*, 105381. DOI: 10.1016/j.resconrec.2020.105381

Nik, V. M., Perera, A. T. D., & Chen, D. (2021). Towards climate resilient urban energy systems: A review. *National Science Review, 8*(3), nwaa134. DOI: 10.1093/nsr/nwaa134 PMID: 34691589

Ning, M. A. (2024). Artificial Intelligence-Driven Decision Support Systems for Sustainable Energy Management in Smart Cities. *International Journal of Advanced Computer Science and Applications, 15*(9).

Nisha Angeline, M., Selvakumar, P., Gupta, M. K., Nazar, N., Elumalai, D., & Bhattacharya, S. (2025). Emerging Materials for Energy Storage and Environmental Applications. In Muduli, K., Moharana, B., Ales, S., & Biswal, D. (Eds.), *Using Computational Intelligence for Sustainable Manufacturing of Advanced Materials* (pp. 53–72). IGI Global Scientific Publishing. DOI: 10.4018/979-8-3693-7974-5.ch003

Noviati, N. D., Maulina, S. D., & Smith, S. (2024). Smart grids: Integrating AI for efficient renewable energy utilization. [ITALIC]. *International Transactions on Artificial Intelligence*, *3*(1), 1–10. DOI: 10.33050/italic.v3i1.644

Nuruzzaman, M., & Rana, S. (2025). IoT-Enabled Condition Monitoring in Power Distribution Systems: A Review of Scada-Based Automation, Real-Time Data Analytics, and Cyber-Physical Security Challenges. *Journal of Sustainable Development and Policy*, *1*(01), 25–43. DOI: 10.63125/pyd1x841

O'Brien, J. (2025). AI's Energy Appetite: Lessons from Leading Hyperscalers on Sustainable Procurement (Doctoral dissertation).

OCDE. (2022). Recomendación del Consejo sobre Principios de alto nivel para la protección del consumidor financiero. OECD Legal Instruments. https://legalinstruments.oecd.org/api/download/?uri=/public/62c8cfd9-65fd-4da9-9bd5-3c9551637814.pdf

OECD. (2022). OECD/INFE Toolkit for Measuring Financial Literacy and Financial Inclusion. OECD Publishing. https://www.oecd.org/en/publications/oecd-infe-toolkit-for-measuring-financial-literacy-and-financial-inclusion-2022_cbc4114f-en.html

OECD. (Organisation for Economic Co-operation and Development). (2011). *Invention and transfer of environmental technologies*. OECD Publishing.

Ohalete, N. C., Aderibigbe, A. O., Ani, E. C., Ohenhen, P. E., Daraojimba, D. O., & Odulaja, B. A. (2023). AI-driven solutions in renewable energy: A review of data science applications in solar and wind energy optimization. *World Journal of Advanced Research and Reviews*, *20*(3), 401–417. DOI: 10.30574/wjarr.2023.20.3.2433

Ojadi, J. O., Odionu, C. S., Onukwulu, E. C., & Owulade, O. A. (2024). AI-Enabled Smart Grid Systems for Energy Efficiency and Carbon Footprint Reduction in Urban Energy Networks. *International Journal of Multidisciplinary Research and Growth Evaluation*, *5*(1), 1549–1566. DOI: 10.54660/.IJMRGE.2024.5.1.1549-1566

Oladapo, B. I., Olawumi, M. A., & Omigbodun, F. T. (2024). Machine learning for optimising renewable energy and grid efficiency. *Atmosphere (Basel)*, *15*(10), 1250. DOI: 10.3390/atmos15101250

Olaoye, G. (2025). Self-Learning Neural Networks in the Cloud: Towards Autonomous AI Systems. *Available at SSRN* 5129553. DOI: 10.2139/ssrn.5129553

Olawade, D. B., Fapohunda, O., Wada, O. Z., Usman, S. O., Ige, A. O., Ajisafe, O., & Oladapo, B. I. (2024). *Smart waste management: A paradigm shift enabled by artificial intelligence*. Waste Management Bulletin.

Olawade, D. B., Wada, O. Z., David-Olawade, A. C., Fapohunda, O., Ige, A. O., & Ling, J. (2024). Artificial Intelligence Potential for Net Zero Sustainability: Current Evidence and Prospects. *Next Sustainability*, *4*, 100041. Advance online publication. DOI: 10.1016/j.nxsust.2024.100041

Omitaomu, O. A., & Niu, H. (2021). Artificial Intelligence Techniques in Smart Grid: A Survey. *Smart Cities*, *4*(2), 548–568. DOI: 10.3390/smartcities4020029

Onukwulu, E. C., Agho, M. O., & Eyo-Udo, N. L. (2023). Developing a Framework for Predictive Analytics in Mitigating Energy Supply Chain Risks. *International Journal of Scholarly Research and Reviews*, *2*(2), 135–155. DOI: 10.56781/ijsrr.2023.2.2.0042

Oudinga, M. (2023). The role of artificial intelligence in enhancing energy management in microgrid systems. *Journal Dimensie Management and Public Sector*, *4*(4), 33–41. DOI: 10.48173/jdmps.v4i4.246

Owino, A. (2023). Challenges of computer vision adoption in the Kenyan agricultural sector and how to solve them: A general perspective. *Advances in Agriculture*, *2023*, 1530629. DOI: 10.1155/2023/1530629

Paiva, S., Ahad, M. A., Tripathi, G., Feroz, N., & Casalino, G. (2021). Enabling Technologies for Urban Smart Mobility: Recent Trends, Opportunities and Challenges. *Sensors (Basel)*, *21*(6), 2143. Advance online publication. DOI: 10.3390/s21062143 PMID: 33803903

Palm, J. (2018). Household installation of solar panels – Motives and barriers in a 10-year perspective. *Energy Policy*, *113*, 1–8. DOI: 10.1016/j.enpol.2017.10.047

Pamucar, D., Deveci, M., Gokasar, I., Tavana, M., & Köppen, M. (2022). A Metaverse Assessment Model for Sustainable Transportation Using Ordinal Priority Approach and Aczel-Alsina Norms. *Technological Forecasting and Social Change*, *182*, 121778. Advance online publication. DOI: 10.1016/j.techfore.2022.121778

Panda, S., Mohanty, S., Rout, P. K., Sahu, B. K., Parida, S. M., Kotb, H., Flah, A., Tostado-Véliz, M., Abdul Samad, B., & Shouran, M. (2022). An insight into the integration of distributed energy resources and energy storage systems with smart distribution networks using demand-side management. *Applied Sciences (Basel, Switzerland)*, *12*(17), 8914. DOI: 10.3390/app12178914

Parekh, R. (2024). Automating the Design Process for Smart Building Technologies. *World Journal of Advanced Research and Reviews, 23*(2), 1213–1234. DOI: 10.30574/wjarr.2024.23.2.2461

Pathak, J., Wikner, A., Fussell, R., Chandra, S., Hunt, B. R., Girvan, M., & Ott, E. (2018). Hybrid forecasting of chaotic processes: Using machine learning in conjunction with a knowledge-based model. *Chaos (Woodbury, N.Y.), 28*(4), 041101. DOI: 10.1063/1.5028373 PMID: 31906641

Patidar, S., Kumar, N., & Jindal, R. (2024). IoT Data Stream Handling, Analysis, Communication and Security Issues: A Systematic Survey. *Wireless Personal Communications*, 1–50. DOI: 10.1007/s11277-024-11177-1

Patterson, D., Gonzalez, J., Le, Q., Liang, C., Munguia, L. M., Rothchild, D., . . . Dean, J. (2021). Carbon emissions and large neural network training. arXiv preprint arXiv:2104.10350.

Pedrycz, W. (2020). Welcome to the exciting world of "green computing and smart environments". *Journal of Smart Environments and Green Computing, 1*(1), 1–2. DOI: 10.20517/jsegc.2020.01

Penzenstadler, B., Raturi, A., Richardson, D., Calero, C., Femmer, H., & Franch, X. (2014, May). Systematic mapping study on software engineering for sustainability (SE4S). In *Proceedings of the 18th international conference on evaluation and assessment in software engineering* (pp. 1-14). DOI: 10.1145/2601248.2601256

Perera, A. T. D., Nik, V. M., Chen, D., Scartezzini, J.-L., & Hong, T. (2020a). Quantifying the impacts of climate change and extreme climate events on energy systems. *Nature Energy, 5*(2), 150–159. DOI: 10.1038/s41560-020-0558-0

Pillai, K. S., Sneha, M. L., Aiswarya, S., Anand, A. B., & Prasad, G. (2023). Municipal Solid Waste Management: A Review of Machine Learning Applications. In *E3S Web of Conferences* (Vol. 455, p. 02018). EDP Sciences. DOI: 10.1051/e3sconf/202345502018

Plume Labs. (2022). AI-Powered Air Quality Forecasting.

Porter, M. E., & Van der Linde, C. (1995). Toward a new conception of the environment-competitiveness relationship. *The Journal of Economic Perspectives, 9*(4), 97–118. DOI: 10.1257/jep.9.4.97

Powroźnik, P., & Szcześniak, P. (2024). Predictive Analytics for Energy Efficiency: Leveraging Machine Learning to Optimize Household Energy Consumption. *Energies, 17*(23), 5866. Advance online publication. DOI: 10.3390/en17235866

Pradhan, N. R., Singh, A. P., Verma, S., & Mohapatra, P. K. (2022). A blockchain based lightweight peer-to-peer energy trading framework for secured high throughput micro-transactions. *Scientific Reports*, *12*(1), 14523. DOI: 10.1038/s41598-022-18603-z PMID: 36008545

PricewaterhouseCoopers (PwC) US. (2025). *2025 AI Business Predictions*. PwC US Tech & Analytics.

Proença, P. F., & Simões, P. (2020). Taco: Trash annotations in context for litter detection. *arXiv preprint arXiv:2003.06975*.

Proskuryakova, L. (2023). The contribution of low-carbon energy technologies to climate resilience. *Climate (Basel)*, *11*(12), 231. DOI: 10.3390/cli11120231

Pudjianto, D., Ramsay, C., & Strbac, G. (2007). Virtual power plant and system integration of distributed energy resources. *IET Renewable Power Generation*, *1*(1), 3–9. DOI: 10.1049/iet-rpg:20060023

Qadir, S. A., Ahmad, F., Al-Wahedi, A. M A B., Iqbal, A., & Ali, A. (2024). Navigating the Complex Realities of Electric Vehicle Adoption: A Comprehensive Study of Government Strategies, Policies, and Incentives. *Energy Strategy Reviews*, *53*, 101379. Advance online publication. DOI: 10.1016/j.esr.2024.101379

Qiu, D., Ye, Y., Papadaskalopoulos, D., & Strbac, G. (2020). A Deep Reinforcement Learning Method for Pricing Electric Vehicles With Discrete Charging Levels. *IEEE Transactions on Industry Applications*, *56*(5), 5901–5912. DOI: 10.1109/TIA.2020.2984614

Qiu, X., Parcollet, T., Fernandez-Marques, J., Gusmao, P. P., Gao, Y., Beutel, D. J., & Lane, N. D. (2023). A first look into the carbon footprint of federated learning. *Journal of Machine Learning Research*, *24*(129), 1–23.

Qudus, L. (2025). Leveraging Artificial Intelligence to Enhance Process Control and Improve Efficiency in Manufacturing Industries. *International Journal of Computer Applications Technology and Research*, *14*(2), 18–38. DOI: 10.7753/IJCATR1402.1002

Radovanović, A., Koningstein, R., Schneider, I., Chen, B., Duarte, A., Roy, B., Xiao, D., Haridasan, M., Hung, P., Care, N., Talukdar, S., Mullen, E., Smith, K., Cottman, M. E., & Cirne, W. (2022). Carbon-aware computing for datacenters. *IEEE Transactions on Power Systems*, *38*(2), 1270–1280. DOI: 10.1109/TPWRS.2022.3173250

Rahayu, B. (2024). Transformative Synergy: Machine Learning and Neural Networks in Environmental Science. *Asian Journal of Basic and Applied Sciences, 10*(1).

Rainforest Foundation US. (2022). AI and Satellite Tech for Indigenous Forest Monitoring.

Rajalakshmi, J., Sumangali, K., Jayanthi, J., & Muthulakshmi, K. (2023). Artificial intelligence with earthworm optimization assisted waste management system for smart cities. *Global NEST Journal, 25,* 190–197.

Rajaperumal, T. A., & Columbus, C. C. (2025). Transforming the electrical grid: The role of AI in advancing smart, sustainable, and secure energy systems. *Energy Informatics, 8*(1), 51. DOI: 10.1186/s42162-024-00461-w

Ramya, J., Yerraguravagari, S. S., Gaikwad, S., & Gupta, R. K. (2024). AI and Machine Learning in Predictive Analytics: Revolutionizing Business Strategies through Big Data Insights. *Library of Progress-Library Science, Information Technology & Computer, 44*(3).

Rana, S. (2025). AI-driven fault detection and predictive maintenance in electrical power systems: A systematic review of data-driven approaches, digital twins, and self-healing grids. *American Journal of Advanced Technology and Engineering Solutions, 1*(01), 258–289. DOI: 10.63125/4p25x993

Rane, N., Choudhary, S., & Rane, J. (2024). "Artificial Intelligence and Machine Learning for Resilient and Sustainable Logistics and Supply Chain Management." *SSRN* Preprint, 8 pages. DOI: DOI: 10.2139/ssrn.4847087

Rao, P. S., Yaqoob, S. I., Ahmed, M. A., Abdinabievna, P. S., Yaseen, S. M., & Arumugam, M. (2023). RETRACTED ARTICLE: Integrated Artificial Intelligence and Predictive Maintenance of Electric Vehicle Components with Optical and Quantum Enhancements. *Optical and Quantum Electronics, 55*(10), 855. Advance online publication. DOI: 10.1007/s11082-023-05135-7

Rashid, H. (2024). Designing Cost-Effective, AI-Powered Demand Forecasting Models for the Manufacturing Sector. *Journal of Artificial Intelligence & Cloud Computing. J Arti Inte & Cloud Comp, 3*(1), 2–5.

Rasp, S., Dueben, P. D., Scher, S., Weyn, J. A., Mouatadid, S., & Thuerey, N. (2020). WeatherBench: a benchmark data set for data-driven weather forecasting. Journal of Advances in Modeling Earth Systems, 12(11), e2020MS002203.

Ravi, A., Saranathan, M., Achuthan, P. H. K., Lavanya, M. C., & Rajini, V. (2022). A Comprehensive review on the current trends in Micro-Phasor Measurement Units. *IOP Conference Series. Materials Science and Engineering, 1258*(1), 012045. DOI: 10.1088/1757-899X/1258/1/012045

Raz, D., Bintz, C., Guetler, V., Tam, A., Katell, M., Dailey, D., . . . Young, M. (2021, July). Face Mis-ID: An interactive pedagogical tool demonstrating disparate accuracy rates in facial recognition. In Proceedings of the 2021 AAAI/ACM Conference on AI, Ethics, and Society (pp. 895-904).

Recycleye. (2024, December) *How AI robots help reduce the cost of waste sorting in MRFs*. https://recycleye.com/how-ai-robots-reduce-cost-waste-sorting/

Rehan, H. (2024). The Future of Electric Vehicles: Navigating the Intersection of AI, Cloud Technology, and Cybersecurity. *International Journal of Scientific Research and Management, 12*(4), 1127–1143. DOI: 10.18535/ijsrm/v12i04.ec04

Rekeraho, A., Cotfas, D. T., Cotfas, P. A., Bălan, T. C., & Tuyishime Emmanuel and Acheampong, R. (2023). Cybersecurity challenges in IoT-based smart renewable energy. In *Research Square*.

Rennings, K. (2000). Redefining innovation — Eco-innovation research and the contribution from ecological economics. *Ecological Economics, 32*(2), 319–332. DOI: 10.1016/S0921-8009(99)00112-3

Reuter, M., Buchwitz, M., & Schneising, O. (2020). Towards monitoring localized CO_2 emissions from space: Co-located regional CO_2 and NO_2 enhancements observed by the OCO-2 and Sentinel-5 Precursor satellites. *Atmospheric Chemistry and Physics, 20*(2), 1027–1040.

Riabova, J. (2024, October). in the Era of Digital Transformation. In Subject-Oriented Business Process Management. *Models for Designing Digital Transformations: 15th International Conference, S-BPM ONE 2024, Weiden, Germany, May 21–22, 2024 Proceedings, 2206*, 189.

Ribeiro, M. T., Singh, S., & Guestrin, C. (2016). “Why should I trust you?”: Explaining the predictions of any classifier. *Proceedings of the 22nd ACM SIGKDD International Conference on Knowledge Discovery and Data Mining*, 1135-1144.

Rodrigues, T. A., Patrikar, J., Oliveira, N. L., Matthews, H. S., Scherer, S., & Samaras, C. (2022). Drone flight data reveal energy and greenhouse gas emissions savings for very small package delivery. *Patterns (New York, N.Y.), 3*(8), 100569. DOI: 10.1016/j.patter.2022.100569 PMID: 36033593

Rogers, E. M. (2003). *Diffusion of innovations* (5th ed.). Free Press.

Rogge, K. S., & Reichardt, K. (2016). Policy mixes for sustainability transitions: An extended concept and framework for analysis. *Research Policy, 45*(8), 1620–1635. DOI: 10.1016/j.respol.2016.04.004

Rolnick, D. (2019). Tackling climate change with machine learning. arXiv:1906.05433.

Rosero-Montalvo, P. D., López-Batista, V. F., & Peluffo-Ordóñez, D. H. (2022). A new data-preprocessing-related taxonomy of sensors for iot applications. *Information (Basel), 13*(5), 241. DOI: 10.3390/info13050241

Rubab, S., Khan, M. M., Uddin, F., Abbas Bangash, Y., & Taqvi, S. A. A. (2022). A study on ai-based waste management strategies for the covid-19 pandemic. *ChemBioEng Reviews, 9*(2), 212–226. DOI: 10.1002/cben.202100044

Safari, A., Daneshvar, M., & Anvari-Moghaddam, A. (2024). Energy intelligence: A systematic review of artificial intelligence for energy management. *Applied Sciences (Basel, Switzerland), 14*(23), 11112. DOI: 10.3390/app142311112

Sakapaji, S. C., & Puthenkalam, J. J. (2023). Harnessing AI for climate-resilient agriculture: Opportunities and challenges. *Ejtas, 1*(6), 1144–1158. DOI: 10.59324/ejtas.2023.1(6).111

Saleh, F., Elhendawi, A., Darwish, A. S., & Farrell, P. (2024). A framework for leveraging the incorporation of AI, BIM, and IoT to achieve smart sustainable cities. *Journal of Intelligent Systems and Internet of Things, 11*(2), 75–84. DOI: 10.54216/JISIoT.110207

Salman, S. K. (2019). Evolution of conventional power systems to smart grids. *2019 54th International Universities Power Engineering Conference (UPEC)*, 1–6.

Samek, W., Wiegand, T., & Müller, K.-R. (2017). *Explainable Artificial Intelligence: Understanding, Visualizing and Interpreting Deep Learning Models*. arXiv.

Sami, K. N., Amin, Z. M. A., & Hassan, R. (2020). Waste management using machine learning and deep learning algorithms. *International Journal on Perceptive and Cognitive Computing, 6*(2), 97–106. DOI: 10.31436/ijpcc.v6i2.165

Sandalow, D. B., McCormick, C., Kucukelbir, A., Friedmann, J., Nachmany, M., Lee, H., ... Ruane, A. C. (2024). Artificial Intelligence for climate change mitigation roadmap. Academic Press.

Sandler, M., Howard, A., Zhu, M., Zhmoginov, A., & Chen, L. C. (2018). Mobilenetv2: Inverted residuals and linear bottlenecks. In *Proceedings of the IEEE conference on computer vision and pattern recognition* (pp. 4510-4520).

Sanh, V., Debut, L., Chaumond, J., & Wolf, T. (2019). DistilBERT, a distilled version of BERT: smaller, faster, cheaper and lighter. arXiv preprint arXiv:1910.01108.

Sardianou, E., & Genoudi, P. (2013). Which factors affect the willingness of consumers to adopt renewable energies? *Renewable Energy*, *57*, 1–4. DOI: 10.1016/j.renene.2013.01.031

Sarker, I. H., Furhad, M. H., & Nowrozy, R. (2021). AI-Driven Cybersecurity: An Overview, Security Intelligence Modeling and Research Directions. *SN Computer Science*, *2*(3), 173. Advance online publication. DOI: 10.1007/s42979-021-00557-0

Sarma, G., & Zabaniotou, A. (2021). Understanding vulnerabilities of renewable energy systems for building their resilience to climate change hazards: Key concepts and assessment approaches. *Renew. Energy Environ. Sustain.*, *6*, 35. DOI: 10.1051/rees/2021035

Satpathy, I., Nayak, A., & Jain, V. (2025). The green city: Sustainable and smart urban living through artificial intelligence. In *Utilizing Technology to Manage Territories* (pp. 273–304). IGI Global.

Schot, J., & Steinmueller, W. E. (2018). Three frames for innovation policy: R&D, systems of innovation and transformative change. *Research Policy*, *47*(9), 1554–1567. DOI: 10.1016/j.respol.2018.08.011

Schwartz, R., Dodge, J., Smith, N. A., & Etzioni, O. (2020). Green ai. *Communications of the ACM*, *63*(12), 54–63. DOI: 10.1145/3381831

Selvaraju, R. R., Cogswell, M., Das, A., Vedantam, R., Parikh, D., & Batra, D. (2017). Grad-CAM: Visual Explanations from Deep Networks via Gradient-Based Localization. In *Proceedings of the IEEE International Conference on Computer Vision (ICCV)* (pp. 618–626). IEEE. https://www.computer.org/csdl/proceedings-article/2017/iccv/3394747/12S4860R1iX

SENER (Secretaría de Energía). (2024). *Energy Statistics and Market Analysis*. Government of Mexico.

Sethumadhavan, R. P. P. S., Selvakumar, P., Kadu, S. G., Chandel, P. S., & Gupta, S. (2025). Environmental Sustainability and Employee Participation. In Şeker, C., Örücü, E., & Kaya, A. (Eds.), *Green Leadership Strategies for Increased Employee Engagement and Performance* (pp. 275–298). IGI Global Scientific Publishing. DOI: 10.4018/979-8-3373-2106-6.ch009

Shadi, M. R., Mirshekali, H., & Shaker, H. R. (2025). Explainable artificial intelligence for energy systems maintenance: A review on concepts, current techniques, challenges, and prospects. *Renewable & Sustainable Energy Reviews*, *216*, 115668. DOI: 10.1016/j.rser.2025.115668

Shafik, W. (2024). Incorporating Artificial Intelligence for Urban and Smart Cities' Sustainability. In *Maintaining a Sustainable World in the Nexus of Environmental Science and AI* (pp. 23-58). IGI Global.

Shahab, S., & Anjum, M. (2022). Solid waste management scenario in india and illegal dump detection using deep learning: An AI approach towards the sustainable waste management. *Sustainability (Basel), 14*(23), 15896. DOI: 10.3390/su142315896

Shahana, A., Hasan, R., Farabi, S. F., Akter, J., Mahmud, M. A., Johora, F. T., & Suzer, G. (2024). AI-Driven Cybersecurity: Balancing Advancements and Safeguards. *Journal of Computer Science and Technology Studies*, *6*(2), 88–97. DOI: 10.32996/jcsts.2024.6.2.9

Shahrabani, M. M. N., & Apanaviciene, R. (2025). Evaluation of smart building integration into a smart city by applying machine learning techniques. *Buildings (Basel, Switzerland), 15*(12), 1–39. DOI: 10.3390/buildings15122031

Shahzad, S., & Jasińska, E. (2024). Renewable revolution: A review of strategic flexibility in future power systems. *Sustainability (Basel), 16*(13), 5454. DOI: 10.3390/su16135454

Shamsuddoha, M., Kashem, M. A., & Nasir, T. (2025). A Review of Transportation 5.0: Advancing Sustainable Mobility Through Intelligent Technology and Renewable Energy. *Future Transportation*, *5*(1), 8. Advance online publication. DOI: 10.3390/futuretransp5010008

Sharifhosseini, S. M., Niknam, T., Taabodi, M. H., Aghajari, H. A., Sheybani, E., Javidi, G., & Pourbehzadi, M. (2024). Investigating Intelligent Forecasting and Optimization in Electrical Power Systems: A Comprehensive Review of Techniques and Applications. *Energies*, *17*(21), 5385. DOI: 10.3390/en17215385

Sharma, N., Litoriya, R., & Sharma, A. (2021). Application and analysis of K-means algorithms on a decision support framework for municipal solid waste management. In *Advanced Machine Learning Technologies and Applications* [Springer Singapore.]. *Proceedings of AMLTA, 2020*, 267–276.

Sharma, R., & Singhal, M. (2024). "Sustainable Transport Using AI and IoT." In Sustainable Development Through Machine Learning, AI and IoT. *Communications in Computer and Information Science*, *2196*, 273–283. DOI: 10.1007/978-3-031-71729-1_25

Sheldon, J., Zhu, W., Abdullah, A., Bhupathiraju, S. H. V., Sugawara, T., Butler, K. R., & Rampazzi, S. (2024, May). Aquasonic: Acoustic manipulation of underwater data center operations and resource management. In *2024 IEEE Symposium on Security and Privacy (SP)* (pp. 331-349). IEEE. DOI: 10.1109/SP54263.2024.00201

Sherstinsky, A. (2020). Fundamentals of recurrent neural network (RNN) and long short-term memory (LSTM) network. *Physica D. Nonlinear Phenomena, 404*, 132306. DOI: 10.1016/j.physd.2019.132306

Shobeiry, S. M. (2024). Ai-enabled modern power systems: Challenges, solutions, and recommendations. In *Artificial Intelligence in the Operation and Control of Digitalized Power Systems* (pp. 19–67). Springer. DOI: 10.1007/978-3-031-69358-8_2

Shreyas Madhav, A. V., Rajaraman, R., Harini, S., & Kiliroor, C. C. (2022). Application of artificial intelligence to enhance collection of E-waste: A potential solution for household WEEE collection and segregation in India. *Waste Management & Research, 40*(7), 1047–1053. DOI: 10.1177/0734242X211052846 PMID: 34726090

Shrimali, G., & Kniefel, J. (2011). Are government policies effective in promoting deployment of renewable electricity resources? *Energy Policy, 39*(9), 4726–4741. DOI: 10.1016/j.enpol.2011.06.055

Shufian, A., Shib, S. K., Dipto, D. R., Rahman, M. T., Hannan, N., & Fattah, S. A. (2025). Fuzzy logic-controlled three-phase dynamic voltage restorer for enhancing voltage stabilization and power quality. *International Journal of Electrical Power & Energy Systems, 166*, 110517. DOI: 10.1016/j.ijepes.2025.110517

Simonyan, K., & Zisserman, A. (2014). Very deep convolutional networks for large-scale image recognition. *arXiv preprint arXiv:1409.1556.*

Song, Y., Wan, C., Hu, X., Qin, H., & Lao, K. (2022). Resilient power grid for smart city. *IEnergy, 1*(3), 325–340. DOI: 10.23919/IEN.2022.0043

Soshinskaya, M., Crijns-Graus, W. H. J., Guerrero, J. M., & Vasquez, J. C. (2014). Microgrids: Experiences, barriers and success factors. *Renewable & Sustainable Energy Reviews, 40*, 659–672. DOI: 10.1016/j.rser.2014.07.198

Souza, A., Jasoria, S., Chakrabarty, B., Bridgwater, A., Lundberg, A., Skogh, F., & Shenoy, P. (2023, October). Casper: Carbon-aware scheduling and provisioning for distributed web services. In *Proceedings of the 14th International Green and Sustainable Computing Conference* (pp. 67-73). DOI: 10.1145/3634769.3634812

Stavrev, S., & Ginchev, D. (2024). Reinforcement learning techniques in optimizing energy systems. *Electronics (Basel), 13*(8), 1459. DOI: 10.3390/electronics13081459

Stern, D. (2024). How AI and ML can help reduce energy usage and carbon emissions in buildings. *Corporate Real Estate Journal, 14*(1), 66–74. DOI: 10.69554/IKXQ8332

Strezoski, L. (2023). Distributed energy resource management systems—DERMS: State of the art and how to move forward. *Wiley Interdisciplinary Reviews. Energy and Environment, 12*(1), e460. DOI: 10.1002/wene.460

Strubell, E., Ganesh, A., & McCallum, A. (2020, April). Energy and policy considerations for modern deep learning research. *Proceedings of the AAAI Conference on Artificial Intelligence, 34*(09), 13693–13696. DOI: 10.1609/aaai.v34i09.7123

Sulaiman, A., Nagu, B., Kaur, G., Karuppaiah, P., Alshahrani, H., Al Reshan, M. S., AlYami, S., & Shaikh, A. (2023). Artificial intelligence-based secured power grid protocol for smart city. *Sensors (Basel), 23*(19), 8016. DOI: 10.3390/s23198016 PMID: 37836846

Sun, N., Miao, Y., Jiang, H., Ding, M., & Zhang, J. (2024). From principles to practice: A deep dive into AI ethics and regulations. arXiv preprint arXiv:2412.04683.

Sun, J., Guan, X., Yuan, S., Guo, Y., Tan, Y., & Gao, Y. (2024). Public health perspectives on green efficiency through smart cities, artificial intelligence for healthcare and low carbon building materials. *Frontiers in Public Health, 12*, 1440049. DOI: 10.3389/fpubh.2024.1440049 PMID: 39741936

Sun, Y., Wan, L., Mangla, S. K., Xu, X., & Song, M. (2024). Uncovering the interactions between the enterprise AI transformation, supply chain concentration, and corporate risk-taking capacity. *IEEE Transactions on Engineering Management, 71*, 11315–11327. DOI: 10.1109/TEM.2024.3411631

Tan, M., & Le, Q. (2019, May). Efficientnet: Rethinking model scaling for convolutional neural networks. In International conference on machine learning (pp. 6105-6114). PMLR.

Tang, Z., Zhang, P., & Krawec, W. O. (2022). Enabling resilient quantum-secured microgrids through software-defined networking. *IEEE Transactions on Quantum Engineering, 3*, 1–11. DOI: 10.1109/TQE.2022.3216530

Tariq, M. U. (2025). AI Tools and Technologies for Management Education: Enhancing Learning Through Innovation. In M. R. & S. R. (Eds.), *Integrating AI and Machine Learning into Business and Management Education* (pp. 105-132). IGI Global Scientific Publishing. DOI: 10.4018/979-8-3373-2150-9.ch005

Tariq, M. U. (2024). Smart Transportation Systems: Paving the Way for Sustainable Urban Mobility. In *Contemporary Solutions for Sustainable Transportation Practices* (pp. 254–283). IGI Global., DOI: 10.4018/979-8-3693-3755-4.ch010

Tariq, M. U. (2025). AI-Driven Benchmarking and Standards in Cybersecurity Insurance. In Alawida, M., Almomani, A., & Alauthman, M. (Eds.), *AI-Driven Cybersecurity Insurance: Innovations in Risk, Governance, and Digital Resilience* (pp. 83–110). IGI Global Scientific Publishing. DOI: 10.4018/979-8-3373-5545-0.ch004

Tariq, M. U. (2025). Bridging the Digital and Human Divide: Fostering Inclusion and Well-Being in Tech-Driven Workplaces. In Achour, Z. (Ed.), *Leading Inclusive Workplaces Through Digital Transformation and Organizational Change* (pp. 139–168). IGI Global Scientific Publishing. DOI: 10.4018/979-8-3373-5127-8.ch005

Tariq, M. U. (2025). Green FinTech-Driving Sustainable Finance and Empowering Underserved Communities. In Masri, M., Sait, M., & Anshari, M. (Eds.), *Promoting Inclusivity and Accessibility With FinTech* (pp. 279–306). IGI Global Scientific Publishing. DOI: 10.4018/979-8-3373-1112-8.ch010

Tariq, M. U. (2025). The Role of AI in Transforming Hospitality Operations-Enhancing Efficiency and Guest Experience. In Talukder, M., & Chowdhury, H. (Eds.), *Smart Operations and Enhancing Guest Experience in the Hospitality Industry* (pp. 415–446). IGI Global Scientific Publishing. DOI: 10.4018/979-8-3373-2145-5.ch016

Thirunagalingam, A., & Selvakumar, P. (2025). Application of Automation and Robotics in the Oil and Gas Industry. In Abdullayev, V., & Khang, A. (Eds.), *Revolutionizing AI and Robotics in the Oil and Gas Industry* (pp. 229–244). IGI Global Scientific Publishing. DOI: 10.4018/979-8-3693-8156-4.ch011

Tschand, A., Rajan, A. T. R., Idgunji, S., Ghosh, A., Holleman, J., Kiraly, C., & Reddi, V. J. (2025, March). MLPerf Power: Benchmarking the Energy Efficiency of Machine Learning Systems from μWatts to MWatts for Sustainable AI. In *2025 IEEE International Symposium on High Performance Computer Architecture (HPCA)* (pp. 1201-1216). IEEE. DOI: 10.1109/HPCA61900.2025.00092

Tuli, E. A., Lee, J.-M., & Kim, D.-S. (2024). Integration of quantum technologies into metaverse: Applications, potentials, and challenges. *IEEE Access : Practical Innovations, Open Solutions*, *12*, 29995–30019. DOI: 10.1109/ACCESS.2024.3366527

Tunde, A. H., Yusuf, S. O., Taiwo, A. I., Ocran, G., Owusu, P., & Paul-Adeleye, A. H. (2024). AI-driven innovations in energy efficiency: Transforming smart buildings and urban areas through technology and digital transformation. *World Journal of Advanced Research and Reviews, 24*(1).

Udayaprasad, P. K., Shreyas, J., Srinidhi, N. N., Dilip Kumar, S. M. N., Dayananda, P., Askar, S. S., & Abouhawwash, M. (2024). Energy Efficient Optimized Routing Technique With Distributed SDN-AI to Large Scale I-IoT Networks. *IEEE Access : Practical Innovations, Open Solutions, 12*, 2742–2759. DOI: 10.1109/ACCESS.2023.3346679

Ukoba, K., Olatunji, K. O., Adeoye, E., Jen, T.-C., & Madyira, D. M. (2024a). Optimizing renewable energy systems through artificial intelligence: Review and future prospects. *Energy & Environment, 35*(7), 3833–3879. DOI: 10.1177/0958305X241256293

Ullah, K., Ahsan, M., Hasanat, S. M., Haris, M., Yousaf, H., Raza, S. F., Tandon, R., Abid, S., & Ullah, Z. (2024). Short-term load forecasting: A comprehensive review and simulation study with CNN-LSTM hybrids approach. *IEEE Access : Practical Innovations, Open Solutions, 12*, 111858–111881. DOI: 10.1109/ACCESS.2024.3440631

Umar, A., Kumar, D., & Ghose, T. (2025). Decentralized energy trading in microgrids: A blockchain-integrated model for efficient power flow with social welfare optimization. *Electrical Engineering, 107*, 2677–2695. DOI: 10.1007/s00202-024-02635-x

UNDP – United Nations Development Programme. (2020). Empowering Local Action for Climate Justice.

Upadhyaya, A. N., Debnath, P., Rani, P., Selvakumar, P., Manjunath, T. C., & Bhattacharya, S. (2025). AI-Based Disaster Prediction and Early Warning Systems. In Mobo, F. (Ed.), *AI and Emerging Technologies for Emergency Response and Smart Cities* (pp. 55–74). IGI Global Scientific Publishing. DOI: 10.4018/979-8-3693-9770-1.ch003

Uzondu, N. C., & Lele, D. D. (2024). Comprehensive analysis of integrating smart grids with renewable energy sources: Technological advancements, economic impacts, and policy frameworks. *Eng. Sci. Technol. j., 5*(7), 2334–2363.

Vaddi, R. S. (2022). A review of UAV-based remote sensing for atmospheric monitoring. *Sensors (Basel), 22*(8), 3080. PMID: 35459066

VanGeet, O., Lintner, W., & Tschudi, B. (2011). *FEMP best practices guide for energy-efficient data center design*. National Renewable Energy Laboratory.

Varone, A., Heilmann, Z., Porruvecchio, G., & Romanino, A. (2024). Solar parking lot management: An IoT platform for smart charging EV fleets, using real-time data and production forecasts. *Renewable & Sustainable Energy Reviews, 189*, 113845. DOI: 10.1016/j.rser.2023.113845

Veefkind, J. P., Aben, I., McMullan, K., Förster, H., de Vries, J., Otter, G., Claas, J., Eskes, H. J., de Haan, J. F., Kleipool, Q., van Weele, M., Hasekamp, O., Hoogeveen, R., Landgraf, J., Snel, R., Tol, P., Ingmann, P., Voors, R., Kruizinga, B., & Levelt, P. F. (2012). TROPOMI on the ESA Sentinel-5 Precursor: A GMES mission for global observations of the atmospheric composition for climate, air quality and ozone layer applications. *Remote Sensing of Environment, 120*, 70–83. DOI: 10.1016/j.rse.2011.09.027

Venkatesh, V., Morris, M. G., Davis, G. B., & Davis, F. D. (2003). User acceptance of information technology: Toward a unified view. *Management Information Systems Quarterly, 27*(3), 425–478. DOI: 10.2307/30036540

Vinuesa, R., Azizpour, H., Leite, I., Balaam, M., Dignum, V., Domisch, S., Felländer, A., Langhans, S. D., Tegmark, M., & Fuso Nerini, F. (2020). The role of artificial intelligence in achieving the Sustainable Development Goals. *Nature Communications, 11*(1), 233. DOI: 10.1038/s41467-019-14108-y PMID: 31932590

Wall, P. J., Saxena, D., & Brown, S. (2021). Artificial intelligence in the Global South (AI4D): Potential and risks. arXiv preprint arXiv:2108.10093.

Wan, A., Chang, Q., Khalil, A.-B., & He, J. (2023). Short-term power load forecasting for combined heat and power using CNN-LSTM enhanced by attention mechanism. *Energy, 282*, 128274. DOI: 10.1016/j.energy.2023.128274

Wang, L. (2025). *Designing pull-based energy system rules for public buildings: An application of modularity*. Academic Press.

Wang, C., Qin, J., Qu, C., Ran, X., Liu, C., & Chen, B. (2021). A smart municipal waste management system based on deep-learning and Internet of Things. *Waste Management (New York, N.Y.), 135*, 20–29. DOI: 10.1016/j.wasman.2021.08.028 PMID: 34461487

Wang, Y., Chen, Q., Hong, T., & Kang, C. (2019). Review of Smart Meter Data Analytics: Applications, Methodologies, and Challenges. *IEEE Transactions on Smart Grid, 10*(3), 3125–3148. DOI: 10.1109/TSG.2018.2818167

Watson, J., Byrnes, R., Ockwell, D., & Stua, M. (2015). Lessons from China: Building technological capabilities for low carbon technology transfer and development. *Climatic Change, 131*(3), 387–399. DOI: 10.1007/s10584-014-1124-1

Wilson, A., & Anwar, M. R. (2024). The future of adaptive machine learning algorithms in high-dimensional data processing. *International Transactions on Artificial Intelligence, 3*(1), 97–107. DOI: 10.33050/italic.v3i1.656

Wiser, R., & Bolinger, M. (2021). *2020 Wind Technologies Market Report*. Lawrence Berkeley National Laboratory.

Wodecki, A. (2020). *Artificial intelligence in management: Self-learning and autonomous systems as key drivers of value creation*. Edward Elgar Publishing. DOI: 10.4337/9781839104954

Wohlin, C. (2014). Guidelines for snowballing in systematic literature studies and a replication in software engineering. *Proceedings of the 18th International Conference on Evaluation and Assessment in Software Engineering*, 1-10. DOI: 10.1145/2601248.2601268

Wooster, M. J., Roberts, G., Perry, G. L. W., & Kaufman, Y. J. (2015). Retrieval of biomass combustion rates and totals from fire radiative power observations: FRP derivation and calibration relationships. *Journal of Geophysical Research, D, Atmospheres, 110*(D24).

World Economic Forum. (2025). *Advancing Responsible AI Innovation: A Playbook*. World Economic Forum in collaboration with Accenture.

World Resources Institute. (2022). Global Forest Watch.

Wu, H., Wang, X., & Fan, Z. (2025). Addressing the sustainable AI trilemma: a case study on LLM agents and RAG. arXiv preprint arXiv:2501.08262.

Wu, T. W., Zhang, H., Peng, W., Lü, F., & He, P. J. (2023). Applications of convolutional neural networks for intelligent waste identification and recycling: A review. *Resources, Conservation and Recycling, 190*, 106813. DOI: 10.1016/j.resconrec.2022.106813

Xu, A., Chang, H., Xu, Y., Li, R., Li, X., & Zhao, Y. (2021). Applying artificial neural networks (ANNs) to solve solid waste-related issues: A critical review. *Waste Management (New York, N.Y.), 124*, 385–402. DOI: 10.1016/j.wasman.2021.02.029 PMID: 33662770

Xu, C. (2021). A review of onboard AI technologies for satellite remote sensing. *IEEE Journal of Selected Topics in Applied Earth Observations and Remote Sensing, 14*, 5197–5211.

Yarally, T., Cruz, L., Feitosa, D., Sallou, J., & Van Deursen, A. (2023). Uncovering energy-efficient practices in deep learning training: Preliminary steps towards green ai. arXiv preprint arXiv:2303.13972. DOI: 10.1109/CAIN58948.2023.00012

Yarbrough, J. (2024). *Revolutionizing Waste Management: AI-Powered Real-time Characterization for Efficient Handling of Non-Recyclable Municipal Solid Waste* (No. NREL/PR-2700-88053). National Renewable Energy Laboratory (NREL), Golden, CO (United States).

Yazmyradov, S., & Lee, H. J. (2024). A Comprehensive Review of AI Security: Threats, Challenges, and Mitigation Strategies. *International Journal of Internet. Broadcasting and Communication, 16*(4), 375–384.

Yin, R. K. (2014). *Case study research: Design and methods* (5th ed.). Sage Publications.

Yitmen, I., Almusaed, A., Hussein, M., & Almssad, A. (2025). AI-Driven Digital Twins for Enhancing Indoor Environmental Quality and Energy Efficiency in Smart Building Systems. *Buildings (Basel, Switzerland), 15*(7), 1030. DOI: 10.3390/buildings15071030

Zanella, A., Bui, N., Castellani, A., Vangelista, L., & Zorzi, M. (2014). Internet of things for smart cities. *IEEE Internet of Things Journal, 1*(1), 22–32. DOI: 10.1109/JIOT.2014.2306328

Zapata, V., Gernaat, D. E. H. J., Yalew Seleshi, G., Santos da Silva, S. R., Iyer, G., Hejazi, M., & van Vuuren, D. P. (2022a). Climate change impacts on the energy system: A model comparison. *Environmental Research Letters, 17*(3), 34036. DOI: 10.1088/1748-9326/ac5141

Zewe, A. (2025). Explained: Generative AI's Environmental Impact. *MIT News*.

Zhang, C., Wu, J., Long, C., & Cheng, M. (2017). Review of Existing Peer-to-Peer Energy Trading Projects. *Energy Procedia, 105*, 2563–2568. DOI: 10.1016/j.egypro.2017.03.737

Zhang, C., Wu, J., Zhou, Y., Cheng, M., & Long, C. (2018). Peer-to-Peer energy trading in a Microgrid. *Applied Energy, 220*, 1–12. DOI: 10.1016/j.apenergy.2018.03.010

Zhang, W., Chen, L., Luo, J., & Liu, J. (2022). A two-stage container management in the cloud for optimizing the load balancing and migration cost. *Future Generation Computer Systems, 135*, 303–314. DOI: 10.1016/j.future.2022.05.002

Zhang, Y., Cheng, C., Cai, H., Jin, X., Jia, Z., Wu, X., Su, H., & Yang, T. (2022). Long-term stochastic model predictive control and efficiency assessment for hydro-wind-solar renewable energy supply system. *Applied Energy, 316*, 119134. DOI: 10.1016/j.apenergy.2022.119134

Zhao, T., Bell, L., Horner, M. W., Sulik, J., & Zhang, J. (2018). Consumer responses towards home energy financial incentives: A survey-based study. *Energy Policy, 114*, 845–857. DOI: 10.1016/j.enpol.2017.12.0

Zohuri, B., & Mossavar-Rahmani, F. (2024). Balancing Sustainability and Innovation the Future of Global Energy Policy. *J Econ Managem Res*, *5*(6), 2–6.

Zong, Z., & Guan, Y. (2024). AI-Driven Intelligent Data Analytics and Predictive Analysis in Industry 4.0: Transforming Knowledge, Innovation, and Efficiency. *Journal of the Knowledge Economy*, *16*(1), 864–903. DOI: 10.1007/s13132-024-02001-z

About the Contributors

Ben Othman Soufiene is an Assistant Professor of computer science at the Applied College, King Faisal University, Saudi Arabia from 2025. He received his Ph.D. degree in computer science from Manouba University in 2016 for his dissertation on "Secure data aggregation in wireless sensor networks. He also holds M.S. degrees from the Monastir University in 2012. My research interests focus on the Internet of Medical Things, Wireless Body Sensor Networks, Wireless Networks, Artificial Intelligence, Machine Learning and Big Data. Dr. Ben Othman has published more than 120 papers at reputed international journals, conferences, and book chapters. He is an Editorial Board Member in the different Journals and Conferences. He serves as an associate editor/academic editor for international journals including IEEE Access, IEEE Sensors, IEEE Internet of Things, Elsevier, Springer, Taylor & Francis, IGI, IET, Telecommunication Computing Electronics and Control, and Wiley. Dr. Ben Othman is a Technical Program Committee Member for more than a dozen of international conferences.

Mohieddine Rahmouni is an Associate Professor of Economics and Quantitative Methods at King Faisal University and the University of Tunis (Higher School of Economic and Commercial Sciences of Tunis). He earned his Ph.D. through a joint program between the University of Montesquieu Bordeaux-IV and the University of Tunis. Dr. Rahmouni's research spans the economics and econometrics of innovation, applied econometrics, productivity, technology, and industrial dynamics, with a focus on the intersections of these fields in driving economic development and industrial transformation.

Kirti Babber is a Research Scholar in Electronics and Communication Engineering. Her work focuses on exploring innovative technologies and developing smart, practical solutions in the field of communication systems and embedded electronics. She is deeply interested in how technology can bridge gaps between research and real-world applications, improving both efficiency and accessibility.

Kirti believes in continuous learning and collaboration, and her research reflects a commitment to innovation, sustainability, and human-centered design.

Omdev Dahiya is working as an Assistant Professor in the School of Computer Science & Engineering, Lovely Professional University, Phagwara, Punjab, India. He holds a Ph.D. in Computer Science & Engineering from Maharshi Dayanand University, Rohtak, Haryana, India. With over 5 years of experience in teaching and research within the field of Computer Science & Engineering, he has published over 45 research articles indexed in Scopus and Web of Science databases. Additionally, he has supervised numerous UG/PG students and is currently overseeing the work of two Ph.D. scholars, with one already awarded their Ph.D. He has been a session chair and technical program committee member at various national and international conferences. He also serves as a potential reviewer for several international journals, including IEEE ACCESS, Heliyon, Recent Advances in Electrical & Electronic Engineering, and Social Sciences & Humanities Open. He is currently working in various domains of computer engineering which includes Data Mining, Learning Analytics, Cloud Computing, and soft computing approaches involving advanced AI/ML techniques.

Syed Hassan Imam Gardezi is a distinguished futurist, thought leader and strategic advisor with a unique background in private equity, venture capital, alternative and institutional investments. This foundation has afforded him a rare, hands-on perspective across diverse industries and global supply chains, where he has directly overseen integrations and transformations driven by contemporary and emerging technologies. His expertise lies in demystifying and deploying technologies like Machine Learning, Data Science, AI, IoT, and blockchains for tangible strategic advantage. This deep, practical experience in both investment and operational technology transformation directly informs his writing and editing. A recognized thought leader and holder of numerous prestigious fellowships and certifications, including a PhD in Management, an MBA and certifications in emerging technologies like AI and Blockchain, he translates complex technological concepts into actionable business strategy.

Priyanka Gotter is currently serving as an Assistant Professor in the School of Computer Science & Engineering at Lovely Professional University, Phagwara, Punjab, India. She holds an M.Tech in Computer Science & Engineering from Guru Nanak Dev University, Amritsar, Punjab, India. She also pursuing Ph.D. in Computer Science & Engineering from Lovely Professional University, Phagwara, Punjab, India. With over 4 years of teaching experience, Ms. Gotter has been actively involved in academic instruction and mentoring at the undergraduate and postgraduate levels. Her areas of interest include Data Structures, Algorithms, Artificial Intelligence, and Machine Learning. She has contributed to curriculum development and has played a key role in fostering academic excellence among her students.

Sumaiya Thaseen earned a bachelor's degree in computer engineering from Madras University. She received her Masters and Ph.D. degrees. She is currently a senior lecturer in Cyber Security in the school of computing, De Montfort University, Dubai She has 18 years of academic and research experience in Vellore Institute of Technology, Vellore, Tamil Nadu, India. She has published more than 40 peer-reviewed research articles in SCI as well Scopus-indexed international journals, in addition to books, book chapters, and numerous conference proceedings. She has collaborated with many researchers from around the world on projects ranging from cyber security with artificial intelligence, IoT, and healthcare. She successfully completed a 0.6 million Full Stack Development consulting project with Perficient Inc in India while she was working in VIT. She was the Subject Matter Expert (SME) for TCS Ion, where she created and reviewed question banks on Computer Science and Cyber Security. She is an active IEEE member, an ACM member, and a lifetime member of the IEEE Computer Society of India (CSI) and the Indian Science Congress Association.

J. Saira Banu is working as a faculty at the School of Computer Science and Engineering, VIT University, Vellore. She completed her Ph.D. in the domain of parallel computing in multicores from VIT University. She has publications in the domain of parallel computing with good citations. She has Fourteen years of teaching experience. Her areas of interest include parallel computing, computer architecture, Image Processing, and Multicore Programming.

M. Rajesh is an Assistant Professor in the Department of Mechanical Engineering at Rathinam Technical Campus, with over 11 years of teaching experience and 1.5years in industry. Specialization in polymer composites, tribology and optimization. He has published more than 10 journal and 15 International conferences articles. An expert in design courses, He integrates academic excellence with practical engineering knowledge to advance mechanical engineering education and research.

M. R. Arun completed his BE, ME and PhD from Anna University, Chennai. He completed PG Diploma (PGDEEM) from Annamalai University, M.Sc. Counselling Psychology from TNOU University, M.A. Public Administration from Madras University and B.Lib.Sc. from Alagappa University. He is in the Teaching Profession for the past 17 years, Published around 115 papers in Journals, Patents and

Conferences, Active participant crossing more than 200 events and internationally networked through various Professional bodies. He is skilled and technically proficient in various simulation Softwares associated to Research in Engineering. He Organized and Coordinated around 15 events in working Institution. Authored more than 15 Books, 20 Book Chapters and he has been technically integrated in various international journals as an Editorial Board Member and Reviewer. His area of interest includes Video Processing, Remote Automation, Wireless Communication, IoT Sensors, Quantum Computing, Machine Learning & Artificial Intelligence.

T.C. Manjunath was born on 6th February 1967. He did his B.E. from RV College of Engg. in 1989 (B'lore Univ), M.E. in Robotics & Automation from Govt.'s LD College of Engg., (Gujarat Univ., Ahmedabad) in 1992 with FCD-First Rank & PhD in Systems & Control Engg. from IIT Bombay in the year 2007, got best research scholar award for the entire IITB. He is currently working as the Dean Research (R & D) and as Professor in Dept. of Computer Science & Engg. with specialization in IoT, Cyber Security, Blockchain Technology in Rajarajeswari College of Engineering, Bengaluru, Karnataka. He has got a teaching (academic), research & administrative experience of more than 35+ years (10 years as Principal & 8 years as HoD, 1 year as Dean Research & rest as faculty) and has published more than 1500+ research papers in various national & international journals & conferences, with 175+ patents, 100+ book-chapters, 25 awards, 100s of Keynote talks and also published 25 text books in various field of science, engineering & technology, completed projects worth Rs. 9 Crores till date & guided 10 PhD scholars in VTU. He is a member of 30 Professional Society Chapters and is a fellow of IE, IETE, AMIEE, Sr. Member of IEEE and a chartered engineer.

N. Subramani is an Assistant Professor in the Department of Mechanical Engineering at Karpagam College of Engineering, Coimbatore, with over 17.5 years of teaching experience and 1 year in industry. Specializing in Natural Fibre Composites and Engineering Design and Optimization. Mr.N.Subramani has published more than 43 journal articles and 15 National/International conferences. An expert in Design and Thermal courses. Mr.N.Subramani integrates academic excellence with practical engineering knowledge to explore the students expertise in connection with the industry needs and recent trends.

P. Selvakumar is currently an Associate Professor in the Department of Science and Humanities at Nehru Institute of Technology, Coimbatore, Tamil Nadu, India. Prior to joining Nehru Institute of Technology, he held positions as an Assistant Professor in the Department of Science and Humanities various engineering colleges in Coimbatore, Tamil Nadu, India, spanning from 2011 to 2023. Dr.P. Selvakumar's academic and research interests lie primarily in the field of phytochemistry. He has made significant contributions to his field, with an impressive record that includes 30 journal publications, 12 patents, and 20 conference papers. Additionally, he has organized 5 seminars and conferences, further demonstrating his commitment to academic and scientific exchange. Throughout his career, Dr. P.Selvakumar has been recognized for his dedication and excellence in teaching, having been honored with several awards, including the Best Teacher Awards. He holds a prominent position in the academic community, contributing actively to research.

Ganesh Kumar R received MCA from Bharathidasan University in 1999, M.E and Ph.D in Computer Science and Engineering from Sathyabama University, Chennai in 2007 and 2015 respectively. He has 26 years of Teaching experience and CCNA Certified Professional. Currently working as Professor in CHRIST (Deemed to be University), School of Engineering and Technology, Kengeri Campus, Bangalore. He has published research papers in International Journals, presented at International Conferences, National and International Patents, Copyrights, Book chapters and Text books. His area of research interest includes Data Science, Cloud Computing, Block Chain Technology and Image processing. Dr. Ganesh Kumar is a Senior Member in IEEE, Fellow-Institute of Engineers (FIE) and also associated with Professional Technical bodies like ISTE, IAENG, Internet Society and Python Software Foundation.

Ravikumar R. Natarajan is an experienced academician with over 12+ years of teaching and research expertise across institutions in India, the UAE, and the Maldives. He currently serves as an Assistant Professor at Marwadi University, Rajkot, specializing in Machine Learning, Recommender Systems, and AI-driven applications in healthcare and security. He holds a Ph.D. in Machine Learning and Recommender Systems from Amity University, Jaipur, a Master of Engineering in CSE from VMRF University, Salem, and a Bachelor of Engineering in CSE from Anna University. He also has a MiniMBA from the Swiss eLearning Institute. Dr. Ravikumar has authored more than 8+ multiple SCIE/Scopus-indexed journal articles, 25+ IEEE/Springer conference papers, and co-authored a book on Artificial Intelligence. He has also published 9 patents and contributed 10+ book chapters in emerging domains such as Federated Learning, Blockchain, and Digital Twins for IGI Global, Wiley and Springer publisher. He actively mentors students through initiatives like the Research Activities Club (RAC) and Student Research Internship Training (SRIT), promoting innovation and research culture.

Pallapati Ravi Kumar is currently serving as an Assistant Professor in the Department of Computer Science and Engineering at Koneru Lakshmaiah Education

Foundation (K L University), located in Vaddeswaram, Guntur, Andhra Pradesh, Bharat. He is dedicated to academic excellence and research in the fields of Deep Learning and Data Science. With a strong commitment to fostering innovation and analytical thinking among students, he actively contributes to both teaching and research activities within the department. His academic and research pursuits focus on developing intelligent systems and data-driven solutions that address real-world challenges through advanced computational techniques.

Carlos Rodríguez-Maillard is a process and transformation leader at RED Ambiental, a leading waste management company in Mexico. His mission is to improve the business strategy across the organization, using his skills in Digital Transformation, problem-solving, change management and business management. He is also a professor at EGADE Business School (Tecnológico de Monterrey). He shares his knowledge in Organizational Performance Management, User-Centered Design, and Project Management classes with MBA and MBM students. He has a PhD in business administration, an MBA, a master's in finance and technology administration and multiple articles and book chapters published in academic and business journals. He is fluent in Spanish, English, and Portuguese and enjoy learning new things and facing new challenges.

Aarthi S is a dedicated Assistant Professor in Computer Science with over six years of teaching experience. She is currently pursuing her Ph.D. at Marwadi University, Gujarat, and holds an M.Phil. and MCA from Bharathiar University, Tamil Nadu. Throughout her academic career, Aarthi has specialized in programming languages such as C++, Java, and C, and contributed to research in recommender systems and AI-related data privacy. She has published journal articles, contributed to book chapters, and actively participated in technical workshops. Known for her student mentorship and engagement in academic events, Aarthi remains committed to fostering innovation and excellence in computer science education.

Amara S A L G Gopala Gupta, is an Assistant Professor at Department of Computer Science and Engineering, KL University, Vaddeswaram, India. Amara S A L G Gopala Gupta holds a Bachelor of Technology in Computer Science and Engineering, degree from Jawaharlal Nehru Technological University, Kakinada, India (2005-2009), and a Master of Technology in Computer Science and Engineering, degree from Jawaharlal Nehru Technological University, Kakinada, India (2010-2012) and a Doctor of Philosophy (PhD) from K L University, Vaddeswaram, India.

Jenifer Shanmugasundaram is a highly motivated computer science enthusiast with a strong passion for research in machine learning, natural language processing, and computer vision. She is particularly interested in exploring how AI can be applied to solve real-world problems in areas such as sustainability, public safety, and mental health. With a forward-thinking mindset and a deep curiosity for emerging technologies, she is committed to advancing responsible and impactful AI solutions through continuous research and experimentation.

Vikas Chand Sharma, an Associate Professor at Chandigarh University and a researcher at Guru Nanak Dev University, Amritsar, brings 13 years of experience in architectural design, urban planning, and research. He specializes in sustainable urban design strategies for historic cities, combining phenomenological approaches with digital mapping. His foundational research on religious spatial patterns and new urbanism was published in 2016. He serves on the Scientific Committee for Contesti. Città, Territori, Progetti and is a member of the Editorial Board for several international journals. His work appears in the International Journal of Construction Education and Research, Journal of Architecture & Urbanism, and Landscape Research.

Muhammad Usman Tariq has more than 17+ year's experience in industry and academia. He is a prolific author of research articles, case studies, book chapters, and academic books, with work spanning a wide range of disciplines. His contributions also include several patents. He is founder and CEO of The Case HQ, a unique repository for courses, narrative and video case studies. Additionally, he is President of Gulf Higher Education (Gulf HE) focused on advancing academic practice. He has been working as a consultant and trainer for industries representing six sigma, quality, health and safety, environmental systems, project management, and information security standards. His work has encompassed sectors in aviation, manufacturing, food, hospitality, education, finance, research, software and transportation. He has diverse and significant experience working with accreditation agencies of ABET, ACBSP, AACSB, WASC, CAA, EFQM and NCEAC. He is a Principal Fellow of Advance HE UK, Senior Fellow of SEDA UK, and Chartered Fellow of CIPD UK other than Senior Member IEEE.

Mythili Thirugnanam, a Professor at VIT Vellore, is a senior academician with over 17+ years of experience in Computer Science and Engineering. She holds a Ph.D. in the field and specializes in Artificial Intelligence, Deep Learning, Medical Image Processing, and Software Engineering. Her research focuses on AI-driven diagnostic and decision-making systems, with notable contributions to nationally and international fund project in collaboration with ISRO and the SPARC-UKIERI program. She has worked on applications such as radiographic defect detection, predictive modeling for material failure, and AI-enhanced imaging. In industry, she collaborates with Samsung R&D on real-time computer vision tasks. Dr. Mythili has published over 50 research articles, delivered 12+ technical talks, and guided multiple Ph.D. scholars in AI and medical imaging. Her work bridges theoretical research with real-world impact across aerospace, healthcare, and smart technologies.

Index

A

C

D

E

S

U

W

www.ingramcontent.com/pod-product-compliance
Lightning Source LLC
LaVergne TN
LVHW062332300926
849161LV00007B/1662